Lecture Notes in Physics

Volume 832

For further volumes:
http://www.springer.com/series/5304

The Lecture Notes in Physics

The series Lecture Notes in Physics (LNP), founded in 1969, reports new developments in physics research and teaching—quickly and informally, but with a high quality and the explicit aim to summarize and communicate current knowledge in an accessible way. Books published in this series are conceived as bridging material between advanced graduate textbooks and the forefront of research and to serve three purposes:

- to be a compact and modern up-to-date source of reference on a well-defined topic
- to serve as an accessible introduction to the field to postgraduate students and nonspecialist researchers from related areas
- to be a source of advanced teaching material for specialized seminars, courses and schools

Both monographs and multi-author volumes will be considered for publication. Edited volumes should, however, consist of a very limited number of contributions only. Proceedings will not be considered for LNP.

Volumes published in LNP are disseminated both in print and in electronic formats, the electronic archive being available at springerlink.com. The series content is indexed, abstracted and referenced by many abstracting and information services, bibliographic networks, subscription agencies, library networks, and consortia.

Proposals should be sent to a member of the Editorial Board, or directly to the managing editor at Springer:

Christian Caron
Springer Heidelberg
Physics Editorial Department I
Tiergartenstrasse 17
69121 Heidelberg/Germany
christian.caron@springer.com

Jean-Pierre Rozelot · Coralie Neiner

Editors

The Pulsations of the Sun and the Stars

 Springer

Editors
Dr. Jean-Pierre Rozelot
Dept. LAGRANGE
Observatoire de la Cote d'Azur
av. Copernic
06130 Grasse
France
e-mail: rozelot@obs-azur.fr

Coralie Neiner
Observatoire de Meudon GEPI
Bâtiment Copernic A, Place Jules Janssen 5
92195 Meudon, Cedex
France
e-mail: Coralie.Neiner@obspm.fr

ISSN 0075-8450

e-ISSN 1616-6361

ISBN 978-3-642-19927-1

e-ISBN 978-3-642-19928-8

DOI 10.1007/978-3-642-19928-8

Springer Heidelberg Dordrecht London New York

Cover design: eStudio Calamar, Berlin/Figueres

Printed on acid-free paper

Springer is part of Springer Science+Business Media (www.springer.com)

Preface

The focus of this book is placed on the physics of pulsations and brings together the knowledge from the Sun and the stars, with a particular emphasis on recent observations and modelling, and on the influence of pulsations of other physical processes.

Oscillations of the Sun have been widely used in the past to understand its interior structure. The extension of similar studies to more distant stars has raised many difficulties despite the strong efforts of the international community over the past decades. However, we are currently witnessing a complete renewal of the methods and models in the field of pulsations of stars due to a large extent to the launches of the MOST and CoRoT satellites in 2003 and 2006 respectively, which have brought results of unprecedented precision on the pulsations of stars of all types. In particular, pulsations make it possible to derive the internal structure of the stars, which still remains a major scientific enigma. In this context and with the first CoRoT results in hand, it seemed interesting to confront the experience of solar astronomers with that of stellar ones. Transposing the results obtained from heliosismology to asterosismology has already been very profitable. No doubt that the Kepler satellite, launched in 2009, and future space missions for asteroseismology will allow us to go even further in the study of pulsations and the modelling of the internal structure of the various stars.

The General Overlook of this Book is as Follows:

A general up-to-date stand-alone introduction on helioseismology is proposed by A. Kosovichev, followed by two sections, one devoted to the Sun and the other to the stars, linked by a transition chapter from heliosismology to asterosismology written by S. Vauclair.

The first section of the book on the Sun is divided into four chapters.

The Sun is the only solar-type star where the dynamics and the magnetism can be studied in detail and the physical process involved is relatively well understood, in particular those which occur at very small scales. This chapter albeit restricted to the quiet solar photosphere describes the properties of the three cellular scales of motions observed at the solar surface: granulation, mesogranulation and su-pergranulation. The intranet work field, not yet clarified, is also tackled and the questions are posed in a new way.

The second chapter focuses on variations of solar activity which are a result of a complex dynamo process in the convection zone. Despite the known general properties of the solar cycles, a reliable forecast of the 11 year sunspot number is still a problem. However, new methods that take into account the dynamics of turbulent magnetic helicity are capable of providing a forecast of the system, and its application permits a good prediction of the sunspot number.

The third part focuses on solar gravity modes, which are mainly trapped inside the radiative region and consequently are able to provide information on the properties of the solar core. Such a topic is of high interest today as we are wondering if the core may rotate at a higher rate than the outer envelope. However, the detection of solar gravity modes still remains a major challenge. The issue discussed here is important since a theoretical determination of mode amplitudes may help to design the track for gravity modes.

The last chapter of this section deals with the rotation, and more precisely with the differential rotation of the Sun and stars. The effects on the outer shape and to first order, and those concerning the apparent oblateness are tackled. Thanks to the advent of interferometry techniques, the stellar shapes can now be measured with a great accuracy. It is shown that the core density and the gravitational moments can be reached.

Then, the recent developments from helioseismology to asteroseismology are presented. The general basis for asteroseismology, the so-called asymptotic theory of stellar oscillations is discussed. Solar-type stars are discussed and examples for which it was possible to derive precise stellar parameters from seismology are presented, focusing on the helium abundance. The potentiality of asteroseismology for a better knowledge of stellar structure and evolution is huge, and many new results are expected in the near future.

The next section, consisting in five chapters, deals with stellar pulsations.

First, the requirements for a self-consistent interpretation of a collection of observables related to rapidly rotating stars are explored. If the star is otherwise static, rapid rotation through the centrifugal force will affect the force balance and hence the structure of the star. Rotation also changes the surface from a spherical to a spheroidal, and possibly in some cases an ellipsoidal, shape. Such changes have been observed even for the Sun, and the confrontation of the theories are of importance.

In the next chapter, the effects of stellar rotation on adiabatic oscillation frequencies of massive stars are presented, together with methods to evaluate them and some of the main results for four specific stars are shown.

The following chapter deals with the extension of the asymptotic theory of stellar oscillations beyond the case of a non-rotating, non-magnetic spherically symmetric star. A recent application to the high-frequency acoustic modes of rapidly rotating stars is presented.

Then, the complete interaction between low-frequency internal gravity waves and differential rotation in stably strongly stratified stellar radiation zones is examined. This includes the modification of the structure of waves and of the angular velocity.

The last chapter is devoted to excitation of solar-like oscillations that have been detected for more than 10 years. The computed mode excitation rates crucially depend not only on the way turbulent convection is described but also on the stratification and the metal abundance of the upper layers of the star. In turn it is shown how the seismic measurements collected so far allow us to infer properties of turbulent convection in stars.

The audience targeted by this book consists of researchers, PhD students, post-docs, and all scientists seeking a complementary culture or scientists evolving toward new research topics.

This book is based on tutorials and discussions on the same topic held at a CNRS school in Saint-Flour (France) in 2008, which has allowed us to give a progress report on the very last solar developments (structure of the solar core for example) and stellar developments (CoRoT results, new stellar models) for a better understanding of stellar pulsations and internal structure in general. Let us remind that a first book titled "The rotation of the Sun and Stars" (LNP 765) resulted from a previous CNRS school held in Obernai (France) in 2007. We hope that this new book the "Pulsations of the Sun and Stars" will provide an interesting sequell for the reader.

The editors sincerely thank the authors for the great quality of their contributions published here. They hope that this new book will help to a better knowledge of the wonderful world which surrounds us.

December 2010   J. P. Rozelot
                                                                          C. Neiner

Contents

Part II   Section 1: The Sun as a Star

Part I
General Overview

A. Kosovichev (Stanford University, USA)

Chapter 1
Advances in Global and Local Helioseismology: An Introductory Review

Alexander G. Kosovichev

Abstract Helioseismology studies the structure and dynamics of the Sun's interior by observing oscillations on the surface. These studies provide information about the physical processes that control the evolution and magnetic activity of the Sun. In recent years, helioseismology has made substantial progress towards the understanding of the physics of solar oscillations and the physical processes inside the Sun, thanks to observational, theoretical and modeling efforts. In addition to global seismology of the Sun based on measurements of global oscillation modes, a new field of local helioseismology, which studies oscillation travel times and local frequency shifts, has been developed. It is capable of providing 3D images of subsurface structures and flows. The basic principles, recent advances and perspectives of global and local helioseismology are reviewed in this article.

1.1 Introduction

In 1926 in his book *The Internal Constitution of the Stars* Sir Eddington [1] wrote: "At first sight it would seem that the deep interior of the Sun and stars is less accessible to scientific investigation than any other region of the universe. Our telescopes may probe farther and farther into the depths of space; but how can we ever obtain certain knowledge of that which is hidden behind substantial barriers? What appliance can pierce through the outer layers of a star and test the conditions within?"

The answer to this question was provided a half a century later by *helioseismology*. Helioseismology studies the conditions inside the Sun by observing and analyzing oscillations and waves on the surface. The solar interior is not transparent to light but it is transparent to acoustic waves. Acoustic (sound) waves on the Sun are excited by

Alexander Kosovichev (✉)
W.W. Hansen Experimental Physics Laboratory, Stanford University, Stanford, CA 94305, USA
e-mail: sasha@sun.stanford.edu

J.-P. Rozelot and C. Neiner (eds.), *The Pulsations of the Sun and the Stars*,
Lecture Notes in Physics 832, DOI: 10.1007/978-3-642-19928-8_1,
© Springer-Verlag Berlin Heidelberg 2011

(a)

(b)

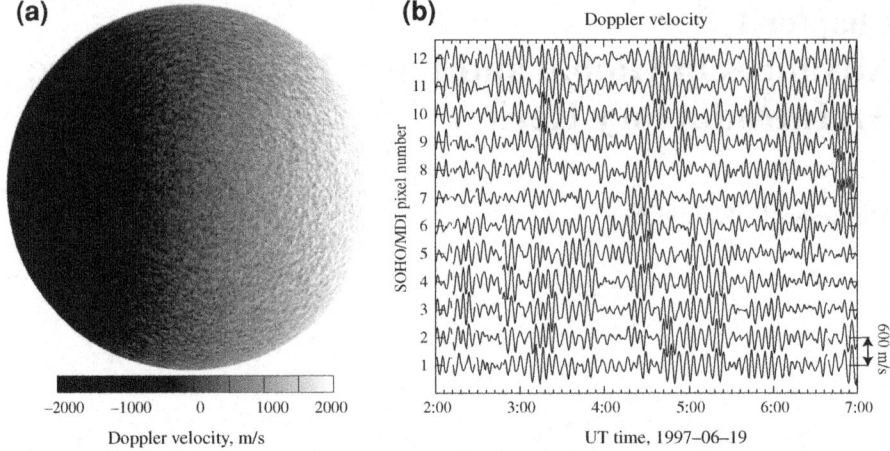

Fig. 1.1 **a** Image of the line-of-sight (Doppler) velocity of the solar surface obtained by the Michelson Doppler Imager (MDI) instrument on board SOHO spacecraft on 1997-06-19, 02:00 UT; **b** Oscillations of the Doppler velocity, measured by MDI at the solar disk center in 12 CCD pixels separated by ~1.4 Mm on the Sun

turbulent convection below the visible surface (photosphere) and travel through the interior with the speed of sound. Some of these waves are trapped inside the Sun and form *resonant oscillation modes*. The travel times of acoustic waves and frequencies of the oscillation modes depend on physical conditions of the internal layers (temperature, density, velocity of mass flows, etc.). By measuring the travel times and frequencies one can obtain information about these condition. This is the basic principle of helioseismology. Conceptually it is very similar to the Earth's seismology. The main difference is that the Earth's seismology studies mostly individual events, earthquakes, while helioseismology is based on the analysis of acoustic noise produced by solar convection. However, recently similar techniques have been applied for ambient noise tomography of Earth's structures. The solar oscillations are observed in variations of intensity of solar images or, more commonly, in the line-of-sight velocity of the surface elements, which is measured from the Doppler shift of spectral lines (Fig. 1.1). Variations caused by these oscillations are very small, much smaller than the noise produced by turbulent convection. Thus, their observation and analysis require special procedures.

Helioseismology is a relatively new discipline of solar physics and astrophysics. It has been developed over the past few decades by a large group of remarkable observers and theorists, and is continued being actively developed. The history of helioseismology has been very fascinating, from the initial discovery of the solar 5-min oscillations and the initial attempts to understand the physical nature and mechanism of these oscillations to detailed diagnostics of the deep interior and subsurface magnetic structures associated with solar activity. This development was not straightforward. As this always happens in science controversial results and ideas provided inspiration for further more detailed studies.

In a brief historical introduction, I describe some key contributions. It is very interesting to follow the line of discoveries that led to our current understanding of the oscillations and to helioseismology techniques. Then, I overview the basic concepts and results of helioseismology. The launch of the Solar Dynamics Observatory (SDO) in 2010 opened a new era in helioseismology. The Helioseismic and Magnetic Imager (HMI) instrument provides uninterrupted high-resolution Doppler-shift and vector magnetogram data over the whole disk. These data will provide a complete information about the solar oscillations and their interaction with solar magnetic fields.

1.2 Brief History of Helioseismology

Solar oscillations were discovered in 1960 by Leighton et al. [2] by analyzing series of Dopplergrams obtained at the Mt. Wilson Observatory. Instead of the expected turbulent behavior of the velocity field they found two distinct classes: large-scale horizontal cellular motions, which they called *supergranulation*, and vertical *quasi-periodic oscillations* with a period of about 300 s (5 min) and a velocity amplitude of about 0.4 km s^{-1}. It turned out that these oscillations are the dominant vertical motion in the lower atmosphere (chromosphere) of the Sun. It is remarkable that they realized the diagnostic potential noting that these oscillations "offer a new means of determining certain local properties of the solar atmosphere, such as the temperature, the vertical temperature gradient, or the mean molecular weight". They also pointed out that the oscillations might be excited in the Sun's granulation layer, and account for a part of the energy transfer from the convection zone into the chromosphere.

This discovery was confirmed by other observers, and for several years it was believed that the oscillations represent transient atmospheric waves excited by *granules*, small convective cells on the solar surface, 1–2 × 10^3 km in size and 8–10 min lifetime. The physical nature of the oscillations at that time was unclear. In particular, the questions whether these oscillations are acoustic or gravity waves, and if they represent traveling or standing waves remained unanswered for almost a decade after the discovery.

Mein [3] applied a 2D Fourier analysis (in time and space) to observational data obtained by John Evans and his colleagues at the Sacramento Peak Observatory in 1962–1965. His idea was to decompose the oscillation velocity field into *normal modes*. He calculated the *oscillation power spectrum* and investigated the relationship between the period and horizontal wavelength (or *frequency–wavenumber diagram*). From this analysis he concluded that the oscillations are acoustic waves that are *stationary* (*evanescent*) in the solar atmosphere. He also made a suggestion that the horizontal structure of the oscillations may be imposed by the convection zone below the surface.

Mein's results were confirmed by Frazier [4] who analyzed high-resolution spectrograms taken at the Kitt Peak National Observatory in 1965. In the wavenumber–frequency diagram he noticed that in addition to the primary 5-min peak there is a secondary lower frequency peak, which was a new puzzle.

This puzzle was solved by Ulrich [5] who following the ideas of Mein and Frazier, calculated the spectrum of standing acoustic waves trapped in a layer below the photosphere. He found that these waves may exist only along discrete line in the wavenumber–frequency $(k - \omega)$ diagram, and that the two peaks observed by Frazier correspond to the first two harmonics (*normal modes*). He formulated the conditions for observing the discrete acoustic modes: observing runs must be longer than 1 h, must cover a sufficiently large region of, at least, 60,000 km in size; the Doppler velocity images must have a spatial resolution of 3,000 km, and be taken at least every 1 min.

At that time the observing runs were very short, typically, 30–40 min. Only in 1974–1975 Deubner [6] was able to obtain three 3-h sets of observations using a magnetograph of the Fraunhofer Institute in Anacapri. He measured Doppler velocities along a $\sim$220, 000 km line on the solar disk by scanning it periodically at 110 s intervals with the scanning steps of about 700 km. The Fourier analysis of these data provided the frequency–wavenumber diagram with three or four *mode ridges* in the oscillation *power spectrum* that represents the squared amplitude of the Fourier components as a function of wavenumber and frequency. Deubner's results provided unambiguous confirmation of the idea that the 5-min oscillations observed on the solar surface represent the standing waves or resonant acoustic modes trapped below the surface. The lowest ridge in the diagram is easily identified as the *surface gravity wave* because its frequencies depend only on the wavenumber and surface gravity. The ridge above is the first acoustic mode, a standing acoustic waves that have one node along the radius. The ridge above this corresponds to the second acoustic modes with two nodes, and so on.

While these observations showed a remarkable qualitative agreement with Ulrich's theoretical prediction, the observed power ridges in the $k - \omega$ diagram were systematically lower than the theoretical mode lines. Soon after, in 1975, Rhodes et al. [7] made independent observations at the vacuum solar telescope at the Sacramento Peak Observatory and confirmed the observational results. They also calculated the theoretical mode frequencies for various solar models, and by comparing these with the observations determined the limits on the depth of the solar convection zone. This, probably, was the first helioseismic inference.

However, it was believed that the acoustic (p)-modes do not provide much information about the solar interior because detailed theoretical calculations of their properties by Ando and Osaki [8] showed that while these mode are determined by interior resonances their amplitudes (*eigenfunctions*) are predominantly concentrated close to the surface. Therefore, the main focus was shifted to observations and analysis of global oscillations of the Sun with periods much longer than 5 min. This task was particularly important for explaining the observed deficit of high-energy *solar neutrinos* [9], which could be either due to a low temperature (or heavy element abundance—low metallicity) in the energy-generating core or due to neutrino oscillations.

In 1975, Hill et al. [10] reported on the detection of oscillations in their measurements of solar oblateness. The periods of these oscillations were between 10 and 40 min. They suggested that the oscillation signals might correspond to global modes of the Sun. Independently, in 1976, two groups, led by Severny at the Crimean Observatory [11] and Isaak at the University of Birmingham [12] found long-period oscillations in global-Sun Doppler velocity signals. The oscillation with a period of 160 min was particularly prominent and stable. The amplitude of this oscillation was estimated close to 2 m/s. Later this oscillation was found in observations at the Wilcox Solar Observatory [13] and at the geographical South Pole [14]. Despite significant efforts to identify this oscillation among the solar resonant modes or find a physical explanation these results remain a mystery. This oscillation lost the amplitude and coherence in the subsequent ground-based measurements and was not found in later observations from SOHO spacecraft [15]. The period of this oscillation was extremely close to 1/9 of a day, and likely was related to terrestrial observing conditions.

Nevertheless, these studies played a very important role in development of helioseismology and emphasized the need for long-term stable and high-accuracy observations from the ground and space. Attempts to detect long-period oscillations (*g-modes*) still continue. However, the focus of helioseismology was shifted to accurate measurements and analysis of the acoustic *p-modes* discovered by Leighton.

The next important step was made in 1979 by the Birmingham group [16]. They observed the Doppler velocity variations integrated over the whole Sun for about 300 h (but typically 8 h a day) at two observatories, Izana, on Tenerife, and Pic du Midi in the Pyrenees. In the power spectrum of 5-min oscillations they detected several equally spaced lines corresponding to global (*low-degree*) acoustic modes, radial, dipole and quadrupole (in terms of the angular degree these are labeled as $\ell = 0, 1$, and 2). Unlike, the previously observed local short-horizontal-wavelength acoustic modes these oscillations propagate into the deep interior and provide information about the structure of the solar core. The estimated frequency spacing between the modes was $67.8\,\mu$Hz. This uniform spacing predicted theoretically by Vandakurov [17] in the framework of a general stellar oscillation theory corresponds to the inverse time that takes for acoustic waves to travel from the surface of the Sun through the center to the opposite side and come back. Thus, the frequency spacing immediately gives an important constraint on the internal structure of the Sun. An initial comparison with the solar models [18, 19] showed that the observed spectrum is consistent with the spectrum of solar models with low metallicity. This result was very exciting because it would provide a solution to the solar neutrino problem. Thus, the determination of solar metallicity (or *heavy element abundance*) became a central problem of helioseismology.

In the same year, 1979, Grec et al. [14] made 5-day continuous measurements at the Amundsen–Scott Station at the South Pole of the global oscillations and confirmed the Birmingham result. Also, they were able to resolve the fine structure of the oscillation spectrum and in addition to the main $67.8\,\mu$Hz spacing (*large frequency separation*) between the strongest peaks of $\ell = 1$ and 2, observe a small $10–16\,\mu$Hz splitting (*small separation*) between the $\ell = 0$ and 2, and $\ell = 1$ and 3 modes. The small separation is mostly sensitive to the central part of the Sun and provides additional diagnostic power.

The comparison of the observed oscillation peaks in the frequency power spectra with the p-mode frequencies calculated for solar models showed that below the surface these oscillations correspond to the standing waves with a large number of nodes along the radius (or *high radial order*). The number of nodes is between 10 and 35, and it was difficult to determine the precise values for the observed modes. This created an uncertainty in the helioseismic determination of the heavy element abundance. Christensen-Dalsgaard and Gough [20] pointed out that while the South Pole and new Birmingham data favor solar models with normal metallicity the low metallicity models cannot be ruled out.

The uncertainty was resolved three years later in 1983 when Duvall and Harvey [21] analyzed the Doppler velocity data measured with a photo-diode array in 200 positions along the North–South direction on the disk, and obtained the diagnostic $k - \omega$ diagram for acoustic modes of degree ℓ, from 1 to 110. This allowed them to connect in the diagnostic diagram the global low-ℓ modes with the high-ℓ modes observed by Deubner. Since the correspondence of the ridges on Deubner's diagram to solar oscillation modes have been determined it was easy to identify the low-ℓ modes by simply counting the ridges corresponding to the low-ℓ frequencies. It turned out that these modes are indeed in the best agreement with the normal metallicity solar model. This result had important implications for the solar neutrino problem because it strongly indicated that the observed deficit of solar neutrinos was not due to a low abundance of heavy elements on the Sun but because of changes in neutrino properties (neutrino oscillations) on their way from the energy-generating core to the Earth. This was later confirmed by direct measurements of solar neutrino properties [22].

It was also important that the definite identification of the observed solar oscillations in terms of normal oscillation modes provided a solid foundation for developing diagnostic methods of helioseismology based on the well-developed mathematical theory of non-radial oscillations of stars [23–25]. This theory provided means for calculating eigenfrequencies and eigenfunctions of normal modes for spherically symmetric stellar models. Mathematically, the problem is reduced to solving a non-linear eigenvalue problem for a fourth-order system of differential equations. This system has two sequences of eigenvalues corresponding to p- and g-modes, and also a degenerate solution, corresponding to f-modes (surface gravity waves). The effects of rotation, asphericity and magnetic fields are usually small and considered by a perturbation theory [26–29].

An important prediction of the oscillation theory is that rotation causes splitting of normal mode frequencies. Without rotation, the normal mode frequencies are degenerate with respect to the azimuthal wavenumber, m, that is, the modes of the angular degree, l, and radial order, n, have the same frequencies irrespective of the azimuthal (longitudinal) wavelength. The stellar rotation removes this degeneracy. Obviously, it does not affect the axisymmetrical ($m=0$) modes, but the frequencies of non-axisymmetrical modes are split. Generally, these modes can be represented as a superposition of two waves running around a star in two opposite directions (prograde and retrograde waves). Without rotation, these modes have the same frequencies and, thus, the same phase speed. In this case, they form a standing wave. However, rotation increases the speed of the prograde wave and decreases the speed of retrograde wave. This results in an increase of the eigenfrequency of the prograde mode, and a frequency decrease of the retrograde mode. This phenomenon is similar to frequency shifts due to the Doppler effect. It is called rotational frequency splitting.

The rotational frequency splitting was first observed by Rhodes, Ulrich and Deubner [30–32]. These measurements provided first evidence that the rotation rate of the Sun is not uniform but increases with depth. The rotational splitting was initially measured for high-degree modes, but then the measurements were extended to the medium- and low-degree range by Duvall and Harvey [33, 34], who made a long continuous series of helioseismology observations at the South Pole. The internal differential rotation law was determined from the data of Brown and Morrow [35]. It was found that the differential latitudinal rotation is confined in the convection zone, and that the radiative interior rotates almost uniformly, and also is slower in the equatorial region than in the convective envelope [36, 37]. Such rotation law was not expected from theories of stellar rotation, which predicted that the stellar cores rotate faster than the envelopes [38]. The knowledge of the Sun's internal rotation law is of particular importance for understanding the dynamo mechanism of magnetic field generation [39].

It became clear that long uninterrupted observations are essential for accurate inferences of the internal structure and rotation of the Sun. Therefore, the observational programs focused on development of global helioseismology networks, GONG [40] and BiSON [41, 42], and also the Solar and Heliospheric Observatory (SOHO) space mission [43]. These projects provided almost continuous coverage for helioseismic observations and also stimulated development of new sophisticated data analysis and inversion techniques.

In addition, the Michelson Doppler Imager (MDI) instrument on SOHO [44] and the GONG+ network, upgraded to higher spatial resolution [45], provided excellent opportunities for developing *local helioseismology*, which provides tools for 3D imaging of the solar interior. The local helioseismology methods are based on measurements of local oscillation properties, such as frequency shift in local areas or variations of travel times.

The idea of using local frequency shifts for inferring the subsurface flows was suggested by Gough and Toomre in 1983 [46]. The method is now called *ring-diagram analysis* [47], because the dispersion relation of solar oscillations forms rings in the horizontal wavenumber plane at a given frequency. It measures shifts of these rings, which are then converted into frequency shifts.

Ten years later, Duvall and his colleagues [48] introduced *time–distance helioseismology* method. In this method, they suggested to measure travel times of acoustic waves from a cross-covariance function of solar oscillations. This function is obtained by cross-correlating oscillation signals observed at two different points on the solar surface for various time lags. When the time lag in the calculations coincides with the travel time of acoustic waves between these points the cross-covariance function shows a maximum. This method provided means for developing *acoustic tomography* techniques [49, 50] for imaging 3D structures and flows with the high-resolution comparable to the oscillation wavelength. These and other methods of local area helioseismology [51, 52] have provided important results on the convective and large-scale flows, and also on the structure and evolution of sunspots and active regions. Their development continues.

The SOHO mission and the GONG network were primarily designed for observing solar oscillation modes of low- and medium-degree, needed for global helioseismology. Local helioseismology requires high-resolution observations of high-degree modes. Because of the telemetry constraints such data are available uninterruptedly from the MDI instrument on SOHO only for 2 months every year. These data provided only snapshots of the subsurface structures and dynamics associated with the solar activity. In order to fully investigate the evolving magnetic activity of the Sun, a new space mission Solar Dynamics Observatory (SDO) was launched on February 11, 2010. It carries the Helioseismic and Magnetic Imager (HMI) instrument, which provides continuous 4096×4096-pixel full-disk images of solar oscillations. These data open new opportunities for investigation the solar interior by local helioseismology [53].

In the modern helioseismology, a very important role is played by *numerical simulations*. Both, global and local helioseismology analyses employ relatively simple analysis the observational data and performing inversions of the fitted frequencies and travel times. For instance, the global helioseismology methods assume that the structures and flows on the Sun are axisymmetrical, and infer only the axisymmetrical components of the sound speed and velocity field. The local helioseismology methods are based on a simplified physics of wave propagation on the Sun. The ring-diagram analysis makes an assumption that the perturbations and flows are horizontally uniform within the area used for calculating the wave dispersion relation, 5–15 heliographic degrees, while a typical size of sunspots is about 1–2°. Most of the time–distance helioseismology inversions are based on a ray-path approximation and ignore the finite wavelength effects that become important at small scales, comparable with the wavelength. Also, all the methods, global and local, do not take into account effects of solar magnetic fields.

Properties of solar oscillations dramatically change in regions of strong magnetic field. In particular, the excitation of oscillations is suppressed in sunspots because the strong magnetic field inhibits convection that drives the oscillations. The magnetic stresses may cause anisotropy of wave speed and lead to transformation of acoustic waves into various MHD type waves. These and other effects have to be investigated and taken into account in the data analysis and inversion procedures. Because of the complexity, these processes can be fully investigated only numerically. The numerical simulations of subsurface solar convection and oscillations were pioneered by Stein and Nordlund [54]. These 3D radiative MHD simulations include all essential physics and provide important insights into the physical processes below the visible surface and also artificial data for helioseismology testing. This type of so-called "realistic" simulations has been used for testing time–distance helioseismology inferences [55], and continues being developed using modern turbulence models [56]. In addition, various aspects of wave propagation and interaction with magnetic fields are studied by solving numerically linearized MHD equations (e.g. [57–59]). The numerical simulations become an important tool for verification and testing of the helioseismology methods and inferences.

1.3 Basic Properties of Solar Oscillations

1.3.1 Oscillation Power Spectrum

The theoretical spectrum of solar oscillation modes shown in Fig. 1.2 covers a wide range of frequencies and angular degrees. It includes oscillations of three types: *acoustic (p) modes, surface gravity (f) modes* and *internal gravity (g) modes*. In this spectrum, the modes are organized in a series of curves corresponding to different overtones of non-radial modes, which are characterized by the number of nodes along the radius (or by the radial order, n). The angular degree, l, of the corresponding spherical harmonics describes the horizontal wave number (or inverse horizontal wavelength). The p-modes cover the frequency range from 0.3 to 5 mHz (or from 3 to 55 min in oscillation periods). The low frequency limit corresponds to the first radial harmonic, and the upper limit is set by the acoustic cut-off frequency of the solar atmosphere. The g-modes frequencies have an upper limit corresponding to the maximum Brunt–Väisälä frequency ($\sim$0.45 mHz) in the radiative zone and occupy the low-frequency part of the spectrum. The intermediate frequency range of 0.3–0.4 mHz at low angular degrees is a region of mixed modes. These modes behave like g-modes in the deep interior and like p-modes in the outer region. The apparent crossings in this diagram are not the actual crossings: the mode branches become close in frequencies but do not cross each other. At these points the mode exchange their properties, and the mode branches are diverted. For instance, the f-mode ridge stays above the g-mode lines. A similar phenomenon is known in quantum mechanics as *avoided crossing*.

So far, only the upper part of the solar oscillation spectrum is observed. The lowest frequencies of detected p- and f-modes are about 1 mHz. At lower frequencies the mode amplitudes decrease below the noise level, and the modes become unobservable. There have been several attempts to identify low-frequency p-modes or even g-modes in the noisy spectrum, but so far these results are not convincing.

The observed power spectrum is shown in Fig. 1.3. The lowest ridge is the f-mode, and the other ridges are p-modes of the radial order, n, starting from $n = 1$. The ridges of the oscillation modes disappear in the convective noise at frequencies below 1 mHz. The power spectrum is obtained from the SOHO/MDI data, representing 1024×1024-pixel images of the line-of-sight (Doppler) velocity of the solar surface taken every minute without interruption. When the oscillations are observed in the integrated solar light ("Sun-as-a-star") then only the modes of low angular

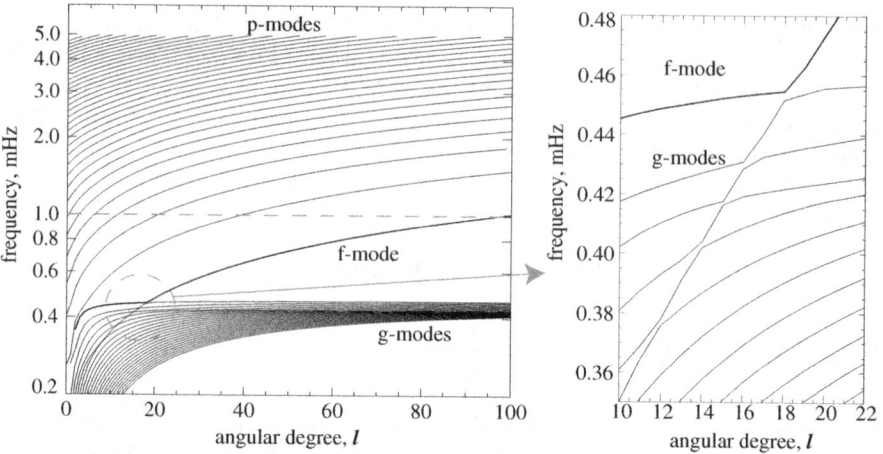

Fig. 1.2 Theoretical frequencies of solar oscillation modes calculated for a standard solar model for the range of angular degree l from 0 to 100, and for the frequency range from 0.2 to 5 mHz. The solid curves connect modes corresponding to the different oscillation overtones (radial orders). The *dashed grey horizontal line* indicate the low-frequency observational limit: only the modes above this line have been reliably observed. The *right panel* shows an area of the avoided crossing of *f*- and *g*-modes (indicated by the *gray dashed circle* in left panel)

Fig. 1.3 Power spectrum obtained from a 6-day long time series of solar oscillation data from the MDI instrument on SOHO in 1996 (ν is the cyclic frequency of the oscillations, l is the angular degree, λ_h is the horizontal wavelength in megameters)

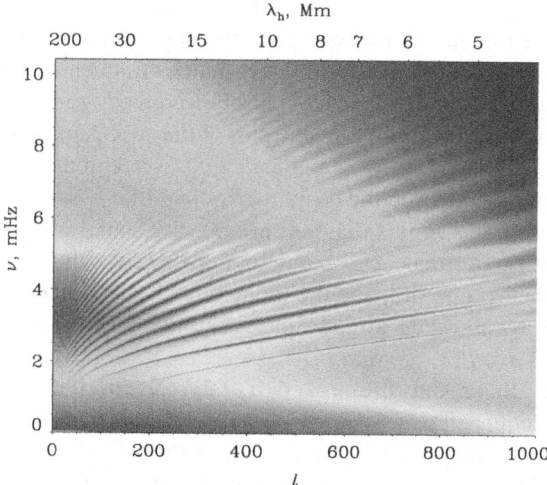

degree are detected in the power spectrum (Fig. 1.4). These modes have a mean period of about 5 min, and represent *p*-modes of high radial order n modes. The n-values of these modes can be determined by tracing in Fig. 1.3 the high-n ridges of the high-degree modes into the low-degree region. This provides unambiguous identification of the low-degree solar modes. Obviously, the mode identification is much more difficult for spatially unresolved oscillations of other stars.

Fig. 1.4 Power spectral density (PSD) of low-degree solar oscillations, obtained from the integrated light observations (Sun-as-a-star) by the GOLF instrument on SOHO, from 11/04/1996 to 08/07/2008

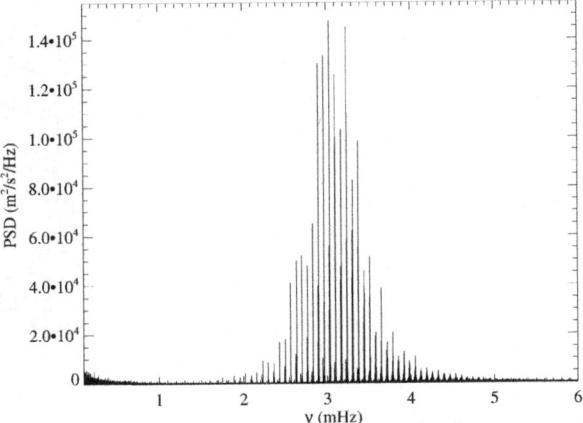

1.3.2 Excitation by Turbulent Convection

Observations and numerical simulations have shown that solar oscillations are driven by turbulent convection in a shallow subsurface layer with a superadiabatic stratification, where convective velocities are the highest. However, details of the stochastic excitation mechanism are not fully established. Solar convection in the superadiabatic layer forms small-scale granulation cells. Analysis of the observations and numerical simulations has shown that sources of solar oscillations are associated with strong downdrafts in dark intergranular lanes [60]. These downdrafts are driven by radiative cooling and may reach near-sonic velocity of several kilometers per second. This process has features of convective collapse [61].

Calculations of the work integral for acoustic modes using the realistic numerical simulations of Stein and Nordlund [62] have shown that the principal contribution to the mode excitation is provided by turbulent Reynolds stresses and that a smaller contribution comes from non-adiabatic pressure fluctuations. Because of the very high Reynolds number of the solar dynamics the numerical modeling requires an accurate description of turbulent dissipation and transport on the numerical subgrid scale. The recent radiative hydrodynamics modeling using the Large-Eddy Simulations (LES) approach and various subgrid scale (SGS) formulations [56] showed that among these formulations the most accurate description in terms of the total amount of the stochastic energy input to the acoustic oscillations is provided by a dynamic Smagorinsky model [63, 64] (Fig. 1.5a).

The observations show that the modal lines in the oscillation power spectrum are not Lorentzians but display a strong asymmetry [67, 68]. Curiously, the asymmetry has the opposite sense in the power spectra calculated from Doppler velocity and intensity oscillations. The asymmetry itself can be easily explained by interference of waves emanated by a localized source [69], but the asymmetry reversal is surprising and indicates on a complicated radiative dynamics of the excitation process. The reversal has been attributed to a *correlated noise* contribution to the observed intensity

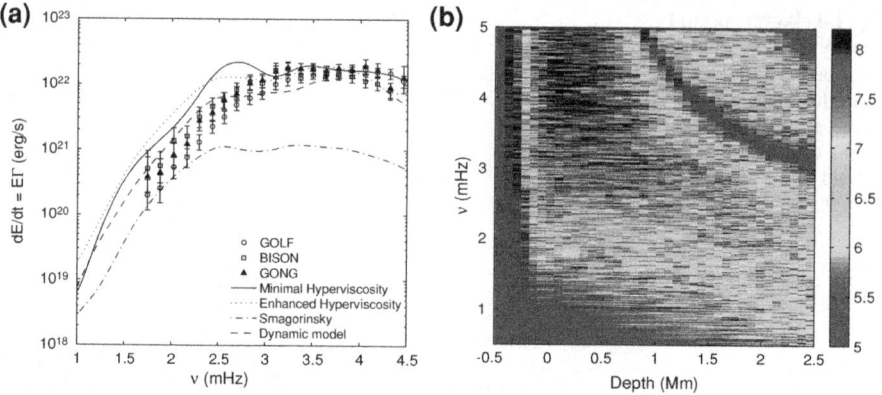

Fig. 1.5　a Comparison of observed and calculated rate of stochastic energy input to modes for the entire solar surface (in erg s^{-1}). *Different curves* show the numerical simulation results obtained for four turbulence models: hyperviscosity (*solid*), enhanced hyperviscosity (*dots*), Smagorinsky (*dash-dots*), and dynamic model (*dashes*). Observed distributions: *circles* SOHO–GOLF, *squares* BISON, and *triangles* GONG for $l = 1$ [65]. **b** Logarithm of the work integrand in units of erg cm^{-2} s^{-1}, as a function of depth and frequency from numerical simulations with the dynamic turbulence model [66]

oscillations [70], but the physics of this effect is still not fully understood. However, it is clear that the line shape of the oscillation modes and the phase-amplitude relations of the velocity and intensity oscillations carry substantial information about the excitation mechanism and, thus, require careful data analysis and modeling.

1.3.3 Line Asymmetry and Pseudo-modes

Figure 1.6 shows the power spectrum for oscillations of the angular degree, $l = 200$, obtained from the SOHO/MDI Doppler velocity and intensity data [70]. The line asymmetry is apparent, particularly, at low frequencies. In the velocity spectrum, there is more power in the low-frequency wings than in the high-frequency wings of the spectral lines. In the intensity spectrum, the distribution of power is reversed. The data also show that the asymmetry varies with frequency. It is the strongest for the *f*-mode and low-frequency *p*-mode peaks. At higher frequencies the peaks become more symmetrical, and extend well above *the acoustic cut-off frequency* (1.51), which is ∼5–5.5 mHz.

Acoustic waves with frequencies below the cut-off frequency are completely reflected by the surface layers because of the steep density gradient. These waves are trapped in the interior, and their frequencies are determined by the resonant conditions, which depend on the solar structure. But the waves with frequencies above the cut-off frequency escape into the solar atmosphere. Above this frequency the power spectrum peaks correspond to so-called "pseudo-modes". These are caused by con-

Fig. 1.6 Power spectra of
$l = 200$ modes obtained
from SOHO/MDI
observations of **a** Doppler
velocity, **b** continuum
intensity [70]

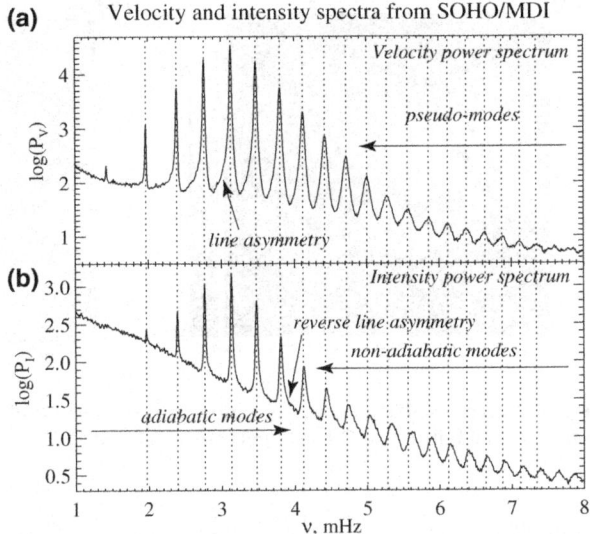

structive interference of acoustic waves excited by the sources located in the granulation layer and traveling upward, and by the waves traveling downward, reflected in the deep interior and arriving back to the surface. Frequencies of these modes are no longer determined by the resonant conditions of the solar structure. They depend on the location and properties of the excitation source ("source resonance"). The pseudo-mode peaks in the velocity and intensity power spectra are shifted relative to each other by almost a half-width. They are also slightly shifted relative to the normal mode peaks although they look like a continuation of the normal-mode ridges in Figs. 1.3 and 1.7. This happens because the excitation sources are located in a shallow subsurface layer, which is very close to the reflection layers of the normal modes. Changes in the frequency distributions below and above the acoustic cut-off frequency can be easily noticed by plotting the frequency differences along the modal ridges.

The asymmetrical profiles of normal-mode peaks are also caused by the localized excitation sources. The interference signal between acoustic waves traveling from the source upwards, and the waves traveling from the source downward and coming back to the surface after the internal reflection depends on the wave frequency. Depending on the multipole type of the source the interference signal can be stronger at frequencies lower or higher than the resonant normal frequencies, thus resulting in asymmetry in the power distribution around the resonant peak. Calculations of Nigam et al. [70] showed that the asymmetry observed in the velocity spectra and the distribution of the pseudo-mode peaks can be explained by a composite source consisting of a monopole term (mass term) and a dipole term (force due to Reynolds stress) located in the zone of superadiabatic convection at a depth of $\simeq 100$ km below the photosphere. In this model, the reversed asymmetry in the intensity power spectra

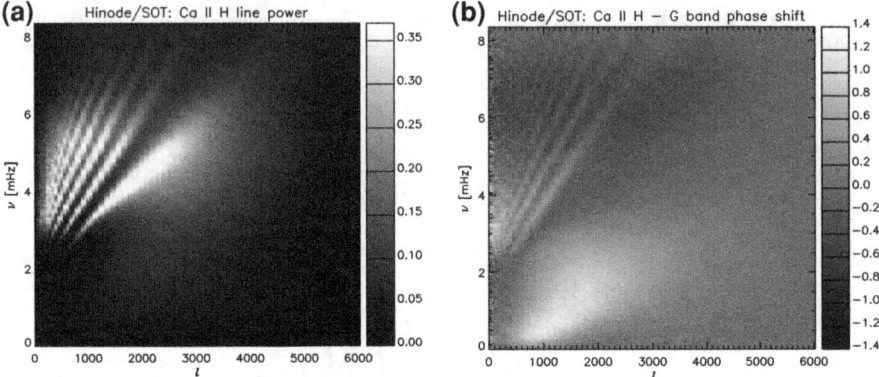

Fig. 1.7 a The oscillation power spectrum from HINODE CaII H line observations. **b** The phase shift between CaII H and G-band (units are in radians) [71]

is explained by effects of a correlated noise added to the oscillation signal through fluctuations of solar radiation during the excitation process. Indeed, if the excitation mechanism is associated with the high-speed turbulent downdrafts in dark lanes of granulation the local darkening contributes to the intensity fluctuations caused by excited waves. The model also explains the shifts of pseudo-mode frequency peaks and their higher amplitude in the intensity spectra. The difference between the correlated and uncorrelated noise is that the correlated noise has some phase coherence with the oscillation signal, while the uncorrelated noise has no coherence.

While this scenario looks plausible and qualitatively explains the main properties of the power spectra, details of the physical processes are still uncertain. In particular, it is unclear whether the correlated noise affects only the intensity signal or both the intensity and velocity. It has been suggested that the velocity signal may have a correlated contribution due to convective overshoot [72]. Attempts to estimate the correlated noise components from the observed spectra have not provided conclusive results [73, 74]. Realistic numerical simulations [75] have reproduced the observed asymmetries and provided an indication that radiation transfer plays a critical role in the asymmetry reversal.

Recent high-resolution observations of solar oscillations simultaneously in two intensity filters, in molecular G-band and CaII H line, from the HINODE space mission [76, 77] revealed significant shifts in frequencies of pseudo-modes observed in the CaII H and G-band intensity oscillations [71]. The phase of the cross-spectrum of these oscillations shows peaks associated with the p-mode lines but no phase shift for the f-mode (Fig. 1.7b). The p-mode properties can be qualitatively reproduced in a simple model with a correlated background if the correlated noise level in the CaII H data is higher than in the G-band data [71]. Perhaps, the same effect can explain also the frequency shift of pseudo-modes. The CaII H line is formed in the lower chromosphere while the G-band signal comes from the photosphere. But how this may lead to different levels of the correlated noise is unclear.

The HINODE results suggest that multi-wavelength observations of solar oscillations, in combination with the traditional intensity-velocity observations, may help to measure the level of the correlated background noise and to determine the type of wave excitation sources on the Sun. This is important for understanding the physical mechanism of the line asymmetry and for developing more accurate models and fitting formulae for determining the mode frequencies [78].

In addition, HINODE provided observations of non-radial acoustic and surface gravity modes of very high angular degree. These observations show that the oscillation ridges are extended up to $l \simeq 4000$ (Fig. 1.7a). In the high-degree range, $l \geq 2500$ frequencies of all oscillations exceed the acoustic cut-off frequency. The line width of these oscillations dramatically increases, probably due to strong scattering on turbulence [79, 80]. Nevertheless, the ridge structure extending up to 8 mHz (the Nyquist frequency of these observations) is quite clear. Although the ridge slope clearly changes at the transition from the normal modes to the pseudo-modes.

1.3.4 Magnetic Effects: Sunspot Oscillations and Acoustic Halos

In general, the main factors causing variations in oscillation properties in magnetic regions, can be divided in two types: direct and indirect. The direct effects are due to additional magnetic restoring forces that can change the wave speed and may transform acoustic waves into different types of MHD waves. The indirect effects are caused by changes in convective and thermodynamic properties in magnetic regions. These include depth-dependent variations of temperature and density, large-scale flows, and changes in wave source distribution and strength. Both direct and indirect effects may be present in observed properties such as oscillation frequencies and travel times, and often cannot be easily disentangled by data analyses, causing confusions and misinterpretations. Also, one should keep in mind that simple models of MHD waves derived for various uniform magnetic configurations and without stratification or with a polytropic stratification may not provide correct explanations to solar phenomena. In this situation, numerical simulations play an important role in investigations of magnetic effects.

Observed changes of oscillation amplitude and frequencies in magnetic regions are often explained as a result of wave scattering and conversion into various MHD modes. However, recent numerical simulations helped us to understand that magnetic fields not only affect the wave dispersion properties but also the excitation mechanism. In fact, changes in excitation properties of turbulent convection in magnetic regions may play a dominant role in observed phenomena.

1.3.4.1 Sunspot Oscillations

For instance, it is well-known that the amplitude of 5-min oscillations is substantially reduced in sunspots. Observations show that more waves are coming into a sunspot

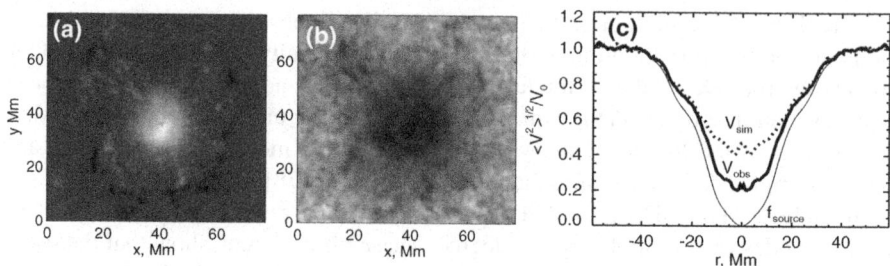

Fig. 1.8 **a** Line-of-sight magnetic field map of a sunspot (AR8243); **b** oscillation amplitude map; **c** profiles of rms oscillation velocities at frequency 3.65 mHz for observations (*thick solid curve*) and simulations (*dashed curve*); the *thin solid curve* shows the distribution of the simulated source strength [83]

than going out of the sunspot area (e.g. [81]). This is often attributed to absorption of acoustic waves in magnetic field due to conversion into slow MHD modes traveling along the field lines (e.g. [82]). However, since convective motions are inhibited by the strong magnetic field of sunspots, the excitation mechanism is also suppressed. 3D numerical simulations of this effect have shown that the reduction of acoustic emissivity can explain at least 50% of the observed power deficit in sunspots (Fig. 1.8) [83].

Another significant contribution comes from the amplitude changes caused by variations in the background conditions. Inhomogeneities in the sound speed may increase or decrease the amplitude of acoustic wave traveling through these inhomogeneities. Numerical simulations of MHD waves using magnetostatic sunspot models show that the amplitude of acoustic waves traveling through a sunspot decreases when the wave is inside the sunspot and then increases when the wave comes out of the sunspot [84]. Simulations with multiple random sources show that these changes in the wave amplitude together with the suppression of acoustic sources can explain most of the observed deficit of the power of 5-min oscillations. Thus, the role of the MHD mode conversion may be insignificant for explaining the power deficit of 5-min photospheric oscillations in sunspots. However, the mode conversion is expected to be significant higher in the solar atmosphere where magnetic forces become dominant.

We should note that while the 5-min oscillations in sunspots come mostly from outside sources there are also 3-min oscillations, which are probably intrinsic oscillations of sunspots. The origin of these oscillations is not yet understood. They are probably excited by a different mechanism operating in strong magnetic field.

HINODE observations added new puzzles to sunspot oscillations. Figure 1.9 shows a sample Ca II H intensity and the relative intensity power maps averaged over 1 mHz intervals in the range from 1 to 7 mHz with logarithmic greyscaling [85]. In the Ca II H power maps, in all the frequency ranges, there is a small area (∼6 arcsec in diameter) near the center of the umbra where the power was suppressed. This type of 'node' has not been reported before. Possibly, the stable high-resolution

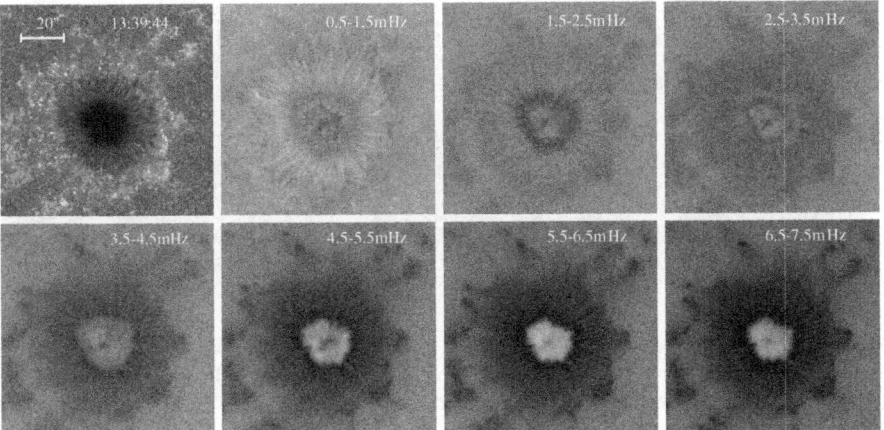

Fig. 1.9 CaII H intensity image from HINODE observations (*top-left*) and the corresponding power maps from CaII H intensity data in five frequency intervals of active region NOAA 10935. The field of view is 100 arcsec square in all the panels. The power is displayed in logarithmic greyscaling [85]

observation made by HINODE/SOT was required to find such a tiny node, although analysis of other sunspots indicates that probably only a particular type of sunspots, e.g., round ones with axisymmetric geometry, exhibit such node-like structure. Above 4 mHz in the Ca II H power maps, power in the umbra is remarkably high. In the power maps averaged over narrower frequency range (0.05 mHz wide, not shown), the region with high power in the umbra seems to be more patchy. This may correspond to elements of umbral flashes, probably caused by overshooting convective elements [86]. The Ca II H power maps show a bright ring in the penumbra at lower frequencies. It probably corresponds to the running penumbral waves. The power spectrum in the umbra has two peaks: one around 3 mHz and the other around 5.5 mHz. The high-frequency peak is caused by the oscillations that excited only in the strong magnetic field of sunspots. The origin of these oscillations is not known yet.

1.3.4.2 Acoustic Halos

In moderate magnetic field regions, such as plages around sunspot regions, observations reveal enhanced emission at high frequencies, 5–7 mHz (with period ∼3 min) [87]. Sometimes this emission is called the "acoustic halo" (Fig. 1.10c). There have been several attempts to explain this effect as a result of wave transformation or scattering in magnetic structures (e.g. [88, 89]). However, numerical simulations show that magnetic field can also change the excitation properties of solar granulation resulting in an enhanced high-frequency emission. In particular, the radiative MHD simulations of solar convection [66] in the presence

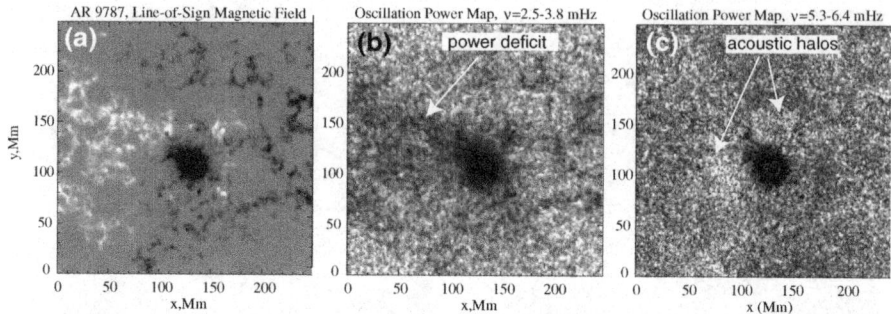

Fig. 1.10 a Line-of-sight magnetic field map of active region NOAA 9787 observed from SOHO/MDI on Jan. 24, 2002 and averaged over a 3-h period; **b** oscillation power map from Doppler velocity measurements for the same period in the frequency 2.5–3.8 mHz; **c** power map for 5.3–6.4 mHz

of vertical magnetic field have shown that the magnetic field significantly changes the structure and dynamics of granulations, and thus the conditions of wave excitation. In magnetic field the granules become smaller, and the turbulence spectrum is shifted towards higher frequencies. This is illustrated in Fig. 1.11, which shows the frequency spectrum of the horizontally averaged vertical velocity. Without a magnetic field the turbulence spectrum declines sharply at frequencies above 5 mHz, but in the presence of magnetic field it develops a plateau. In the plateau region characteristic peaks (corresponding to the "pseudo-modes") appear in the spectrum for moderate magnetic field strength of about 300–600 G. These peaks may explain the effect of the "acoustic halo". Of course, more detailed theoretical and observational studies are required to confirm this mechanism. In particular, multi-wavelength observations of solar oscillations at several different heights would be important. Investigation of the excitation mechanism in magnetic regions is also important for interpretation of the variations of the frequency spectrum of low-degree modes on the Sun, and for asteroseismic diagnostics of stellar activity.

1.3.5 Impulsive Excitation: Sunquakes

"Sunquakes", the helioseismic response to solar flares, are caused by strong localized hydrodynamic impacts in the photosphere during the flare impulsive phase. The helioseismic waves have been observed directly as expanding circular-shaped ripples in SOHO/MDI Dopplergrams [90] (Fig. 1.12).

These waves can be detected in Dopplergram movies and as a characteristic ridge in time–distance diagrams (Fig. 1.13a), [90–93], or indirectly by calculating integrated acoustic emission [94–96]. Solar flares are sources of high-temperature plasma and strong hydrodynamic motions in the solar atmosphere. Perhaps, in all flares such

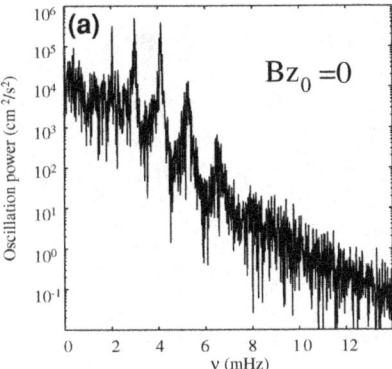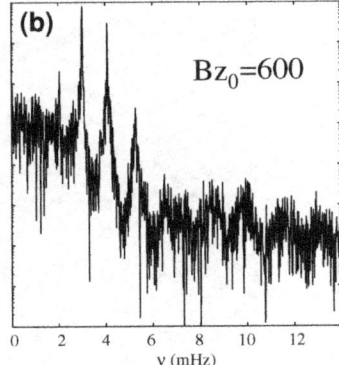

Fig. 1.11 Power spectra of the horizontally averaged vertical velocity at the visible surface for different initial vertical magnetic fields. The peaks on the top of the smooth background spectrum of turbulent convection represent oscillation modes: the sharp asymmetric peaks below 6 mHz are resonant normal modes, while the broader peaks above 6 mHz, which become stronger in magnetic regions, correspond to pseudo-modes [66]

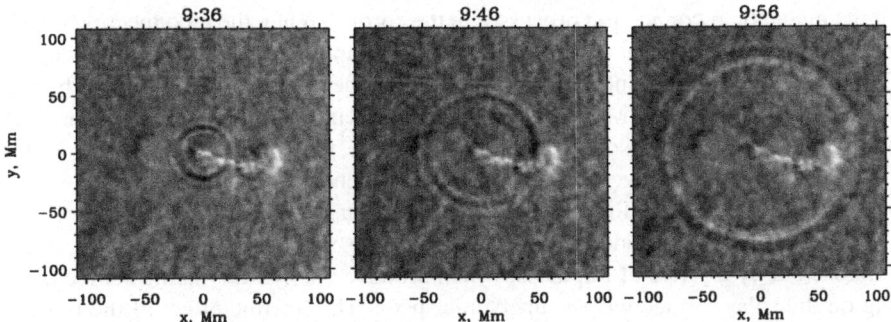

Fig. 1.12 Observations of the seismic response ("sunquakes") of the solar flare of 9 July, 1996, showing a sequence of Doppler-velocity images, taken by the SOHO/MDI instrument. The signal of expanding ripples is enhanced by a factor 4 in the these images

perturbations generate acoustic waves traveling through the interior. However, only in some flares is the impact sufficiently localized and strong to produce the seismic waves with the amplitude above the convection noise level. It has been established in the initial July 9, 1996, flare observations [90] that the hydrodynamic impact follows the hard X-ray flux impulse, and hence, the impact of high-energy electrons.

A characteristic feature of the seismic response in this flare and several others [91–93] is anisotropy of the wave front: the observed wave amplitude is much stronger in one direction than in the others. In particular, the seismic waves excited during the flares of 16 July, 2004, and 15 January, 2005, had the greatest amplitude in the direction of the expanding flare ribbons (Fig. 1.14). The wave anisotropy can be

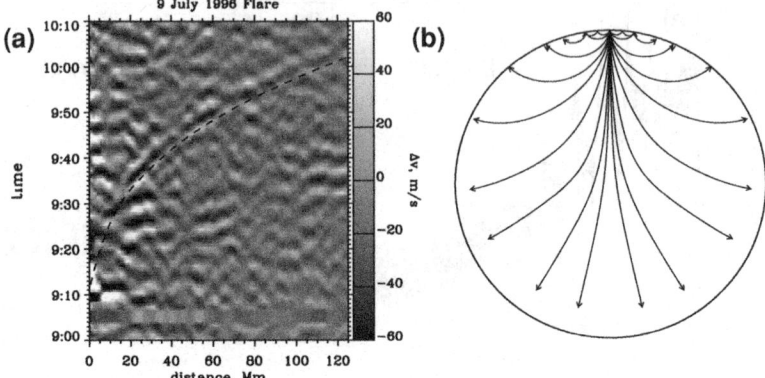

Fig. 1.13 a The time–distance diagram of the seismic response to the solar flare of 9 July, 1996. **b** Illustration of acoustic ray paths of the flare-excited waves traveling through the Sun

attributed to the moving source of the hydrodynamic impact, which is located in the flare ribbons [91, 93, 97]. The motion of flare ribbons is often interpreted as a result of the magnetic reconnection processes in the corona. When the reconnection region moves up it involves higher magnetic loops, the footpoints of which are further apart. The motion of the footpoints of impact of the high-energy particles is particularly well observed in the SOHO /MDI magnetograms showing magnetic transients moving with supersonic speed in some cases [92]. Of course, there might be other reasons for the anisotropy of the wave front, such as inhomogeneities in temperature, magnetic field and plasma flows. However, the source motion seems to be a key factor.

Therefore, we conclude that the seismic wave was generated not by a single impulse but by a series of impulses, which produce the hydrodynamic source moving on the solar surface with a supersonic speed. The seismic effect of the moving source can be easily calculated by convolving the wave Green's function with a moving source function. The result of these calculations is a strong anisotropic wavefront, qualitatively similar to the observations [97]. Curiously, this effect is quite similar to the anisotropy of seismic waves on Earth, when the earthquake rupture moves along the fault. Thus, taking into account the effects of multiple impulses of accelerated electrons and the moving source is very important for sunquake theories. The impulsive sunquake oscillations provide unique information about the interaction of acoustic waves with sunspots. Thus, these effects must be studied in more detail.

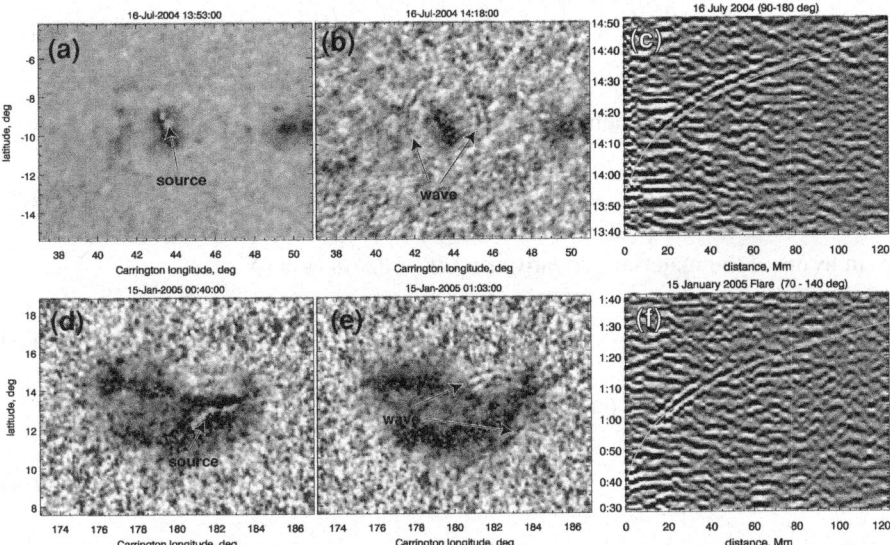

Fig. 1.14 Observations of the seismic response of the Sun ("sunquakes") to two solar flares: **a–c** X3 of 16 July, 2004, and **d–f** X1 flare of 15 January, 2005. The *left panels* show a superposition of MDI *white-light* images of the active regions and locations of the sources of the seismic waves determined from MDI Dopplergrams, the middle column shows the seismic waves, and the right panels show the time–distance diagrams of these events. The *thin yellow curves* in the right panels represent a theoretical time–distance relation for helioseismic waves for the standard solar model [93]

1.4 Global Helioseismology

1.4.1 Basic Equations

A simple theoretical model of solar oscillations can be derived using the following assumptions:

1. linearity: $v/c \ll 1$, where v is velocity of oscillating elements, c is the speed of sound;
2. adiabaticity: $dS/dt = 0$, where S is the specific entropy;
3. spherical symmetry of the background state;
4. magnetic forces and Reynolds stresses are negligible.

The basic governing equations are derived from the conservation of mass, momentum, energy and the Newton's gravity law. The conservation of mass (continuity equation) assumes that the rate of mass change in a fluid element of volume V is equal to the mass flux through the surface of this element (area A):

$$\frac{\partial}{\partial t} \int_V \rho dV = - \int_A \rho \boldsymbol{v} d\boldsymbol{a} = - \int_V \nabla(\rho \boldsymbol{v}) dV, \tag{1.1}$$

where ρ is the density. Then,

$$\frac{\partial \rho}{\partial t} + \nabla(\rho \boldsymbol{v}) = 0, \tag{1.2}$$

or in terms of the material derivative $d\rho/dt = \partial\rho/\partial t + \boldsymbol{v} \cdot \nabla\rho$:

$$\frac{d\rho}{dt} + \rho\nabla\boldsymbol{v} = 0. \tag{1.3}$$

The momentum equation (conservation of momentum of a fluid element) is:

$$\rho\frac{d\boldsymbol{v}}{dt} = -\nabla P + \rho \boldsymbol{g}, \tag{1.4}$$

where P is pressure, $\boldsymbol{g}$ is the gravity acceleration, which can be expressed in terms of gravitational potential Φ: $\boldsymbol{g} = \nabla\Phi$, $d\boldsymbol{v}/dt = \partial\boldsymbol{v}/\partial t + \boldsymbol{v} \cdot \nabla\boldsymbol{v}$ is the material derivative for the velocity vector. The adiabaticity equation (conservation of energy) for a fluid element is:

$$\frac{dS}{dt} = \frac{d}{dt}\left(\frac{P}{\rho^\gamma}\right) = 0, \tag{1.5}$$

or

$$\frac{dP}{dt} = c^2\frac{d\rho}{dt}, \tag{1.6}$$

where $c^2 = \gamma P/\rho$ is the squared adiabatic sound speed. The gravitational potential is calculated from the Poisson equation:

$$\nabla^2\Phi = 4\pi G\rho. \tag{1.7}$$

Now, we consider small perturbations of a stationary spherically symmetrical star in hydrostatic equilibrium:

$$v_0 = 0, \quad \rho = \rho_0(r), \quad P = P_0(r).$$

If $\boldsymbol{\xi}(t)$ is a vector of displacement of a fluid element then velocity $\boldsymbol{v}$ of this element:

$$\boldsymbol{v} = \frac{d\boldsymbol{\xi}}{dt} \approx \frac{\partial\boldsymbol{\xi}}{\partial t}. \tag{1.8}$$

Perturbations of scalar variables, ρ, P, Φ can be of two general types: Eulerian (denoted with prime symbol) at a fixed position $\boldsymbol{r}$:

$$\rho(\mathbf{r}, t) = \rho_0(r) + \rho'(\mathbf{r}, t),$$

and Lagrangian, measured in the moving element (denoted with δ):

$$\delta\rho(\mathbf{r} + \boldsymbol{\xi}) = \rho_0(r) + \delta\rho(\mathbf{r}, t). \tag{1.9}$$

The Eulerian and Lagrangian perturbations are related to each other:

$$\delta\rho = \rho' + (\boldsymbol{\xi} \cdot \nabla\rho_0) = \rho' + (\boldsymbol{\xi} \cdot \mathbf{e}_r)\frac{d\rho_0}{dr} = \rho' + \xi_r\frac{d\rho_0}{dr}, \tag{1.10}$$

where $\mathbf{e}_r$ is the radial unit vector.

In terms of the Eulerian perturbations and the displacement vector, $\boldsymbol{\xi}$, the linearized mass, momentum and energy equations can be expressed in the following form:

$$\rho' + \nabla(\rho_0\boldsymbol{\xi}) = 0, \tag{1.11}$$

$$\rho_0\frac{\partial \boldsymbol{v}}{\partial t} = -\nabla P' - g_0\mathbf{e}_r\rho' + \rho_0\nabla\Phi', \tag{1.12}$$

$$P' + \xi_r\frac{d P_0}{dr} = c_0^2\left(\rho' + \xi_r\frac{d\rho_0}{dr}\right), \tag{1.13}$$

$$\nabla^2\Phi' = 4\pi G\rho'. \tag{1.14}$$

The equations of solar oscillations can be further simplified by neglecting the perturbations of the gravitational potential, which gives relatively small corrections to theoretical oscillation frequencies. This is so-called Cowling approximation: $\Phi' = 0$.

Now, we consider the linearized equations in the spherical coordinate system, r, θ, ϕ. In this system, the displacement vector has the following form:

$$\boldsymbol{\xi} = \xi_r\mathbf{e}_r + \xi_\theta\mathbf{e}_\theta + \xi_\phi\mathbf{e}_\phi \equiv \xi_r\mathbf{e}_r + \boldsymbol{\xi}_h, \tag{1.15}$$

where $\boldsymbol{\xi}_h = \xi_\theta\mathbf{e}_\theta + \xi_\phi\mathbf{e}_\phi$ is the horizontal component of displacement. Also, we use the equation for divergence of the displacement (called dilatation):

$$\begin{aligned}
\nabla\boldsymbol{\xi} \equiv \mathrm{div}\boldsymbol{\xi} &= \frac{1}{r^2}\frac{\partial}{\partial r}(r^2\xi_r) + \frac{1}{r\sin\theta}\frac{\partial}{\partial\theta}(\sin\theta\xi_\theta) + \frac{1}{r\sin\theta}\frac{\partial\xi_\phi}{\partial\phi} \\
&= \frac{1}{r^2}\frac{\partial}{\partial r}(r^2\xi_r) + \frac{1}{r}\nabla_h\boldsymbol{\xi}_h.
\end{aligned} \tag{1.16}$$

We consider periodic perturbations with frequency $\omega : \boldsymbol{\xi} \propto \exp(i\omega t), \ldots$ Here, ω is the angular frequency measured in rad/s; it relates to the cyclic frequency, ν, which measures the number of oscillation cycles per second, as: $\omega = 2\pi\nu$.

Then, in the Cowling approximation, we obtain the following system of the linearized equations (omitting subscript 0 for unperturbed variables):

$$\rho' + \frac{1}{r^2} \frac{\partial}{\partial r}(r^2 \rho \xi_r) + \frac{\rho}{r} \nabla_h \boldsymbol{\xi}_h = 0, \tag{1.17}$$

$$-\omega^2 \rho \xi_r = -\frac{\partial P'}{\partial r} + g\rho', \tag{1.18}$$

$$-\omega^2 \rho \boldsymbol{\xi}_h = -\frac{1}{r} \nabla_h P', \tag{1.19}$$

$$\rho' = \frac{1}{c^2} P' + \frac{\rho N^2}{g} \xi_r, \tag{1.20}$$

where

$$N^2 = g \left(\frac{1}{\gamma P} \frac{dP}{dr} - \frac{1}{\rho} \frac{d\rho}{dr} \right) \tag{1.21}$$

is *the Brunt–Väisälä (or buoyancy) frequency.*

For the boundary conditions, we assume that the solution is regular at the Sun's center. This corresponds to the zero displacement, $\xi_r = 0$ at $r = 0$, for all oscillation modes except of the dipole modes of angular degree $l = 1$. In the dipole-mode oscillations the center of a star oscillates (but not the center of mass), and the boundary condition at the center is replaced by a regularity condition. At the surface, we assume that the Lagrangian pressure perturbation is zero: $\delta P = 0$ at $r = R$. This is equivalent to the absence of external forces. Also, we assume that the solution is regular at the poles $\theta = 0, \pi$.

We seek a solution of (1.17–1.20) by separation of the radial and angular variables in the following form:

$$\rho'(r, \theta, \phi) = \rho'(r) \cdot f(\theta, \phi), \tag{1.22}$$

$$P'(r, \theta, \phi) = P'(r) \cdot f(\theta, \phi), \tag{1.23}$$

$$\xi_r(r, \theta, \phi) = \xi_r(r) \cdot f(\theta, \phi), \tag{1.24}$$

$$\boldsymbol{\xi}_h(r, \theta, \phi) = \xi_h(r) \nabla_h f(\theta, \phi). \tag{1.25}$$

Then, in the continuity equation:

$$\left[\rho' + \frac{1}{r^2} \frac{\partial}{\partial r}(r^2 \rho \xi_r) \right] f(\theta, \phi) + \frac{\rho}{r} \xi_h \nabla_h^2 f = 0. \tag{1.26}$$

the radial and angular variables can be separated if

$$\nabla_h^2 f = \alpha f, \tag{1.27}$$

where α is a constant.

It is well-known that this equation has a non-zero solution regular at the poles $(\theta = 0, \pi)$ only when

$$\alpha = -l(l+1), \tag{1.28}$$

where l is an integer. This non-zero solution is:

$$f(\theta, \phi) = Y_l^m(\theta, \phi) \propto P_l^m(\theta)e^{im\phi}, \tag{1.29}$$

where $P_l^m(\theta)$ is the associated Legendre function of angular degree l and order m.

Then, the continuity equation for the radial dependence of the Eulerian density perturbation, $\rho'(r)$, takes the form:

$$\rho' + \frac{1}{r^2}\frac{\partial}{\partial r}\left(r^2 \rho \xi_r\right) - \frac{l(l+1)}{r^2}\rho \xi_h = 0. \tag{1.30}$$

The horizontal component of displacement ξ_h can be determined from the horizontal component of the momentum equation:

$$-\omega^2 \rho \xi_h(r) = -\frac{1}{r}P'(r), \tag{1.31}$$

or

$$\xi_h = \frac{1}{\omega^2 \rho r}P'. \tag{1.32}$$

Substituting this into the continuity equation (1.30) we get:

$$\rho\frac{d\xi_r}{dr} + \xi_h\frac{d\rho}{dr} + \frac{2}{r}\rho\xi_r + \frac{P'}{c^2} + \frac{\rho N^2}{g}\xi_r - \frac{L^2}{r^2\omega^2\rho}P' = 0, \tag{1.33}$$

where we define $L^2 = l(l+1)$.

Using the hydrostatic equation for the background (unperturbed) state, $dP/dr = -g\rho$, we finally obtain:

$$\frac{d\xi_r}{dr} + \frac{2}{r}\xi_r - \frac{g}{c^2}\xi_r + \left(1 - \frac{L^2 c^2}{r^2\omega^2}\right)\frac{P'}{\rho c^2} = 0, \tag{1.34}$$

or

$$\frac{d\xi_r}{dr} + \frac{2}{r}\xi_r - \frac{g}{c^2}\xi_r + \left(1 - \frac{S_l^2}{\omega^2}\right)\frac{P'}{\rho c^2} = 0, \tag{1.35}$$

where

$$S_l^2 = \frac{L^2 c^2}{r^2} \tag{1.36}$$

is the *Lamb frequency*.

Similarly, for the momentum equation we obtain:

$$\frac{dP'}{dr} + \frac{g}{c^2} P' + (N^2 - \omega^2)\rho\xi_r = 0. \tag{1.37}$$

The inner boundary condition at the Sun's center is:

$$\xi_r = 0, \tag{1.38}$$

or a regularity condition for $l = 1$.

The outer boundary condition at the surface $(r = R)$ is:

$$\delta P = P' + \frac{dP}{dr}\xi_r = 0. \tag{1.39}$$

Applying the hydrostatic equation, we get:

$$P' - g\rho\xi_r = 0. \tag{1.40}$$

Using the horizontal component of the momentum equation: $P' = \omega^2\rho r\xi_h$, the outer boundary condition (1.40) can be written in the following form:

$$\frac{\xi_h}{\xi_r} = \frac{g}{\omega^2 r}, \tag{1.41}$$

that is, the ratio of the horizontal and radial components of displacement is inverse proportional to the squared oscillation frequency. However, observations show that this relation is only approximate, presumably, because of the external force caused by the solar atmosphere.

Equations (1.35) and (1.37) with boundary conditions (1.38–1.40) constitute an eigenvalue problem for solar oscillation modes. This eigenvalue problem can be solved numerically for any solar or stellar model. The solution gives the frequencies, ω_{nl}, and the radial eigenfunctions, $\xi_r^{(n,l)}(r)$ and $P'^{(n,l)}(r)$, of the normal modes.

The radial eigenfunctions multiplied by the angular eigenfunctions (1.22–1.25) represented by the spherical harmonics (1.29) give 3D oscillation eigenfunctions of the normal modes, e.g.:

$$\xi_r(r, \theta, \phi, \omega) = \xi_r^{(n,l)}(r)Y_l^m(\theta, \phi). \tag{1.42}$$

Examples of such two eigenfunctions for p- and g-modes are shown in Fig. 1.15. It illustrates the typical behavior of the modes: the p-modes are concentrated (have the strongest amplitude) in the outer layers of the Sun, and g-modes are mostly confined in the central region.

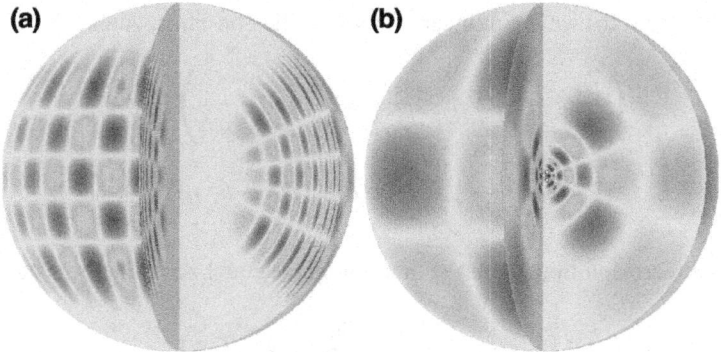

Fig. 1.15 Eigenfunctions (1.42) of two normal oscillation modes of the Sun: **a** p-mode of angular degree $l = 20$, angular degree $m = 16$, and radial order $n = 16$, **b** g-mode of $l = 5$, $m = 3$, and $n = 5$. *Red* and *blue–green* colors correspond to positive and negative values of ξ_r

1.4.2 JWKB Solution

The basic properties of the oscillation modes can be investigated analytically using an asymptotic approximation. In this approximation, we assume that only density $\rho(r)$ varies significantly among the solar properties in the oscillation equations, and seek for an oscillatory solution in the JWKB form:

$$\xi_r = A\rho^{-1/2}e^{ik_r r}, \tag{1.43}$$

$$P' = B\rho^{1/2}e^{ik_r r}, \tag{1.44}$$

where the *radial wavenumber* k_r is a slowly varying function of r; A and B are constants.

Then, substituting these in (1.35) and (1.37) we obtain:

$$\frac{d\xi_r}{dr} = -A\rho^{-1/2}\left(-ik_r + \frac{1}{H}\right)e^{ik_r r}, \tag{1.45}$$

$$\frac{dP'}{dr} = -B\rho^{1/2}\left(-ik_r - \frac{1}{H}\right)e^{ik_r r}, \tag{1.46}$$

where

$$H = \left(\frac{d\log\rho}{dr}\right)^{-1}, \tag{1.47}$$

is the *density scale height*.

From (1.45, 1.46) we get a linear system for the constants, A and B:

$$\left(-ik_r + \frac{1}{H}\right) A - \frac{g}{c^2} A + \frac{1}{c^2}\left(1 - \frac{S_l^2}{\omega^2}\right) B = 0, \qquad (1.48)$$

$$\left(-ik_r - \frac{1}{H}\right) B + \frac{g}{c^2} B + (N^2 - \omega^2) A = 0. \qquad (1.49)$$

It has a non-zero solution when the determinant is equal zero, that is, when

$$k_r^2 = \frac{\omega^2 - \omega_c^2}{c^2} + \frac{S_l^2}{c^2\omega^2}\left(N^2 - \omega^2\right), \qquad (1.50)$$

where

$$\omega_c = \frac{c}{2H} \qquad (1.51)$$

is *the acoustic cut-off frequency*. Here, we used the relation: $N^2 = g/H - g^2/c^2$.

The frequencies of solar modes depend on the sound speed, c, and three characteristic frequencies: acoustic cut-off frequency, ω_c (1.51), Lamb frequency, S_l (1.36), and Brunt–Väisälä frequency, N (1.21). These frequencies calculated for a standard solar model are shown in Fig. 1.16. The acoustic cut-off and Brunt–Väisälä frequencies depend only on the solar structure, but the lamb frequency depends also on the mode angular degree, l. This diagram is very useful for determining the regions of mode propagation. The waves propagate in the regions where the radial wavenumber is real, that $k_r^2 > 0$. If $k_r^2 < 0$ then the waves exponentially decay with distance (become 'evanescent'). The characteristic frequencies define the boundaries of the propagation regions, also called the wave *turning points*. The region of propagation for p- and g-modes are indicated in Fig. 1.16, and are discussed in the following sections.

We define a horizontal wavenumber as

$$k_h \equiv \frac{L}{r}, \qquad (1.52)$$

where $L = \sqrt{l(l+1)}$. This definition follows from the angular part of the wave equation (1.27):

$$\frac{1}{r^2}\nabla_h^2 Y_l^m + \frac{l(l+1)}{r^2} Y_l^m = 0, \qquad (1.53)$$

where ∇_h is the horizontal component of gradient. It can be rewritten in terms of horizontal wavenumber: k_h, $\frac{1}{r^2}\nabla_h^2 Y_l^m + k_h^2 Y_l^m = 0$ if $k_h^2 = l(l+1)/r^2$.

In terms of k_h the Lamb frequency is $S_l = k_h c$, and (1.50) takes the form:

$$k_r^2 = \frac{\omega^2 - \omega_c^2}{c^2} + k_h^2\left(\frac{N^2}{\omega^2} - 1\right), \qquad (1.54)$$

Fig. 1.16 Buoyancy (Brunt–Väisälä) frequency N (*thick curve*), acoustic cut-off frequency, ω_c (*thin curve*) and Lamb frequency S_l for $l = 1, 5, 20, 50,$ and 100 (*dashed curves*) vs. fractional radius r/R for a standard solar model. The *horizontal lines with arrows* indicate the trapping regions for a g mode with frequency $\nu = 0.2$ mHz, and for a sample of five p modes: $l = 1, \nu = 1$ mHz; $l = 5,$ $\nu = 2$ mHz; $l = 20, \nu = 3$ mHz; $l = 50, \nu = 4$ mHz; $l = 100, \nu = 5$ mHz

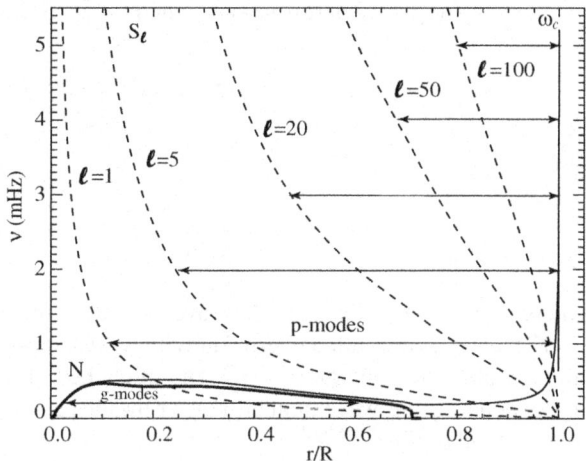

The frequencies of normal modes are determined for the Bohr quantization rule (resonant condition):

$$\int_{r_1}^{r_2} k_r \, dr = \pi(n + \alpha), \tag{1.55}$$

where r_1 and r_2 are the radii of the inner and outer turning points where $k_r = 0$, n is the radial order - integer number, and α is a phase shift which depends on properties of the reflecting boundaries.

1.4.3 Dispersion Relations for p- and g-modes

For high-frequency oscillations, when $\omega^2 \gg N^2$, the dispersion relation (1.54) can be written as:

$$k_r^2 = \frac{\omega^2 - \omega_c^2}{c^2} - \frac{S_l^2}{c^2} = \frac{\omega^2 - \omega_c^2}{c^2} - k_h^2. \tag{1.56}$$

Then, we obtain:

$$\omega^2 = \omega_c^2 + (k_r^2 + k_h^2)c^2 \equiv \omega_c^2 + k^2 c^2. \tag{1.57}$$

This is the dispersion relation for acoustic (p) modes, ω_c is the acoustic cut-off frequency. The waves with frequencies less than ω_c (or wavelength $\lambda > 4\pi H$) do not propagate. These waves exponentially decay, and are called 'evanescent'.

For low-frequency perturbations, when $\omega^2 \ll S_l^2$, one gets:

$$k_r^2 = \frac{S_l^2}{c^2\omega^2}(N^2 - \omega^2) = \frac{k_h^2}{\omega^2}(N^2 - \omega^2), \tag{1.58}$$

and

$$\omega^2 = \frac{k_h^2 N^2}{k_r^2} \equiv N^2 \cos^2\theta, \tag{1.59}$$

where θ is the angle between the wavevector, k, and horizontal surface.

These waves are called internal gravity waves or g-modes. They propagate mostly horizontally, and only if $\omega^2 < N^2$. The frequency of the internal gravity waves does not depend on the wavenumber, but on the direction of propagation. These waves are evanescent if $\omega^2 > N^2$.

1.4.4 Frequencies of p- and g-modes

Now, we use the Bohr quantization rule (1.55) and the dispersion relations for the p- and g-modes (1.57, 1.58) to derive the mode frequencies.

p-modes: The modes propagate in the region where $k_r^2 > 0$; and the radii of the turning points, r_1 and r_2, are determined from the relation $k_r^2 = 0$:

$$\omega^2 = \omega_c^2 + \frac{L^2 c^2}{r^2} = 0. \tag{1.60}$$

The acoustic cut-off frequency is only significant near the Sun's surface. The lower turning point is located in the interior where $\omega_c \ll \omega$ (Fig. 1.16). Then, at the lower turning point, $r = r_1$: $\omega \approx Lc/r$, or

$$\frac{c(r_1)}{r_1} = \frac{\omega}{L} \tag{1.61}$$

represents the equation for the radius of the lower turning point, r_1. The upper turning point is determined by the acoustic frequency term: $\omega_c(r_2) \approx \omega$. Since $\omega_c(r)$ is a steep function of r near the surface, then

$$r_2 \approx R. \tag{1.62}$$

The p-mode propagation region is illustrated in Fig. 1.16. Thus, the resonant condition for the p-modes is:

$$\int_{r_1}^{R} \sqrt{\frac{\omega^2}{c^2} - \frac{L^2}{r^2}}\, dr = \pi(n + \alpha). \tag{1.63}$$

Fig. 1.17 Spectrum of normal modes calculated for the standard solar model. The *thick gray curve* shows *f*-mode. Labels p_1–p_{33} mark *p*-modes of the radial order $n = 1, \ldots, 33$

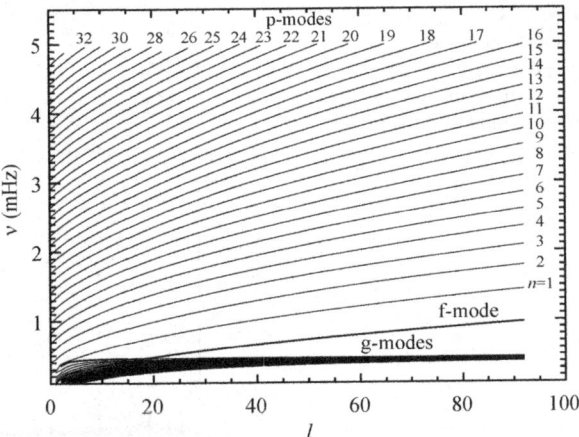

In the case of the low-degree "global" modes, for which $l \ll n$, the lower turning point is almost at the center, $r_1 \approx 0$, and we obtain [17]:

$$\omega \approx \frac{\pi (n + L/2 + \alpha)}{\int_0^R dr/c}. \tag{1.64}$$

This relation shows that the spectrum of low-degree *p*-modes is approximately equidistant with the frequency spacing:

$$\Delta\nu = \left(4 \int\limits_0^R \frac{dr}{c} \right)^{-1}. \tag{1.65}$$

This corresponds very well to the observational power spectrum shown in Fig. 1.4. According to this relation, the frequencies of mode pairs, (n, l) and $(n - 1, l + 2)$, coincide. However, calculations to the second-order approximation shows that the frequencies in these pairs are separated by the amount [98, 99]:

$$\delta\nu_{nl} = \nu_{nl} - \nu_{n-1,l+2} \approx -(4l + 6)\frac{\Delta\nu}{4\pi^2\nu_{nl}} \int\limits_0^R \frac{dc}{dr}\frac{dr}{r}. \tag{1.66}$$

This is the so-called "small separation". For the Sun, $\Delta\nu \approx 136\,\mu$Hz, and $\delta\nu \approx 9\,\mu$ Hz. The *l*-*ν* diagram for the *p*-modes is illustrated in Fig. 1.17.

g-modes: The turning points, $k_r = 0$, are determined from (1.58):

$$N(r) = \omega. \tag{1.67}$$

In the propagation region, $k_r > 0$, (see Fig. 1.16), far from the turning points $(N \gg \omega)$:

Fig. 1.18 Periods of solar oscillation modes in the angular degree range, $l = 0$–10. Labels g_1–g_6 mark g-modes of the radial order $n = 1, \ldots, 6$

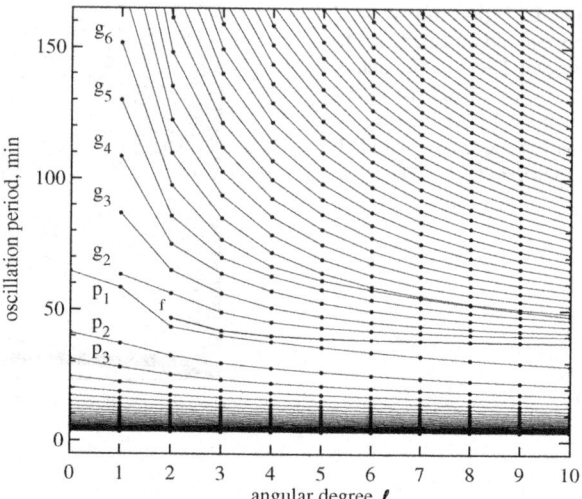

$$k_r \approx \frac{LN}{r\omega}. \tag{1.68}$$

Then, from the resonant condition:

$$\int_{r_1}^{r_2} \frac{L}{\omega} N \frac{dr}{r} = \pi(n + \alpha) \tag{1.69}$$

we find an asymptotic formula for the g-mode frequencies:

$$\omega \approx \frac{L \int_{r_1}^{r_2} N \frac{dr}{r}}{\pi(n + \alpha)}. \tag{1.70}$$

It follows that for a given l value the oscillation periods form a regular equally spaced pattern:

$$P = \frac{2\pi}{\omega} = \frac{\pi(n + \alpha)}{L \int_{r_1}^{r_2} N \frac{dr}{r}}. \tag{1.71}$$

The distribution of numerically calculated g-mode periods is shown in Fig. 1.18.

1.4.5 Asymptotic Ray-path Approximation

The asymptotic approximation provides an important representation of solar oscillations in terms of the ray theory. Consider the wave path equation in the ray approximation:

$$\frac{\partial \boldsymbol{r}}{\partial t} = \frac{\partial \omega}{\partial \boldsymbol{k}}. \tag{1.72}$$

Then, the radial and angular components of this equation are:

$$\frac{dr}{dt} = \frac{\partial \omega}{\partial k_r}, \tag{1.73}$$

$$r\frac{d\theta}{dt} = \frac{\partial \omega}{\partial k_h}. \tag{1.74}$$

Using the dispersion relation for acoustic (p) modes:

$$\omega^2 = c^2(k_r^2 + k_h^2), \tag{1.75}$$

in which we neglected the ω_c term (it can be neglected everywhere except near the upper turning point, R), we get

$$dt = \frac{dr}{c\left(1 - k_h^2 c^2/\omega^2\right)^{1/2}}. \tag{1.76}$$

From this we find the travel time from the lower turning point to the surface.

The equation for the acoustic ray path is given by the ratio of equations (1.74) and (1.76):

$$r\frac{d\theta}{dr} = \left(\frac{\partial \omega}{\partial k_h}\right) \Big/ \left(\frac{\partial \omega}{\partial k_r}\right) = \frac{k_h}{k_r}, \tag{1.77}$$

or

$$r\frac{d\theta}{dr} = \frac{k_h}{k_r} = \frac{L/r}{\sqrt{\omega^2/c^2 - L^2/r^2}}. \tag{1.78}$$

For any given values of ω and l, and initial coordinates, r and θ, this equation gives trajectories of ray paths of p-modes inside the Sun. The ray paths calculated for two solar p-modes are shown in Fig. 1.19a. They illustrate an important property that the acoustic waves excited by a source near the solar surface travel into the interior and come back to surface. The distance, Δ, between the surface points for one skip can be calculated as the integral:

$$\Delta = 2\int_{r_1}^{R} d\theta = 2\int_{r_1}^{R} \frac{L/r}{\sqrt{\omega^2/c^2 - L^2/r^2}} dr \equiv 2\int_{r_1}^{R} \frac{c/r}{\sqrt{\omega^2/L^2 - c^2/r^2}} dr. \tag{1.79}$$

The corresponding travel time is calculated by integrating equation (1.76):

$$\tau = 2\int_{r_1}^{R} dt = \int_{r_1}^{R} \frac{dr}{c\left(1 - k_h^2 c^2/\omega^2\right)^{1/2}} \equiv \int_{r_1}^{R} \frac{dr}{c\left(1 - L^2 c^2/r^2\omega^2\right)^{1/2}}. \tag{1.80}$$

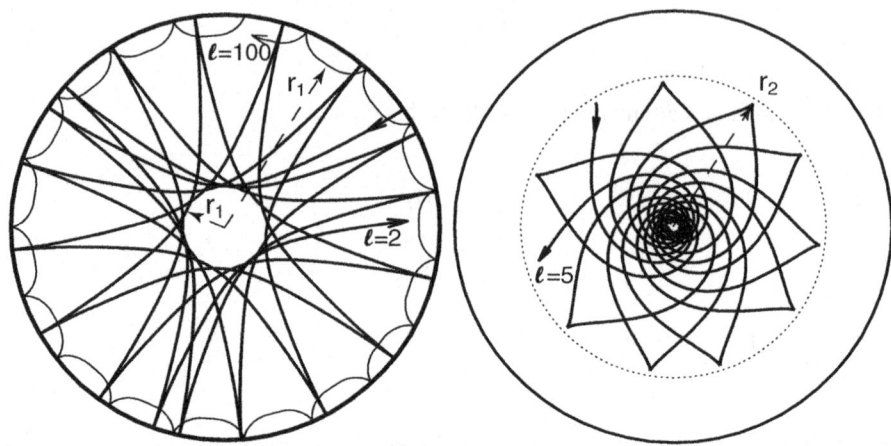

Fig. 1.19 Ray paths for **a** two solar p-modes of angular degree $l = 2$, frequency $\nu = 1429.4$ μHz (*thick curve*), and $l = 100$, $\nu = 3357.5\mu$Hz (*thin curve*); **b** g-mode of $l = 5$, $\nu = 192.6$ μHz (the *dotted curve* indicates the base of the convection zone). The lower turning points, r_1 of the p-modes are shown by *arrows*. The upper turning points of these modes are close to the surface and not shown. For the g-mode, the upper turning point, r_2, is shown by *arrow*. The inner turning point is close to the center and not shown

These equations give a *time–distance* relation, $\tau - \Delta$, for acoustic waves traveling between two surface points through the solar interior. The ray representation of the solar modes and the time–distance relation provided a motivation for developing *time–distance helioseismology* (Sect. 1.7), a local helioseismology method [48].

The ray paths for g-modes are calculated similarly. For the g-modes, the dispersion relation is:

$$\omega^2 = \frac{k_h^2 N^2}{k_r^2 + k_h^2}. \tag{1.81}$$

Then, the corresponding ray path equation is:

$$r\frac{d\theta}{dr} = -\frac{k_r}{k_h} = -\sqrt{\frac{N^2}{\omega^2} - 1}. \tag{1.82}$$

The solution for a g-mode of $l = 5$, $\nu = 192.6\,\mu$Hz is shown in Fig. 1.19b. Note that the g-mode travels mostly in the central region. Therefore, the frequencies of g-modes are mostly sensitive to the central conditions.

1.4.6 Duvall's Law

The solar p-modes, observed in the period range of 3–8 min, can be considered as high-frequency modes and described by the asymptotic theory quite accurately. Consider the resonant condition (1.63) for p-modes:

$$\int_{r_1}^{R} \left(\frac{\omega^2}{c^2} - \frac{L^2}{r^2} \right)^{1/2} dr = \pi(n + \alpha), \tag{1.83}$$

Dividing both sides by ω we get:

$$\int_{r_1}^{R} \left(\frac{r^2}{c^2} - \frac{L^2}{\omega^2} \right)^{1/2} \frac{dr}{r} = \frac{\pi(n + \alpha)}{\omega}. \tag{1.84}$$

Since the lower integral limit, r_1 depends only on the ratio L/ω, then the whole left-hand side is a function of only one parameter, L/ω, that is:

$$F \left(\frac{L}{\omega} \right) = \frac{\pi(n + \alpha)}{\omega}. \tag{1.85}$$

This relation represents the so-called Duvall's law [100]. It means that a 2D dispersion relation $\omega = \omega(n, l)$ is reduced to the 1D relation between two ratios L/ω and $(n + \alpha)/\omega$. With an appropriate choice of parameter α (e.g. 1.5) these ratios can be easily calculated from a table of observed solar frequencies. An example of such calculations, shown in Fig. 1.20, illustrates that the Duvall's law holds quite well for the observed solar modes. The short bottom branch that separates from the main curve corresponds to f-modes.

1.4.7 Asymptotic Sound–Speed Inversion

The Duvall's law demonstrates that the asymptotic theory provides a rather accurate description of the observed solar p-modes. Thus, it can be used for solving the inverse problem of helioseismology: determination of the internal properties from the observed frequencies. Theoretically, the internal structure of the Sun is described by the stellar evolution theory [101]. This theory calculates the thermodynamic structure of the Sun during the evolution on the Main Sequence. The evolutionary model of the current age $\approx 4.6 \times 10^9$ years, is called the standard solar model. Helioseismology provides estimates of the interior properties, such as the sound–speed profiles, that can be compared with the predictions of the standard model.

Our goal is to find corrections to a solar model from the observed frequency differences between the Sun and the model using the asymptotic formula for the Duvall's law [102].

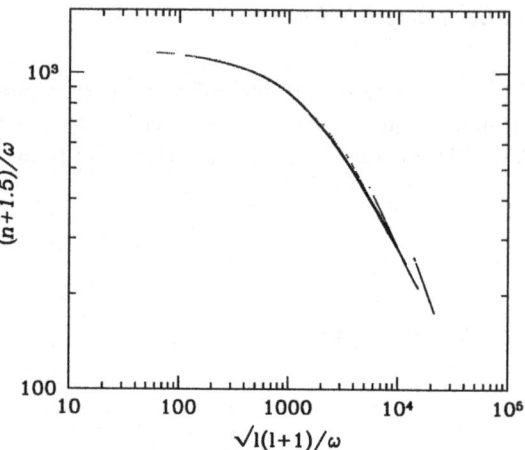

Fig. 1.20 The observed Duvall's law relation for modes of $l = 0$–250

We consider a small perturbation of the sound–speed, $c \rightarrow c + \Delta c$, and the corresponding perturbation of frequency: $\omega \rightarrow \omega + \Delta \omega$. Then, from (1.84) we obtain:

$$\int_{r_t}^{R} \left[\frac{(\omega + \Delta\omega)^2}{(c + \Delta c)^2} - \frac{L^2}{r^2} \right]^{1/2} dr = \pi(n + \alpha). \tag{1.86}$$

Expanding this in terms of $\Delta c/c$ and $\Delta\omega/\omega$ and keeping only the first-order terms we get:

$$\frac{\Delta\omega}{\omega} \int_{r_t}^{R} \frac{dr}{c \left(1 - L^2 c^2/r^2\omega^2\right)^{1/2}} = \int_{r_t}^{R} \frac{\Delta c}{c} \frac{dr}{c \left(1 - L^2 c^2/r^2\omega^2\right)^{1/2}}. \tag{1.87}$$

If we introduce a new variable:

$$T = \int_{r_t}^{R} \frac{dr}{c \left(1 - L^2 c^2/r^2\omega^2\right)^{1/2}}, \tag{1.88}$$

then

$$\frac{\Delta\omega}{\omega} = \frac{1}{T} \int_{r_t}^{R} \frac{\Delta c}{c} \frac{dr}{c \left(1 - L^2 c^2/r^2\omega^2\right)^{1/2}}. \tag{1.89}$$

This equation has a simple physical interpretation: T is the travel time of acoustic waves to travel along the acoustic ray path between the lower and upper turning points (Fig. 1.19). The right-hand side integral is an average of the sound–speed perturbations along this ray path (compare with (1.80)).

Equation (1.89) can be reduced to *the Abel integral equation* by making a substitution of variables. The new variables are:

$$x = \frac{\omega^2}{L^2}, \tag{1.90}$$

$$y = \frac{c^2}{r^2}, \tag{1.91}$$

where x is a measured quantity, and y is associated with the sound–speed distribution of an unperturbed solar model.

Then, we obtain an equation for x and y:

$$F(x) = \int_0^x \frac{f(y)dy}{\sqrt{x-y}}, \tag{1.92}$$

where

$$F(x) = T\frac{\Delta\omega}{\omega}\frac{1}{\sqrt{x}},$$

$$f(y) = \frac{\Delta c}{c}\frac{1}{2y^{3/2}\left(\frac{d\log c}{d\log r}+1\right)}.$$

To solve for $f(y)$ we multiply both sides of (1.19) by $dx/\sqrt{z-x}$ and integrate with respect to x from 0 to z:

$$\int_0^z \frac{F(x)dx}{\sqrt{z-x}} = \int_0^z \frac{dx}{\sqrt{z-x}} \int_0^x \frac{f(y)dy}{\sqrt{x-y}}$$

$$= \int_0^x f(y)dy \int_y^z \frac{dx}{\sqrt{(z-x)(x-y)}}.$$

Here we changed the order of integration.

Note that

$$\int_y^z \frac{dx}{\sqrt{(z-x)(x-y)}} = \pi,$$

then

$$\int_0^z \frac{F(x)dx}{\sqrt{z-x}} = \pi \int_0^x f(y)dy.$$

Differentiating with respect to x, we obtain the final solution:

$$f(y) = \frac{1}{\pi} \frac{d}{dx} \int_0^z \frac{F(x)dx}{\sqrt{z-x}}. \tag{1.93}$$

Then, from $f(y)$ we find the sound–speed correction $\Delta c/c$.

This method based on linearization of the asymptotic Abel integral is called "differential asymptotic sound–speed inversion" [102]. It provides estimates of the sound–speed deviations from a reference solar model.

Alternatively, the sound–speed profile inside the Sun can be found from a solution of the Abel obtained by differentiating the Duvall's law equation (1.84) with respect to variable $y = L/\omega$. Then, this equation can be solved analytically. The solution provides an implicit relationship between the solar radius and sound speed [103]:

$$\ln(r/R) = \int_{r/c}^{R/c_s} \frac{dF}{dy} \left(y^2 - \frac{r^2}{c^2}\right)^{-1/2} dy, \tag{1.94}$$

where c_s is the sound speed at the solar surface $r = R$. The calculation of the derivative, dF/dy, is essentially differentiation of a smooth function approximating the Duvall's law, that is differentiating $\pi(n + \alpha)/\omega$ with respect to L/ω. Both these quantities are obtained from the observed frequency table, $\omega(n, l)$.

The first inversion result using this approach was published by Christensen-Dalsgaard et al. [102]. These technique can be generalized by including the Brunt–Väisälä frequency term in the p-mode dispersion relation, and also taking into account the frequency dependence of the phase shift, α [36]. The results show that this inversion procedure provides a good agreement with the solar models, used for testing, except the central core, where the asymptotic and Cowling approximations become inaccurate.

Figure 1.21 shows the inversion results [104] for the p-mode frequencies measured by Duvall et al. [105]. The deviation of the sound speed from a standard solar model is about 1%. Later, the agreement between the solar model and the helioseismic inversions was improved by using more precise opacity tables and including element diffusion in the model calculations [101]. Also, a more accurate inversion method was developed by using a perturbation theory based on a variational principle for the normal mode frequencies (Sect. 1.5).

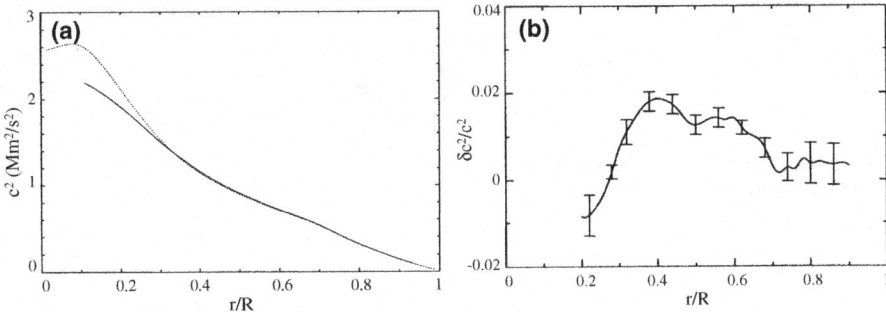

Fig. 1.21 a Result of the asymptotic sound inversion (*solid curve*) [104] for the *p*-mode frequencies [105]. It confirmed the standard solar model (model 1) [106] (*dots*). The large discrepancy in the central region is due to inaccuracy of the data and the asymptotic approximation. **b** The relative difference in the squared sound speed between the asymptotic inversions of the observed and theoretical frequencies

1.4.8 Surface Gravity Waves (f-mode)

The surface gravity (*f*-mode) waves are similar in nature to the surface ocean waves. They are driven by the buoyancy force, and exist because of the sharp density decrease at the solar surface. These waves are missing in the JWKB solution. These waves propagate at the surface boundary where Lagrangian pressure perturbation $\delta P \sim 0$.

To investigate these waves we consider the oscillation equations in terms of δP by making use of the relation between the Eulerian and Lagrangian variables (1.10):

$$P' = \delta P + g\rho\xi_r.$$

The oscillation equations (1.35) and (1.37) in terms of ξ_r and δP are:

$$\frac{d\xi_r}{dr} - \frac{L^2 g}{\omega^2 r^2}\xi_r + \left(1 - \frac{L^2 c^2}{\omega^2 r^2}\right)\frac{\delta P}{\rho c^2} = 0, \qquad (1.95)$$

$$\frac{d\delta P}{dr} + \frac{L^2 g}{\omega^2 r^2}\delta P - \frac{g\rho f}{r}\xi_r = 0, \qquad (1.96)$$

where

$$f \approx \frac{\omega^2 r}{g} - \frac{L^2 g}{\omega^2 r}. \qquad (1.97)$$

These equations have a peculiar solution:

$$\delta P = 0, \quad f = 0.$$

For this solution:

$$\omega^2 = \frac{Lg}{R} = k_h g \tag{1.98}$$

the dispersion relation for the f-mode.

The eigenfunction equation:

$$\frac{d\xi_r}{dr} - \frac{L}{r}\xi_r = 0 \tag{1.99}$$

has a solution

$$\xi_r \propto e^{k_h(r-R)} \tag{1.100}$$

exponentially decaying with depth.

These waves are similar in nature to water waves which have the same dispersion relation: $\omega^2 = gk_h$. The f-mode waves are incompressible: $\nabla \cdot v = 0$. These waves are not sensitive to the sound speed but are sensitive to the density gradient at the solar surface. They are used for measurements of the 'seismic radius' of the Sun.

1.4.9 The Seismic Radius

The frequencies of f-modes are:

$$\omega^2 = gk_h \equiv \frac{GM}{R^2}\frac{L}{R} \equiv L\frac{GM}{R^3}. \tag{1.101}$$

If the frequencies are determined in observations for given l, then we can define the 'seismic radius', R, as

$$R = \left(\frac{LGM}{\omega^2}\right)^{1/3}. \tag{1.102}$$

The procedure of measuring the solar seismic radius is simple [107]. The lower curve in Fig. 1.22a shows the relative difference between the f-mode frequencies of $l = 88$–250 calculated for a standard solar model (model S) and the frequencies obtained from the SOHO/MDI observations. This difference shows that the model frequencies are systematically, by $\approx 6.6 \times 10^{-4}$, lower than the observed frequencies. Then from (1.101):

$$\frac{\Delta R}{R} = -\frac{2}{3}\frac{\Delta v}{v} \approx 4.4 \times 10^{-4}, \tag{1.103}$$

This means that the seismic radius is approximately equal to 695.68 Mm, which is about 0.3 Mm less than the standard radius, 695.99 Mm, used for calibrating the

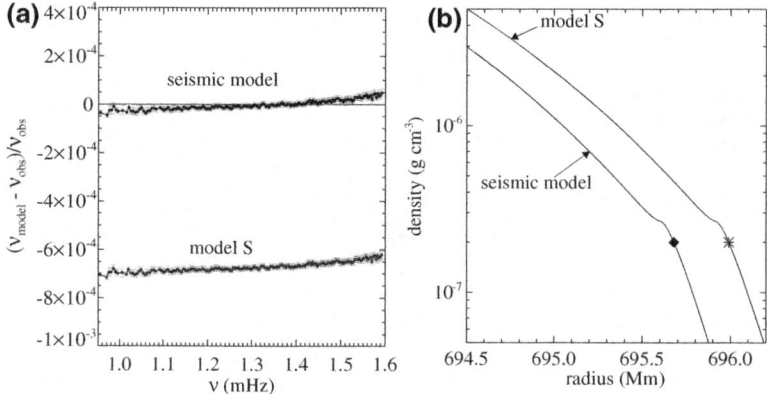

Fig. 1.22 **a** Relative differences between the *f*-mode frequencies of $l = 88$–250 computed for a standard solar model (model S) and the observed frequencies. The 'seismic model' frequencies are obtained by scaling the frequencies of model S with factor 1.00066, which corresponds to scaling down the model radius with $(1.00066)^{2/3} \approx 1.00044$. The error bars are 3σ error estimates of the observed frequencies. **b** Density as a function of radius near the surface for the standard and seismic models. The *star* indicates the photospheric radius. The *diamond* shows the seismic radius, 695.68 Mm

model calculation. This radius is usually measured astrometrically as a position of the inflection point in the solar limb profile. However, in the model calculations it is considered as a radius where the optical depth of continuum radiation is equal 1. The difference between this radius and the radius of the inflection point can explain the discrepancy between the model and seismic radius.

Figure 1.22b illustrates the density profiles in the standard solar model (model S [101]) and a 'seismic' model, calibrated to the seismic radius. The *f*-mode frequencies of the seismic model match the observations.

Since the *f*-mode frequencies provide an accurate estimate of the seismic radius, then it is interesting to investigate the variations of the solar radius during the solar activity cycle, which are important for understanding physical mechanisms of solar variability (e.g. [108]). Figure 1.23 shows the *f*-mode frequency variations during the solar cycle 23, in 1997–2004, relative to the *f*-mode frequencies observed in 1996 during the solar minimum [109].

The results show a systematic increase of the *f*-mode frequency with the increased solar activity, which means a decrease of the seismic radius. However, the variations of the *f*-mode frequencies are not constant as this is expected from (1.103) for a simple homologous change of the solar structure. A detailed investigation of these variations showed that the frequency dependence can be explained if the variations of the solar structure are not homologous and if the deeper subsurface layers expand but the shallower layers shrink with the increased solar activity [109, 110].

Fig. 1.23 Average relative frequency differences for f-mode $\langle \delta \nu / \nu \rangle$ as a function of $\langle \nu \rangle$, averaged frequencies binned every 20μHz. The reference year is 1996

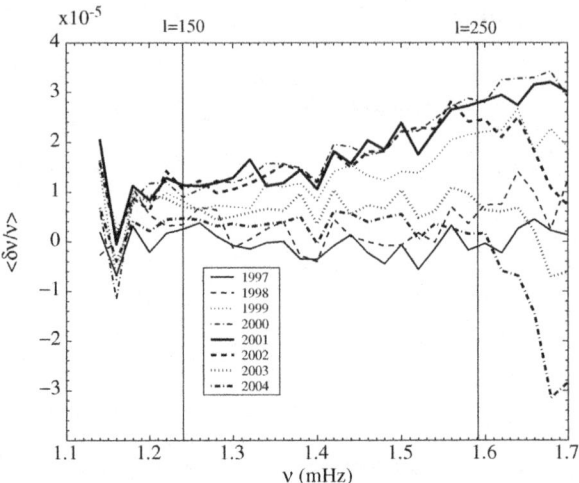

1.5 General Helioseismic Inverse Problem

In the asymptotic (high-frequency or short-wavelength) approximation (1.84), the oscillation frequencies depend only on the sound–speed profile. This dependence is expressed in terms of the Abel integral equation (1.89), which can be solved analytically.

In a general case, the relation between the frequencies and internal properties is more complicated, the frequencies depend not only on the sound speed, but also on other internal properties, and there is no analytical solution. Generally, the frequencies determined from the oscillation equations (1.35) and (1.37) depend on the density, $\rho(r)$, the pressure, $P(r)$, and the adiabatic exponent, $\gamma(r)$. However, ρ and P are not independent and related to each other through the hydrostatic equation:

$$\frac{dP}{dr} = -g\rho, \qquad (1.104)$$

where $g = Gm/r^2$, $m = 4\pi \int_0^r \rho r'^2 dr'$. Therefore, only two thermodynamic (hydrostatic) properties of the Sun are independent, e.g. pairs of (ρ, γ), (P, γ), or their combinations: $(P/\rho, \gamma)$, (c^2, γ), (c^2, ρ) etc.

The general inverse problem of helioseismology is formulated in terms of small corrections to the standard solar model because the differences between the Sun and the standard model are typically 1% or less. When necessary the corrections can be applied repeatedly using an iterative procedure.

1.5.1 Variational Principle

We consider the oscillation equations as a formal operator equation in terms of the vector displacement, ξ:

$$\omega^2 \xi = \mathcal{L}(\xi), \tag{1.105}$$

where $\mathcal{L}$ in the general case is an integro-differential operator. If we multiply this equation by ξ^* and integrate over the mass of the Sun, we get:

$$\omega^2 \int_V \rho \xi^* \cdot \xi dV = \int_V \xi^* \cdot \mathcal{L}\xi \rho dV, \tag{1.106}$$

where ρ is the model density, V is the solar volume.

Then, the oscillation frequencies can be determined as a ratio of two integrals:

$$\omega^2 = \frac{\int_V \xi^* \cdot \mathcal{L}\xi \rho dV}{\int_V \rho \xi^* \cdot \xi dV}. \tag{1.107}$$

The frequencies are expressed in terms of eigenfunctions ξ and the solar properties represented by coefficients of the operator $\mathcal{L}$. For small perturbations of solar parameters the frequency change will depend on these perturbations and the corresponding perturbations of the eigenfunctions, e.g.

$$\delta\omega^2 = \Psi[\delta\rho, \delta\gamma, \delta\xi]. \tag{1.108}$$

The variational principle states that the perturbations of the eigenfunctions constitute second-order corrections, that is, to the first-order approximation the frequency variations depend only on variations of the model properties:

$$\delta\omega^2 \approx \Psi[\delta\rho, \delta\gamma]. \tag{1.109}$$

The variational principle allows us to neglect the perturbation of the eigenfunctions in the first-order perturbation theory. This was first established by Rayleigh. Thus, (1.107) is called the Rayleigh's Quotient, and the variational principle is called the Rayleigh's Principle. The original formulation of this principle is: for an oscillatory system the kinetic energy averaged over a period is equal to the averaged potential energy. In our case, the left-hand side of (1.106) is proportional to the mean kinetic energy, and the right-hand side is proportional to the potential energy of solar oscillations.

1.5.2 Perturbation Theory

We consider a small perturbation of operator $\mathcal{L}$ caused by variations of the solar structure properties:

$$\mathcal{L}(\xi) = \mathcal{L}_0(\xi) + \mathcal{L}_1(\xi).$$

Then, the corresponding frequency perturbations are determined from the following equation:

$$\delta\omega^2 = \frac{\int_V \xi^* \cdot \mathcal{L}_1 \xi \rho dV}{\int_V \rho \xi^* \cdot \xi dV},$$

or

$$\frac{\delta\omega}{\omega} = \frac{1}{2\omega_0 I} \int_V \xi^* \cdot \mathcal{L}_1 \xi \rho dV, \qquad (1.110)$$

where

$$I = \int_V \rho \xi^* \cdot \xi dV \qquad (1.111)$$

is so-called *mode inertia* or *mode mass*. The *mode energy* is $E = I\omega_0^2 a^2$, where a is the amplitude of the surface displacement. The mode eigenfunctions are usually normalized such that $\xi_r(R) = 1$.

Using explicit formulations for operator $\mathcal{L}_1$, equation (1.110) can be reduced to a system of integral equations for a chosen pair of independent variables [111–114], e.g. for (ρ, γ)

$$\frac{\delta\omega^{(n,l)}}{\omega^{(n,l)}} = \int_0^R K_{\rho,\gamma}^{(n,l)} \frac{\delta\rho}{\rho} dr + \int_0^R K_{\gamma,\rho}^{(n,l)} \frac{\delta\gamma}{\gamma} dr, \qquad (1.112)$$

where $K_{\rho,\gamma}^{(n,l)}(r)$ and $K_{\gamma,\rho}^{(n,l)}(r)$ are sensitivity (or 'seismic') kernels. They are calculated using the initial solar model parameters, ρ_0, P_0, γ, and the oscillation eigenfunctions for these model, ξ.

1.5.3 Kernel Transformations

The sensitivity kernels for various pairs of solar parameters can be obtained by using the relations among these parameters, which follows from the equations of solar structure('stellar evolution theory').

A general procedure for calculating the sensitivity kernels developed by Kosovichev [114] can be illustrated in an operator form. Consider two pairs of solar variables, X and Y, e.g.

$$X = \left(\frac{\delta\rho}{\rho}, \frac{\delta\gamma}{\gamma}\right); \quad Y = \left(\frac{\delta u}{u}, \frac{\delta Y}{Y}\right),$$

where $u = P/\rho Y$ is the helium abundance.

The linearized structure equations (the hydrostatic equilibrium equation and the equation of state) that relate these variables can be written symbolically:

$$\mathcal{A}X = Y. \tag{1.113}$$

Let K_X and K_Y be the sensitivity kernels for X and Y, then the frequency perturbation is:

$$\frac{\delta\omega}{\omega} = \int_0^R K_X \cdot X dr \equiv \langle K_X \cdot X \rangle, \tag{1.114}$$

where $\langle \cdot \rangle$ denotes the inner product. Similarly,

$$\frac{\delta\omega}{\omega} = \langle K_Y \cdot Y \rangle. \tag{1.115}$$

Then from (1.114) and (1.115) we obtain the following relation:

$$\langle K_Y \cdot Y \rangle = \langle K_Y \cdot \mathcal{A}X \rangle = \langle \mathcal{A}^* K_Y \cdot X \rangle, \tag{1.116}$$

where $\mathcal{A}^*$ is an adjoint operator. This operator is adjoint to the stellar structure operator, $\mathcal{A}$. The second part of (1.116) represent a formal definition of this operator.

From (1.114) and (1.116) we get:

$$\langle \mathcal{A}^* K_Y \cdot X \rangle = \langle K_X \cdot X \rangle.$$

This equation is valid for any X only if

$$\mathcal{A}^* K_Y = K_X. \tag{1.117}$$

This means that the equation for the sensitivity kernels is adjoint to the stellar structure equations. The explicit formulation of the adjoint equations for the sensitivity kernels for various pairs of variables is given in [114].

Examples of the sensitivity kernels for solar properties are shown in Fig. 1.24. Figure 1.25 illustrates the difference in sensitivities of the p- and g-modes. The frequencies of solar p-modes are mostly sensitive to properties of the outer layers of the Sun while the frequencies of g-modes have the greatest sensitivity to the parameters of the solar core.

Fig. 1.24 Sensitivity kernels for the acoustic mode of the angular degree, $l = 10$, and the radial order, $n = 6$. $K_{\rho,\gamma}$ is the kernel for density, ρ, at constant adiabatic exponent, γ; $K_{c^2,\rho}$ is the kernel for the squared sound speed, c^2, at constant ρ; $K_{u,Y}$ is the kernel for function u, the ratio pressure, p, to density at constant helium abundance, Y; and $K_{A^*,\gamma}$ is the kernel for the parameter of convective stability, $A^* = rN^2/g$, at constant γ

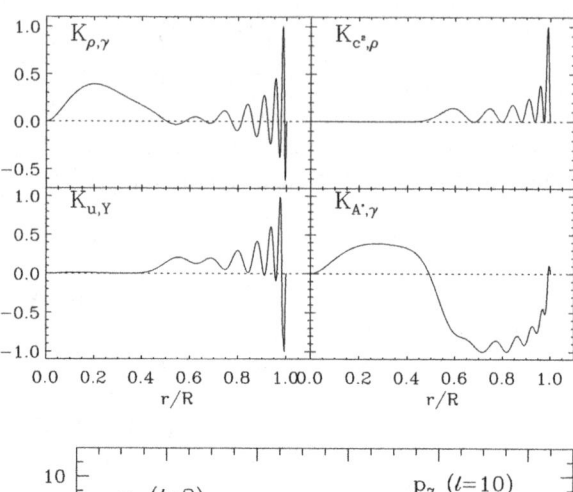

Fig. 1.25 Sensitivity kernels for p- and g-modes for $u = P/\rho$ and helium abundance Y

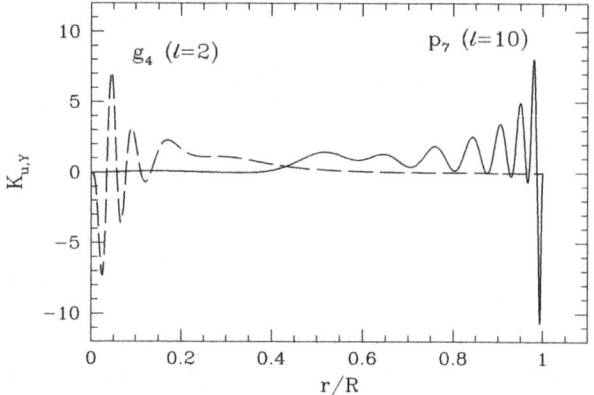

1.5.4 Solution of Inverse Problem

The variational formulation provides us with a system integral equations (1.112) for a set of observed mode frequencies. Typically, the number of observed frequencies, $N \simeq 2000$. Thus, we have a problem of determining two functions from this finite set of measurements. In general, it is impossible to determine these functions precisely. We can always find some rapidly oscillating functions, $f(r)$, such that being added to the unknowns, $\delta\rho/\rho$ and $\delta\gamma/\gamma$, do not change the values of the integrals, e.g.

$$\int_0^R K_{\rho,\gamma}^{(n,l)}(r)f(r)dr = 0.$$

Such problems without a unique solution are called "ill-posed". The general approach is to find a smooth solution that satisfies the integral equations (1.112) by applying some smoothness constraints to the unknown functions. This is called a *regularization procedure*.

There are two basic methods for solution of the helioseismic inverse problem:

1. Optimally Localized Averages (OLA) method (Backus–Gilbert method) [115];
2. Regularized Least-Squares (RLS) method (Tikhonov method) [116].

1.5.5 Optimally Localized Averages Method

The idea of the OLA method is to find a linear combination of data such as the corresponding linear combination of the sensitivity kernels for one unknown has an isolated peak at a given radial point, r_0, (resembling a δ-function), and the combination for the other unknown is close to zero. Then, this linear combination provides an estimate for the first unknown at r_0.

Indeed, consider a linear combination of (1.112) with some unknown coefficient $a^{(n,l)}$:

$$\sum a^{(n,l)} \frac{\delta\omega^{(n,l)}}{\omega^{(n,l)}} = \int_0^R \sum a^{(n,l)} K_{\rho,\gamma}^{(n,l)} \frac{\delta\rho}{\rho} dr + \int_0^R \sum a^{(n,l)} K_{\gamma,\rho}^{(n,l)} \frac{\delta\gamma}{\gamma} dr. \quad (1.118)$$

If in the first term the linear combination of the kernels is close to a δ-function at $r = r_0$, that is

$$\sum a^{(n,l)} K_{\rho,\gamma}^{(n,l)}(r) \simeq \delta(r - r_0), \quad (1.119)$$

and the linear combination in the second term vanishes:

$$\sum a^{(n,l)} K_{\gamma,\rho}^{(n,l)}(r) \simeq 0, \quad (1.120)$$

then (1.118) gives an estimate of the density perturbation, $\delta\rho/\rho$, at $r = r_0$:

$$\sum a^{(n,l)} \frac{\delta\omega^{(n,l)}}{\omega^{(n,l)}} \approx \int_0^R \delta(r - r_0) \frac{\delta\rho}{\rho} dr = \overline{\left(\frac{\delta\rho}{\rho}\right)}_{r_0}. \quad (1.121)$$

Of course, the coefficients, $a^{(n,l)}$, of (1.121) must be calculated from conditions (1.119) and (1.120) for various target radii r_0.

The functions,

$$\sum a^{(n,l)} K_{\rho,\gamma}^{(n,l)}(r) \equiv A(r_0, r), \quad (1.122)$$

$$\sum a^{(n,l)} K_{\gamma,\rho}^{(n,l)}(r) \equiv B(r_0, r), \quad (1.123)$$

are called the *averaging kernels*. They play a fundamental role in the helioseismic inverse theory for determining the resolving power of helioseismic data.

Fig. 1.26 A sample of the optimally localized averaging kernels for the structure function, u, the ratio of pressure, P, to density, ρ, $u = P/\rho$. The second, eliminated, parameter in these kernels is the helium abundance, Y

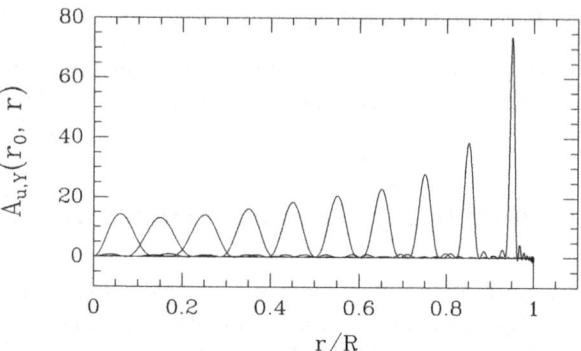

The coefficients, $a^{n,l}$, are determined by minimizing a quadratic form:

$$M(r_0, A, \alpha, \beta) = \int_0^R J(r_0, r) \, [A(r_0, r)]^2 \, dr$$

$$+ \beta \int_0^R [B(r_0, r)]^2 \, dr + \alpha \sum_{i,j} E_{n,l;n',l'} a^{n,l} a^{n',l'}, \tag{1.124}$$

where function $J(r_0, r) = 12(r - r_0)^2$ provides a localization of the averaging kernels $A(r, r_0)$ at $r = r_0$, $E_{n,l;n',l'}$ is a covariance matrix of observational errors, α and β are *regularization parameters*. The first integral in (1.124) represents the Backus–Gilbert criterion of localization for $A(r_0, r)$; the second term minimizes the contribution from $B(r_0, r)$, thus, effectively eliminating the second unknown function ($\delta\gamma/\gamma$ in this case); and the last term minimizes the errors. A practical minimization algorithm is presented in [114]. An example of the averaging kernels is shown in Fig. 1.26.

1.5.6 Inversion Results for Solar Structure

As an example, consider the results of inversion of the recent data obtained from the MDI instrument on board the SOHO space observatory. The data represent 2176 frequencies of solar oscillations of the angular degree, l, from 0 to 250. These frequencies were obtained by fitting peaks in the oscillation power spectra from a 360-day observing run, between May 1, 1996 and April 25, 1997.

Figure 1.27 shows the relative frequency difference, $\delta\omega/\omega$, between the observed frequencies and the corresponding frequencies calculated for the standard model [101]. The frequency difference is scaled with a factor $Q \equiv I(\omega)/I_0(\omega)$, where $I(\omega)$ is the mode inertia, and $I_0(\omega)$ is the mode inertia of radial modes ($l = 0$), calculated at the same frequency.

Fig. 1.27 The relative frequency difference, scaled with the relative mode inertia factor, $Q = I/I_0$ (1.111), between the Sun and the standard solar model

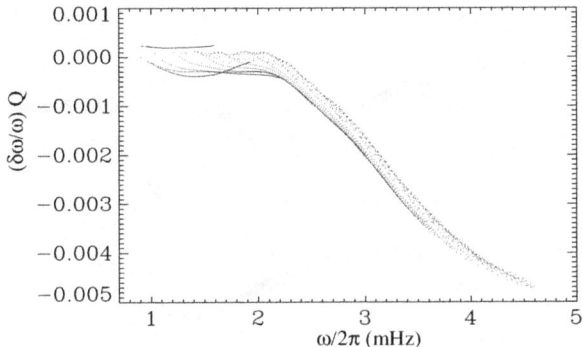

This scaled frequency difference depends mainly on the frequency alone meaning that most of the difference between the Sun and the reference solar model is in the near-surface layers. Physically, this follows from the fact that the p-modes of different l behave similarly near the surface where they propagate almost vertically. This behavior is illustrated by the p-mode ray paths in Fig. 1.19a, which become almost radial near the surface. In the inversion procedure, this frequency dependence is eliminated by adding an additional "surface term" in (1.112) [114]. However, there is also a significant scatter along the general frequency trend. This scatter is due to the variations of the structure in the deep interior, and it is the basic task of the inversion methods to uncover the variations.

First, we test the inversion procedure by considering the frequency difference for two solar models and trying to recover the differences between model properties. Results of the test inversion (Fig. 1.28) show good agreement with the actual differences. However, the sharp variations, like a peak in the parameter of convective stability, $A^* \equiv rN^2/g$, at the base of the convection zone, are smoothed. Also, the inner 5% of the Sun and the subsurface layers (outer 2–3%) are not resolved.

Then, we apply this procedure to the real solar data. The results (Fig. 1.29) show that the differences between the inferred structure and the reference solar model (model S) are quite small, generally less than 1%. The small differences provide a justification for the linearization procedure, based on the variational principle. This also means that the modern standard model of the Sun [101] provides an accurate description of the solar properties compared to the earlier solar model [106], used for the asymptotic inversions(Fig. 1.21). A significant improvement in the solar modeling was achieved by using more accurate radiative opacity data and by including the effects of gravitational settling of heavy elements and element diffusion. However, recent spectroscopic estimates of the heavy element abundance on the Sun, based on radiative hydrodynamics simulations of solar convection, indicated that the heavy element abundance on the Sun may be lower than the value used in the standard model [117]. The solar model with a low heavy element abundance do not agree with the helioseismology measurements (e.g. [118]). This problem in the solar modeling has not been resolved. Thus, the helioseismic inferences of the solar

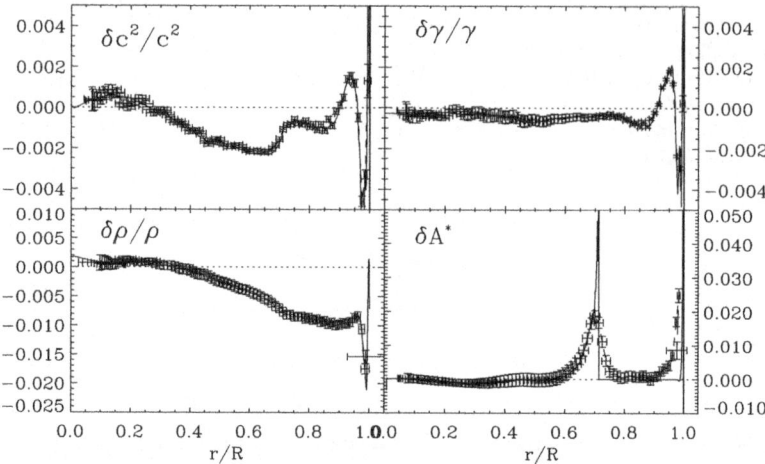

Fig. 1.28 The results of test inversions (points with the error bars, connected with *dashed curves*) of frequency differences between two solar models for the squared sound speed, c^2, the adiabatic exponent, γ, the density, ρ, and the parameter of convective stability, A^*. The *solid curves* show the actual differences between the two models. Random Gaussian noise was added to the frequencies of a test solar model. The *vertical bars* show the formal error estimates, the *horizontal bars* show the characteristic width of the localized averaging kernels. The central points of the averages are plotted at the centers of gravity of the averaging kernels

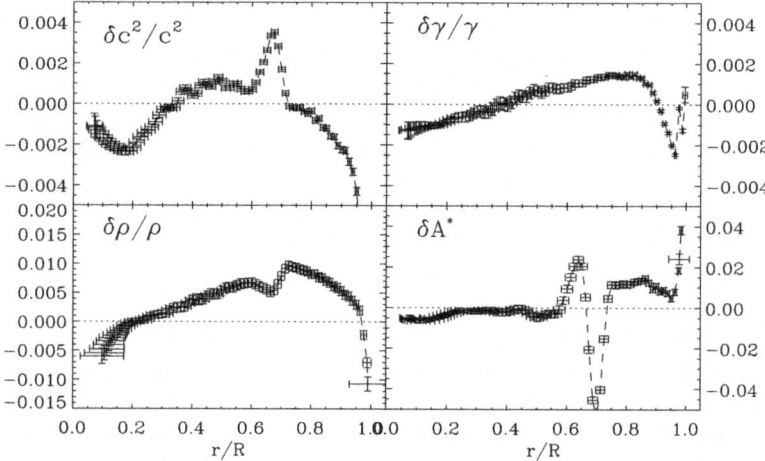

Fig. 1.29 The relative differences between the Sun and the standard solar model [101] in the squared sound speed, c^2, the adiabatic exponent, γ, the density, ρ, and the parameter of convective stability, A^*, inferred from the solar frequencies determined from the 360-day series of SOHO/MDI data

structure lead to better understanding of the structure and evolution of the star, and have important applications in other fields of astrophysics.

The prominent peak of the squared sound speed, $\delta c^2/c^2$, at the base of the convection zone, $r/R \approx 0.7$, indicates on additional mixing which may be caused by rotational shear flows or by convective overshoot. The variation in the sound speed in the energy-generating core at $r/R < 0.2$ might be also caused by a partial mixing.

The monotonic decrease of the adiabatic exponent, γ, in the core was recently explained by the relativistic corrections to the equation of state [119]. Near surface variations of γ, in the zones of ionization of helium and hydrogen, and below these zones, are most likely caused by deficiencies in the theoretical models of the weakly coupled plasma employed in the equation of state calculations [120].

The monotonic decrease of the squared sound speed variation in the convection zone ($r/R > 0.7$) is partly due to an error in the solar seismic radius used to calibrate the standard model [107], and partly due to the inaccurate description of the subsurface layers by the standard solar model, based on the mixing-length convection theory.

1.5.7 Regularized Least-Squares Method

The Regularized Least-Squares (RLS) method [116] is based on minimization of the quantity

$$
\mathcal{E} \equiv \sum_{n,l} \frac{1}{\sigma_{n,l}^2} \left[\frac{\delta\omega^{(n,l)}}{\omega^{(n,l)}} - \int_0^R \left(K_{(f,g)}^{(n,l)} \frac{\delta f}{f} + K_{(g,f)}^{(n,l)} \frac{\delta g}{g} \right) dr \right]^2
$$
$$
+ \int_0^R \left[\alpha_1 \left(L_1 \frac{\delta f}{f} \right)^2 + \alpha_2 \left(L_2 \frac{\delta g}{g} \right)^2 \right] dr, \qquad (1.125)
$$

in which the unknown structure correction functions, $\delta f/f$ and $\delta g/g$, are both represented by piece-wise linear functions or by cubic splines. The second integral specifies smoothness constraints for the unknown functions, in which L_1 and L_2 are linear differential operators, e.g. $L_{1,2} = d^2/d^2r$; σ_i are error estimates of the relative frequency differences.

In this inversion method, the estimates of the structure corrections are, once again, linear combinations of the frequency differences obtained from observations, and corresponding averaging kernels exist too. However, unlike the OLA kernels $A(r_0; r)$, the RLS averaging kernels may have negative sidelobes and significant peaks near the surface, thus making interpretation of the inversion results to some extent ambiguous. Nevertheless, it works well in most cases, and may provide a higher resolution compared to the OLA method.

1.5.8 Inversions for Solar Rotation

The eigenfrequencies of a spherically-symmetrical static star are degenerate with respect to the azimuthal number m. Rotation breaks the symmetry and splits each mode of radial order, n, and angular degree, l, into $(2l + 1)$ components of $m = -l, \ldots, l$ (*mode multiplets*). The rotational frequency splitting can be computed using a more general variational principle derived by Lynden-Bell and Ostriker [121]. From this variational principle, one can obtain mode frequencies ω_{nlm} relative to the degenerate frequency ω_{nl} of the non-rotating star:

$$\Delta\omega_{nlm} \equiv \omega_{nlm} - \omega_{nl} = \frac{1}{I_{nl}} \int_V \left[m\boldsymbol{\xi} \cdot \boldsymbol{\xi}^* + i e_\Omega (\boldsymbol{\xi} \times \boldsymbol{\xi}^*) \right] \Omega \rho dV, \qquad (1.126)$$

where e_Ω is the unit vector defining the rotation axis, and $\Omega = \Omega(r, \theta)$ is the angular velocity which is a function of radius r and co-latitude θ, and I_{nl} is the mode inertia.

Equation (1.126) can be rewritten as a 2D integral equation for $\Omega(r, \theta)$:

$$\Delta\omega_{nlm} = \int_0^R \int_0^\pi K_{nlm}^{(\Omega)}(r, \theta) \Omega(r, \theta) d\theta dr. \qquad (1.127)$$

where $K_{nlm}^{(\Omega)}(r, \theta)$ represent the rotational splitting kernels:

$$K_{nlm}^{(\Omega)}(r, \theta) = \frac{m}{I_{nl}} 4\pi \rho r^2 \left\{ (\xi_{nl}^2 - 2\xi_{nl}\eta_{nl})(P_l^m)^2 + \eta_{nl}^2 \left[\left(\frac{d P_l^m}{d\theta} \right)^2 \right. \right.$$
$$\left. \left. - 2P_l^m \frac{d P_l^m}{d\theta} \frac{\cos\theta}{\sin\theta} + \frac{m^2}{\sin^2\theta} (P_l^m)^2 \right] \right\} \sin\theta. \qquad (1.128)$$

Here ξ_{nl} and η_{nl} are the radial and horizontal components of eigenfunctions of the mean spherically symmetric structure of the Sun, $P_l^m(\theta)$ is an associated normalized Legendre function ($\int_0^\pi (P_l^m)^2 \sin\theta d\theta = 1$). The kernels are symmetric relative to the equator, $\theta = \pi/2$. Therefore, the frequency splittings are sensitive only to the symmetric component of rotation in the first approximation. The non-symmetric component can, in principle, be determined from the second-order correction to the frequency splitting, or from local helioseismic techniques, such as time–distance seismology.

For a given set of observed frequency splitting, $\Delta\omega_{nlm}$, (1.127) constitutes a 2D linear inverse problem for the angular velocity, $\Omega(r, \theta)$, which can be solved by the OLA or RLS techniques.

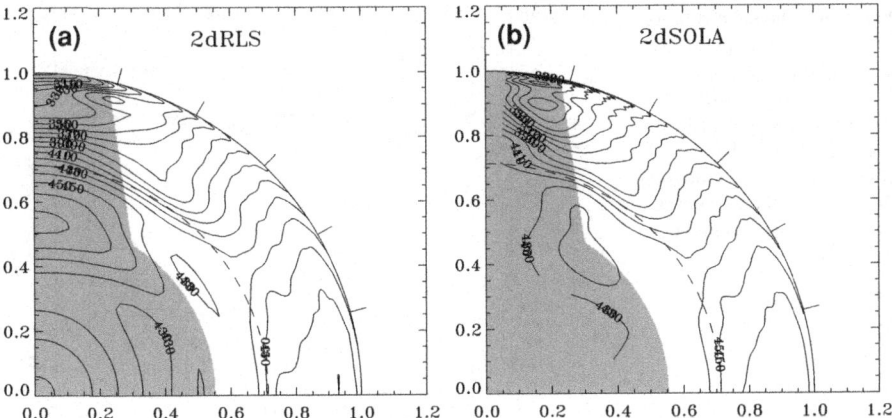

Fig. 1.30 *Contour lines* of the rotation rate (in nHz) inside the Sun obtained by inverting the rotational frequency splittings from a 144-day observing run from SOHO MDI by the RLS and SOLA methods. The *shaded areas* are the areas where the localized averaging kernels substantially deviate from the target positions

1.5.9 Results for Solar Rotation

As an example, we present the inversion results for solar rotation obtained from SOHO data. The frequency splitting data were obtained from the 144-day MDI time series by Schou for $j = 1, \ldots, 36$ and $1 \leq l \leq 250$ [122]. The total number of measurements in this data set was $M = 37366$.

Figure 1.30 shows results of inversion of the SOI-MDI data by the two methods. The results are generally in good agreement in most of the area where good averaging kernels were obtained. However, the results differ in the high-latitude region. In particular, a prominent feature of the RLS inversion at coordinates (0.2, 0.95) in Fig. 1.30a, which can be interpreted as a 'polar jet', is barely visible in Fig. 1.30b, showing the OLA inversion of the same data. Therefore, obtaining reliable inversion results in this region and also in the shaded area is one of the main current goals of helioseismology. This can be achieved by obtaining more accurate measurements of rotational frequency splitting and improving inversion techniques. Of course, the radical improvement can be made by observing the polar regions of the Sun. These measurements can be done by using spacecraft with an orbit highly inclined to the ecliptic plane, such as a proposed Solar Polar Imager (SPI) and POLARIS missions [123].

The most characteristic feature of solar rotation is the differential rotation of the convection zone, which occupies the outer 30% of the solar radius. While the radiative core rotates almost uniformly, the equatorial regions of the convection zone rotate significantly faster than the polar regions. The main interest in understanding the role of the Sun's internal rotation is the dynamo process of generation of solar magnetic fields and the origin of the 11-year sunspot cycle. The results of these

Fig. 1.31 The solar rotation rate as a function of radius at three latitudes. The *horizontal lines* indicate the rotation rate of the surface magnetic flux at the end of solar cycle 22 ("old magnetic flux") and at the beginning of cycle 23 ("new magnetic flux") [126]

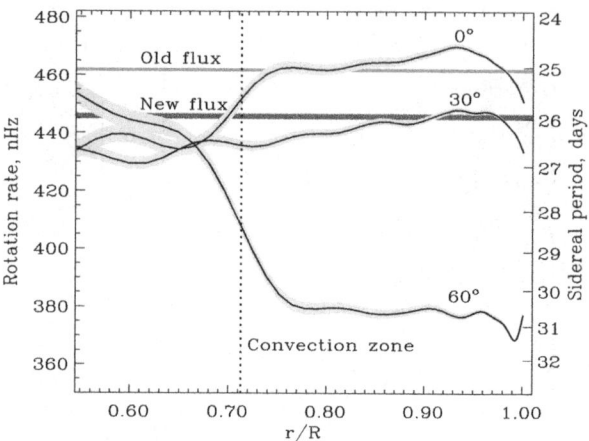

measurements (Fig. 1.31a) reveal two radial shear layers at the bottom of the convection zone (so-called tachocline) and in the upper convective boundary layer. A common assumption is that the solar dynamo operates in the tachocline area (interface dynamo) where it is easier to explain storage of magnetic flux than in the upper convection zone because of the flux buoyancy. However, there are theoretical and observational difficulties with this concept. First, the magnetic field in the tachocline must be quite strong, ∼60–160 kG, to sustain the action of the Coriolis force transporting the emerging flux tubes into high-latitude regions [124]. The magnetic energy of such field is above the equipartition level of the turbulent energy. Second, the back-reaction such strong field should suppress turbulent motions affecting the Reynolds stresses. Since these turbulent stresses support the differential rotation one should expect significant changes in the rotation rate in the tachocline. However, no significant variations with the 11-year solar cycle are detected. Third, magnetic fields often tend to emerge in compact regions on the solar surface during long periods lasting several solar rotations. This effect is known as "complexes of activity" or "active longitudes". However, the helioseismology observations show that the rotation rate of the solar tachocline is significantly lower than the surface rotation rate. Thus, magnetic flux emerging from the tachocline should be spread over longitudes (with new flux lagging the previously emerged flux) whether it remains connected to the dynamo region or disconnected. It is well-known that sunspots rotate faster than surrounding plasma. This means that the magnetic field of sunspots is anchored in subsurface layers. Observations show that the rotation rate of magnetic flux matches the internal plasma rotation in the upper shear layer (Fig. 1.31) indicating that this layer is playing an important role in the solar dynamo, and causing a shift in the dynamo paradigm [125].

Variations in solar rotation clearly related to the 11-year sunspot cycle are observed in the upper convection zone. These are so-called "torsional oscillations" which represent bands of slower and faster rotation, migrating towards the equator as the

Fig. 1.32 **a** Migration of the subsurface zonal flows with latitude during solar cycle 23 from SOHO/MDI data [127]. *Red* shows zones of faster rotation, *green* and *blue* show slower rotation. **b** Variations of the zonal flows with depth and latitude during the first 4 years after the solar minimum. [128]

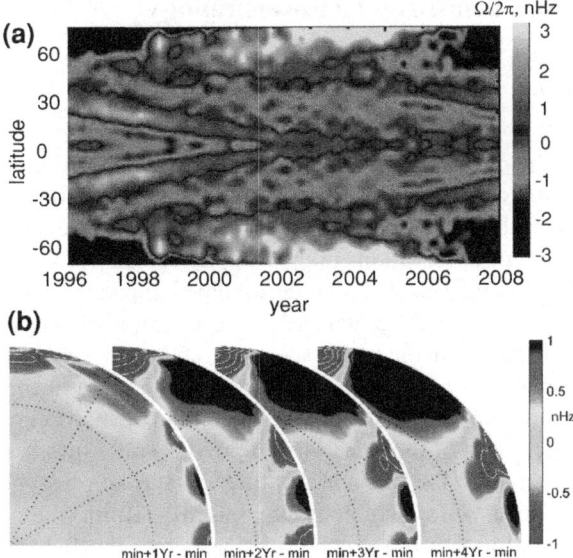

solar cycle progresses (Fig. 1.32). The torsional oscillations were first discovered on the Sun's surface [129], and then were found in the upper convection zone by helioseismology [130, 131]. The depth of these evolving zonal flows is not yet established. However, there are indications that they may be persistent through most of the convection zone, at least, at high latitudes [128]. The physical mechanism is not understood. Nevertheless, it is clear that these zonal flows are closely related to the internal dynamo mechanism that produces toroidal magnetic field. On the solar surface, this field forms sunspots and active regions which tend to appear in the areas of shear flows at the outer (relative to the equator) part of the faster bands. Thus, the torsional flows are an important key to understanding the solar dynamo, and one of the challenges is to establish their precise depth and detect corresponding variations in the thermodynamic structure of the convection zone. Recent modeling of the torsional oscillations by the Lorentz force feedback on differential rotation showed that the poleward-propagating high-latitude branch of the torsional oscillations can be explained as a response of the coupled differential rotation/meridional flow system to periodic forcing in midlatitudes of either mechanical (Lorentz force) or thermal nature [132]. However, the main equatorward-propagating branches cannot be explained by the Lorentz force, but maybe driven by thermal perturbations caused by magnetic field [133]. It is intriguing that starting from 2002, during the solar maximum, the helioseismology observations show new branches of "torsional oscillations" migrating from about 45° latitude towards the equator (Fig. 1.32a). They indicate the start of the next solar cycle, number 24, in the interior, and are obviously related to magnetic processes inside the Sun. However, magnetic field of the new cycle appeared on the surface only in 2008.

1.6 Local-area Helioseismology

1.6.1 Basic Principles

In the previous sections we discussed methods of global helioseismology, which are based on inversions of accurately measured frequencies and frequency splitting of normal oscillation modes of the Sun. The frequencies are measured from long time series of observations of the Doppler velocity of the solar disk. These time series are much longer than the mode lifetimes, typically, two or three 36-day-long 'GONG months', that is 72 or 108 days. The long time series allow us to resolve individual mode peaks in the power spectrum, and accurately measure the frequencies and other parameters of these modes. However, because of the long integration times global helioseismology cannot capture the fast evolution of magnetic activity in subsurface layers of the Sun. Also, it provides only information about the axisymmetrical structure of the Sun and the differential rotation (zonal flows).

Local helioseismology attempts to determine the subsurface structure and dynamics of the Sun in local areas by analyzing local characteristics of solar oscillations, such as frequency and phase shifts and variations in wave travel times. This is a relatively new and rapidly growing field. It takes advantage of high-resolution observations of solar oscillations, currently available from the GONG+ helioseismology network and the space mission SOHO, and are anticipated from the SDO mission.

1.6.2 Ring-diagram Analysis

Local helioseismology was pioneered by Gough and Toomre [46] who first proposed to measure oscillation frequencies of solar modes as a function of the wavevector, $\omega(\mathbf{k})$, (the dispersion relation) in local areas, and use these measurements for diagnostics of the local flows and thermodynamic properties. They noticed that subsurface variations of temperature cause change in the frequencies, and that subsurface flows result in distortion of the dispersion relation because of the advection effect.

This idea was implemented by Hill [47] in the form of a ring-diagram analysis. The name of this technique comes from the ring appearance of the 3D dispersion relation, $\omega = \omega(k_x, k_y)$, in the (k_x, k_y) plane, where k_x and k_y are x- and y-components of the wave vector, $\mathbf{k}$ (Fig. 1.33). The ridges in the vertical cuts represent the same mode ridges as in Fig. 1.3, corresponding to the normal oscillation modes of different radial orders n.

In the presence of a horizontal flow field, $\mathbf{U} = (U_x, U_y)$ the dispersion relation has the form:

$$\omega = \omega_0(k) + \mathbf{k} \cdot \mathbf{U} \equiv \omega_0 + (U_x k_x + U_y k_y), \qquad (1.129)$$

Fig. 1.33 Three-dimensional
power spectrum of solar
oscillations, $P(k_x, k_y, \omega)$.
The vertical panels with *blue
background* show the mode
ridge structure similar to the
global oscillation spectrum
shown in Fig. 1.3. The
horizontal cut with
transparent background
shows the ring structure of
the power spectrum at a
given frequency(courtesy of
Amara Graps)

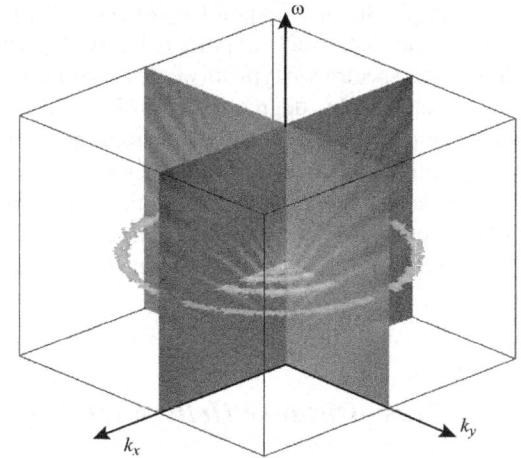

where $\omega_0(k)$ is the symmetrical part of the dispersion relation in the (k_x, k_y)-plane. It
depends only on the magnitude of the wave vector, k. The power spectrum, $P(\omega, k)$,
for each k is fitted with a Lorentzian profile [134]:

$$P(\omega, k) = \frac{A}{(\omega - \omega_0 + k_x U_x + k_y U_y)^2 + \Gamma^2} + \frac{b_0}{k^3}, \qquad (1.130)$$

where A, ω_0, Γ, and b_0 are respectively the amplitude, central frequency, line width
and a background noise parameter.

In some realizations, the fitting formula includes the line asymmetry (Sect. 1.3).
Also, the central frequency can be fitted by assuming a power-law relation: $\omega_0 = ck^p$,
where c and p are constants [47, 135]. This relationship is valid for a polytropic
adiabatic stratification, where $p = 1/2$ [46]. If the flow velocity changes with depth
then the parameter, U, represent a velocity, averaged with the depth with a weighting
factor proportional to the kinetic energy density of the waves, $\rho\boldsymbol{\xi} \cdot \boldsymbol{\xi}$ [136]:

$$U = \frac{\int u(z)\rho\boldsymbol{\xi} \cdot \boldsymbol{\xi} dz}{\int \rho\boldsymbol{\xi} \cdot \boldsymbol{\xi} dz}, \qquad (1.131)$$

where $\boldsymbol{\xi}(z) = (\xi_r, \xi_h)$ is the wave amplitude, given by the mode displacement eigenfunctions (1.15). The integral is taken over the entire extent of the solar envelope.
Equation (1.131) is solved by the RLS or OLA techniques (Sect. 1.5).

The ring-diagram method has provided important results about the structure and
evolution of large-scale and meridional flows and dynamics of active regions [127,
134, 137–139]. In particular, large-scale patterns of subsurface flows converging
around magnetic active regions were discovered [138]. These flows cause variations
of the mean meridional circulation with the solar cycle [134], which may affect
transport of magnetic flux of decaying active regions from low latitudes to the polar
regions, and thus change the duration and magnitude of the solar cycles [140].

However, the ring-diagram technique in the present formulation has limitations in terms of the spatial and temporal resolution and the depth coverage. The local oscillation power spectra are typically calculated for regions with the horizontal size covering 15 heliographic degrees ($\simeq 180$ Mm). This is significantly larger than the typical size of supergranulation and active regions ($\simeq 30$ Mm). There have been attempts to increase the resolution by doing the measurements in overlapping regions (so-called "dense-packed diagrams"). However, since such measurements are not independent, their resolution is unclear. The measurements of the power spectra calculated for smaller regions (2–4° in size) increase the spatial resolution but decrease the depth coverage [141].

1.6.3 Time–Distance Helioseismology (Solar Tomography)

Further developments of local seismology led to the idea to perform measurements of local wave distortions in the time–distance space instead of the traditional frequency–wavenumber Fourier space [48]. In this case, the wave distortions can be measured as perturbations of wave travel times. However, because of the stochastic nature of solar waves it is impossible to track individual wave fronts. Instead, it was suggested to use a cross-covariance (time–distance) function that provides a statistical measure of the wave distortion. Indeed, by cross-correlating solar oscillation signals at two points one may expect that the main contribution to this cross-correlation will be from the waves traveling between these points along the acoustic ray paths [142, 143]. Thus, the cross-covariance function calculated for oscillation signals measured at two points separated by a distance, Δ, for various time lags, τ, has a peak when the time lag is equal to the travel time of acoustic waves between these points. Physically, the cross-covariance function corresponds to the Green's function of the wave equation, representing the wave signal from a point source. Of course, in reality, because of the finite wavelength effects, non-uniform distribution of acoustic sources, and complicated wave interaction with turbulence and magnetic fields the interpretation of the travel-time measurements is extremely challenging. Various approximations are used to relate the observed perturbations of the travel times to the internal properties such as sound–speed perturbations and flow velocities. We discuss the basic principles and the current status of the time–distance helioseismology method in Sect. 1.7.

1.6.4 Acoustic Holography and Imaging

The acoustic holography [144] and acoustic imaging [51] techniques are developed on the principles of day-light imaging by collecting over large areas on the solar surface coherent acoustic signals emitted from selected target points of the interior. The idea is that the signals constructed this way contain information about objects located below the surface because of wave absorption or scattering at the target points. The phases of individual signals are calculated by using the time–distance relation, $\tau(\Delta)$, for acoustic waves traveling along the ray paths. The constructed signals, $\psi_{\text{out,in}}(t)$, are calculated using the following relation [145]:

$$\psi_{\text{out,in}}(t) = \sum_{\tau_1}^{\tau_2} W \overline{\psi}(\Delta, t \pm \tau), \qquad (1.132)$$

where $\overline{\psi}(\Delta, t + \tau)$ is the azimuthal-averaged signal at a distance Δ from a target point at time $t \pm \tau(\Delta)$. The summation variable τ is equally spaced in the interval (τ_1, τ_2); and the weighting factor, $W \propto (\sin \Delta / \tau^2)^{1/2}$, describes the geometrical spreading of acoustic waves with distance. The positive sign in (1.132) corresponds to ψ_{out} constructed with waves traveling outward from a target point ("egression signal" [144]), while the negative sign provides ψ_{in} constructed with the incoming waves ("ingression signal").

The amplitude and phase of the constructed signals contain information about subsurface perturbation. A practical approach to extract this is to cross-correlate the outgoing and incoming signals [146, 147]:

$$C(t) = \int \psi_{\text{in}}(t') \psi_{\text{out}}(t' + t) dt', \qquad (1.133)$$

and then to measure time shifts of this function for various target positions relative to the corresponding quiet Sun values. These measurements correspond to the travel-time variations obtained by time–distance helioseismology [148, 149]. Further analysis of the travel-time variations is similar to the time–distance helioseismology method [50]. The advantages and disadvantages of the time–distance helioseismology and acoustic holography/imaging are not clear. Both approaches are being tested using various types of artificial data and applied for measuring subsurface structures and flows. Most of the current inferences of subsurface structures and flows have been obtained using the time–distance approach [48, 50]. The time–distance helioseismology method, also called *solar tomography* is described in more detail in the following section.

1.7 Solar Tomography

1.7.1 Time–distance Diagram

Solar acoustic waves (*p*-modes) are excited by turbulent convection near the solar surface and travel through the interior with the speed of sound. Because the sound speed increases with depth the waves are refracted and reappear on the surface at some distance from the source. The wave propagation is illustrated in Fig. 1.34.

Fig. 1.34 A cross-section
diagram through the solar
interior showing a sample of
wave paths inside the Sun

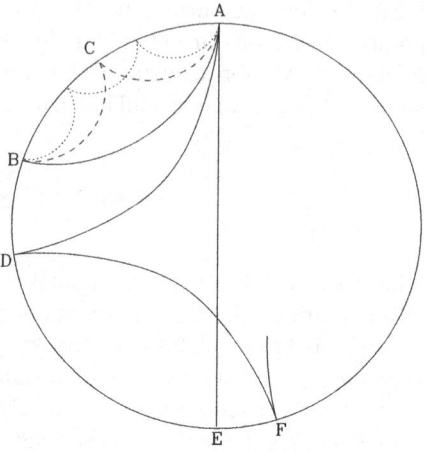

Waves excited at point A will reappear at the surface points B, C, D, E, F, and others
after propagating along the ray paths indicated by the curves connecting these points.

The basic idea of *time–distance helioseismology*, or *helioseismic tomography*, is
to measure the acoustic travel time between different points on the solar surface, and
then to use these measurements for inferring variations of wave-speed perturbations
and flow velocities in the interior by inversion [48]. This idea is similar to seismology
of Earth. However, unlike in Earth, the solar waves are generated stochastically by
numerous acoustic sources in a subsurface layer of turbulent convection.

Therefore, the wave travel time is determined from the cross-covariance function,
$\Psi(\tau, \Delta)$, of the oscillation signal, $f(t, r)$:

$$\Psi(\tau, \Delta) = \int_0^T f(t, r_1) f^*(t + \tau, r_2) dt, \tag{1.134}$$

where Δ is the horizontal distance between two points with coordinates r_1 and
r_2, τ is the lag time, and T is the total time of the observations. The normalized
cross-covariance function is called cross-correlation. The time–distance analysis is
based on non-normalized cross-covariance. Because of the stochastic nature of solar
oscillations, function Ψ must be averaged over some areas to achieve a good signal-
to-noise ratio sufficient for measuring the travel times. The oscillation signal, $f(t, r)$,
is measured from the Doppler shift or intensity of a spectral line. A typical cross-
covariance function obtained from full-disk solar observations of the Doppler shift
shown in Fig. 1.35a displays a set of ridges. The ridges correspond to acoustic wave
packets traveling between two points on the surface directly through the interior or
with intermediate reflections (bounces) from the surface as illustrated in Fig. 1.34.

The waves originating at point A may reach point B directly (solid curve) forming
the first-bounce ridge, or after one bounce at point C (dashed curve) forming the

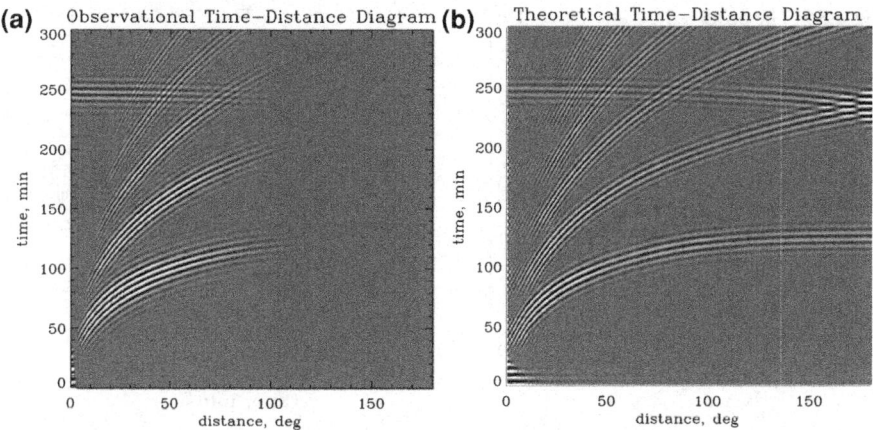

Fig. 1.35 The observational **a** and theoretical **b** cross-covariance functions (time–distance diagrams) as a function of distance on the solar surface, Δ, and the delay time, τ. The lowest set of ridges ('first bounce') corresponds to acoustic waves propagated to the distance, Δ, without additional reflections from the solar surface. The second from the bottom ridge ('second bounce') is produced by the waves arriving to the same distance after one reflection from the surface, and the third ridge ('third bounce') results from the waves arriving after two bounces from the surface. The backward ridge at $\tau \approx 250$ min is a continuation of the second-bounce ridge due to the choice of the angular distance range from 0 to 180° (that is, the counterclockwise distance ADF in Fig. 1.34 is substituted with the clockwise distance AF). Because of foreshortening close to the solar limb the observational cross-covariance function covers only $\sim 110°$ of distance

second-bounce ridge, or after two bounces (dotted curve)—the third-bounce ridge and so on. Because the sound speed is higher in the deeper layers the direct waves arrive first, followed by the second-bounce and higher-bounce waves.

The cross-covariance function represents a *time–distance diagram*, or a solar 'seismogram'. Fig. 1.36 shows the cross-covariance signal as a function of time for the travel distance, Δ, of 30°. It consists of three wave packets corresponding to the first, second and third bounces. Ideally, like in Earth seismology, the seismogram can be inverted to infer the structure and flows using a wave theory. However, in practice, modeling the wave fronts is a computationally intensive task. Therefore, the analysis is performed by measuring and inverting the phase and group travel times of the wave packets employing various approximations, the most simple and powerful of which is the ray-path approximation.

Generally, the observed solar oscillation signal corresponds to displacement or pressure perturbation, and can be represented in terms of the normal modes eigenfunctions. Therefore, the cross-covariance function also can be expressed in terms of the normal modes. In addition, it can be represented as a superposition of traveling wave packets, as we show in the next subsection [50]. An example of the theoretical cross-covariance function calculated using normal p-modes of the standard solar model is shown in Fig. 1.35b. This model reproduces the observational cross-covariance function very well in the observed range of distances, from 0 to

Fig. 1.36 The observed
cross-covariance signal as a
function of time at the
distance of 30°

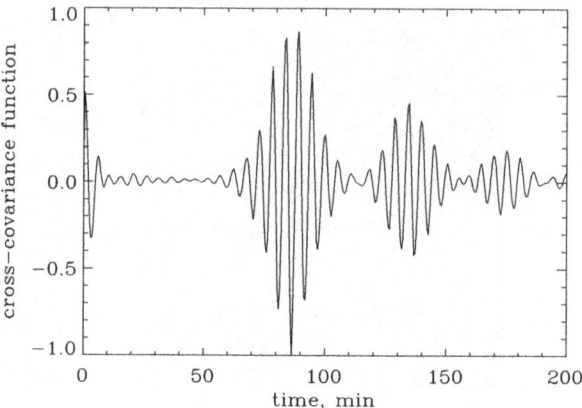

90°. The theoretical model was calculated for larger distances than the corresponding observational diagram in Fig. 1.35a, including points on the far side of the Sun,
which is not accessible for measurements. A backward propagating ridge originating
from the second-bounce ridge at 180° is a geometrical effect due to the choice of the
range of the angular distance from 0 to 180°. In the theoretical diagram (Fig. 1.35b)
one can notice a very weak backward ridge between 30 and 70° and at 120 min. This
ridge is due to reflection from the boundary between the convection and radiative
zones. However, this signal has not been detected in observations.

1.7.2 Wave Travel Times

For simplicity we consider solar oscillation signals observed not far from the disk
center and describe these in terms of the radial displacement neglecting the horizontal
displacement. The general theory was developed by Nigam and Kosovichev [150].
In the simple case, the solar oscillation signal can be represented in terms of the
radial eigenfunctions (1.42):

$$f(t, r, \theta, \phi) = \sum_{nlm} a_{nlm} \xi_r^{(n,l,m)}(r, \theta, \phi) \exp(i\omega_{nlm}t + i\phi_{nlm}), \qquad (1.135)$$

where n, l and m are the radial order, angular degree and angular order of a normal
mode respectively, $\xi_{nlm}(r, \theta, \phi)$ is a mode eigenfunction in the spherical coordinates,
r, θ and ϕ, ω_{nlm} is the eigenfrequency, and ϕ_{nlm} is an initial phase of the mode. Using
(1.135), we calculate the cross-covariance function, and express it as a superposition
of traveling wave packets. Such a representation is important for interpretation of
the time–distance data. A similar correspondence between the normal modes and the
wave packets has been discussed for surface oscillations in Earth's seismology [151]
and also for ocean waves [152].

To simplify the analysis, we consider the spherically symmetrical case. In this case, the mode eigenfrequencies do not depend on the azimuthal order m. For a radially stratified sphere, the eigenfunctions can be represented in terms of spherical harmonics $Y_{lm}(\theta, \phi)$ (1.42):

$$\xi_r^{(n,l,m)}(r, \theta, \phi) = \xi_r^{(n,l)}(r) Y_{lm}(\theta, \phi), \qquad (1.136)$$

where $\xi_r^{(n,l)}(r)$ is the radial eigenfunction [153].

Using, the convolution theorem [154] we express the cross-covariance function in terms of a Fourier intergral:

$$\Psi(\tau, \Delta) = \int\limits_{-\infty}^{\infty} F(\omega, r_1) F^*(\omega, r_2) \exp(i\omega\tau) d\omega, \qquad (1.137)$$

where $F(\omega, r)$ is Fourier transform of the oscillation signal $f(t, r)$.

The oscillation signal is considered as band-limited and filtered to select a p-mode frequency range using a Gaussian transfer function:

$$G(\omega) = \exp\left[-\frac{1}{2} \left(\frac{\omega - \omega_0}{\delta\omega} \right)^2 \right], \qquad (1.138)$$

where ω is the cyclic frequency, ω_0 is the central frequency and $\delta\omega$ is the characteristic bandwidth of the filter. The cross-covariance function in Fig. 1.35 displays three sets of ridges which correspond to the first, second and third bounces of acoustic wave packets from the surface.

The time series used in our analysis are considerably longer than the travel time τ, therefore, we can neglect the effect of the window function, and represent $F(\omega, r)$ in the form

$$F(\omega, r, \theta, \phi) \approx A \sum_{nlm} \xi_r^{(n,l)}(r) Y_{lm}(\theta, \phi) \delta(\omega - \omega_{nl}) \exp\left[-\frac{1}{2} \left(\frac{\omega - \omega_0}{\delta\omega} \right)^2 \right], \qquad (1.139)$$

where $\delta(x)$ is the delta-function, ω_{nl} are frequencies of the normal modes, and A is the amplitude of the Gaussian envelope of the amplitude spectrum at $\omega = \omega_0$. In addition, we assume the normalization conditions: $\xi_r^{(n,l)}(R) = 1$, $a_{nl} = AG(\omega_{nl})$. Then, the cross-covariance function is

$$\Psi(\tau, \Delta) = A^2 \sum_{nl} \exp\left[-\left(\frac{\omega_{nl} - \omega_0}{\delta\omega} \right)^2 + i\omega_{nl}\tau \right] \sum_{m=-l}^{l} Y_{lm}(\theta_1, \phi_1) Y_{lm}^*(\theta_2, \phi_2), \qquad (1.140)$$

where θ_1, ϕ_1 and θ_2, ϕ_2 are the spherical heliographic coordinates of the two observational points. The sum of the spherical function products is:

$$\sum_{m=-l}^{l} Y_{lm}(\theta_1, \phi_1) Y_{lm}^*(\theta_2, \phi_2) = \alpha_l P_l(\cos \Delta), \tag{1.141}$$

where $P_l(\cos \Delta)$ is the Legendre polynomial, Δ is the angular distance between points 1 and 2 along the great circle on the sphere, $\cos \Delta = \cos \theta_1 \cos \theta_2 + \sin \theta_1 \sin \theta_2 \cos(\phi_2 - \phi_1)$, and $\alpha_l = \sqrt{4\pi/(2l+1)}$. Then, the cross-covariance function is:

$$\Psi(\tau, \Delta) \approx A^2 \sum_{nl} \alpha_l P_l(\cos \Delta) \exp\left[-\left(\frac{\omega_{nl} - \omega_0}{\delta\omega}\right)^2 + i\omega_{nl}\tau\right]. \tag{1.142}$$

For large values of $l\Delta$, but when Δ is small,

$$P_l(\cos \Delta) \simeq \sqrt{\frac{2}{\pi L \Delta}} \cos\left(L\Delta - \frac{\pi}{4}\right). \tag{1.143}$$

Thus,

$$\Psi(\tau, \Delta) = A^2 \sum_{nl} \frac{2}{L\sqrt{\Delta}} \exp\left[-\frac{(\omega_{nl} - \omega_0)^2}{\delta\omega^2}\right] \cos(\omega_{nl}\tau) \cos(L\Delta). \tag{1.144}$$

Now the double sum can be reduced to a convenient sum of integrals if we regroup the modes so that the outer sum is over the ratio $v = \omega_{nl}/L$ and the inner sum is over ω_{nl}.

According to the ray-path theory, the travel distance Δ of an acoustic wave is determined by the ratio v, which represents the horizontal angular phase velocity ($v = \omega_{nl}/L \equiv (\omega_{nl}/k_h)/r$). Because of the band-limited nature of the function G, only values of L which are close to $L_0 \equiv \omega_0/v$ contribute to the sum. We consider the relation L vs ω_{nl} as a continuous function along the mode ridges (Fig. 1.3), and expand L near the central frequency ω_0 :

$$L \simeq L_0 + \frac{\partial L}{\partial \omega_{nl}}(\omega_{nl} - \omega_0) = \frac{\omega_0}{v} + \frac{\omega_{nl} - \omega_0}{u}, \tag{1.145}$$

where $u \equiv \partial \omega_{nl}/\partial L$. Furthermore,

$$\cos(\omega_{nl}\tau) \cos(L\Delta) = \cos\left[\left(\tau - \frac{\Delta}{u}\right)\omega_{nl} + \left(\frac{1}{u} - \frac{1}{v}\right)\Delta\omega_0\right], \tag{1.146}$$

and the other term is identical except that τ has been replaced with $-\tau$ (negative time lag). The result is that the double sum in (1.144) becomes

$$\Psi(\tau, \Delta) \simeq A^2 \sum_{v} \frac{2}{L_0\sqrt{\Delta}} \sum_{\omega_{nl}} \exp\left[-\frac{(\omega - \omega_0)^2}{\delta\omega^2}\right] \cos\left[\left(\pm\tau - \frac{\Delta}{u}\right) + \left(\frac{1}{u} - \frac{1}{v}\right)\Delta\omega_0\right]. \tag{1.147}$$

The inner sum can be approximated by an integral, considering ω_{nl} as a continuous variable along the mode ridges:

$$\int_{-\infty}^{\infty} d\omega \exp\left[-\frac{(\omega - \omega_0)^2}{\delta\omega^2}\right] \cos\left[\left(\tau - \frac{\Delta}{u}\right)\omega - \left(\frac{1}{u} - \frac{1}{v}\right)\Delta\omega_0\right] =$$

$$\sqrt{\pi\delta\omega^2} \exp\left[-\frac{\delta\omega^2}{4}\left(\tau - \frac{\Delta}{u}\right)^2\right] \cos\left[\omega_0\left(\tau - \frac{\Delta}{v}\right)\right]. \tag{1.148}$$

The integration limits reflect the fact that the amplitude function $G(\omega)$ is essentially zero for very large and very small frequencies. Finally, the cross-covariance is expressed in the following form [50]:

$$\Psi(\tau, \Delta) = B \sum_{v} \cos\left[\omega_0\left(\tau - \tau_{\text{ph}}\right)\right] \exp\left[-\frac{\delta\omega^2}{4}\left(\tau - \tau_{\text{gr}}\right)^2\right], \tag{1.149}$$

where B is constant, $\tau_{\text{ph}} = \Delta/v$ and $\tau_{\text{gr}} = \Delta/u$ are the phase and group travel times. Equation (1.149) has the form of a Gabor wavelet. The phase and group travel times are measured by fitting individual terms of equation (1.149) to the observed cross-covariance function using a least-squares technique.

1.7.3 Deep- and Surface-Focus Measurement Schemes

As we have pointed out the travel-time measurements require averaging of the cross-covariance function in order to obtain a good signal-to-noise ratio. Two typical schemes of the spatial averaging suggested by Duvall [155] are shown in Fig. 1.37.

For the so-called 'surface-focusing' scheme (Fig. 1.37a) the measured travel times are mostly sensitive to the near surface condition at the central point where the ray paths are focused. However, by measuring the travel times for several distances and applying an inversion procedure it is possible to infer the distribution of the variations of the wave speed and flow velocities with depth. The averaging also can be done in such a way that the 'focus' point is located beneath the surface. An example of the 'deep-focusing' scheme is shown in Fig. 1.37b. In this case the travel times are more sensitive to deep structures but still inversions are required for correct interpretation.

1.7.4 Sensitivity Kernels: Ray-path Approximation

The travel-time inversion procedures are based on theoretical relations between the travel-time variations and interior properties constituting the forward problem of local helioseismology. Similarly to global helioseismology, these relations are expressed

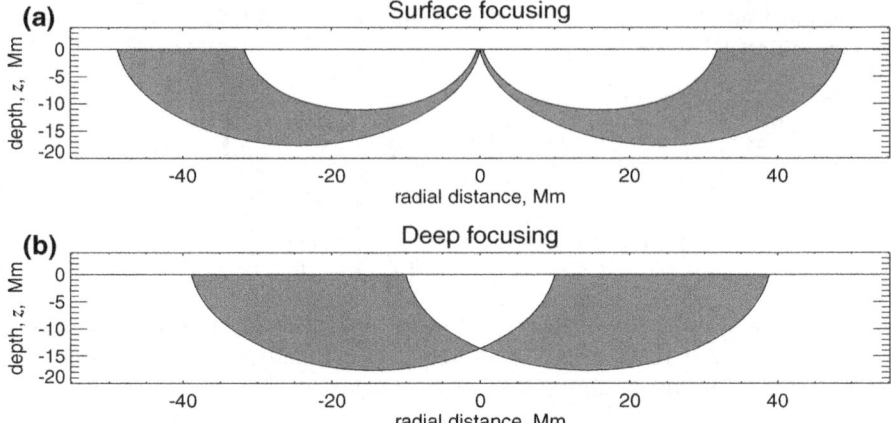

Fig. 1.37 The regions of ray propagation (*shaded areas*) as a function of depth, z, and the radial distance, Δ, from a point on the surface for two observing schemes: 'surface focusing' (**a**) and 'deep focusing' (**b**). The rays are also averaged over a circular regions on the surface, forming 3D figures of revolution

in the form of linear integral equations with sensitivity kernels. Two basic types of the sensitivity kernels have been used: ray-path kernels [50] and Born-approximation kernels [156–158]. The ray-path kernels are based on a simple and generally robust theoretical ray approximation, but they do not take into account finite wavelength effects and thus are not sufficiently accurate for diagnostics of small-scale structures. For reliable inferences it is important to use both these kernels.

In the ray approximation, the travel times are sensitive only to the perturbations along the ray paths given by Hamilton's equations (1.72). The variations of the phase travel time obey the Fermat's Principle:

$$\delta\tau = \frac{1}{\omega} \int_{\Gamma} \delta k \, dr, \tag{1.150}$$

where δk is the perturbation of the wave vector, k, due to the structural inhomogeneities and flows along the unperturbed ray path, Γ. Using the dispersion relation for acoustic waves in the convection zone the travel-time variations can be expressed in terms of the sound–speed, magnetic field strength and flow velocity.

The dispersion relation for magnetoacoustic waves in the convection zone is

$$(\omega - k \cdot U)^2 = \omega_c^2 + k^2 c_f^2, \tag{1.151}$$

where U is the flow velocity, ω_c is the acoustic cut-off frequency, $c_f^2 = \frac{1}{2}\left(c^2 + c_A^2 + \sqrt{(c^2 + c_A^2)^2 - 4c^2(k \cdot c_A)^2/k^2}\right)$ is the fast magnetoacoustic speed, $c_A = B/\sqrt{4\pi\rho}$ is the vector Alfvén velocity, B is the magnetic field strength,

c is the adiabatic sound speed, and ρ is the plasma density. If we assume that, in the unperturbed state $U = B = 0$, then, to the first-order approximation

$$\delta\tau = -\int_\Gamma \left[\frac{(n \cdot U)}{c^2} + \frac{\delta c}{c} S + \left(\frac{\delta\omega_c}{\omega_c}\right) \frac{\omega_c^2}{\omega^2 c^2 S} + \frac{1}{2}\left(\frac{c_A^2}{c^2} - \frac{(k \cdot c_A)^2}{k^2 c^2}\right) S \right] ds.$$

(1.152)

where n is a unit vector tangent to the ray, $S = k/\omega$ is the phase slowness.

Then, we separate the effects of flows and structural perturbations by measuring the travel times of acoustic waves traveling in opposite directions along the same ray path, and calculating the difference, τ_{diff} and the mean, τ_{mean}, of these reciprocal travel times:

$$\delta\tau_{\text{diff}} = -2\int_\Gamma \frac{(n \cdot U)}{c^2} ds;$$

(1.153)

$$\delta\tau_{\text{mean}} = -\int_\Gamma \left[\frac{\delta c}{c} S + \left(\frac{\delta\omega_c}{\omega_c}\right) \frac{\omega_c^2}{\omega^2 c^2 S} + \frac{1}{2}\left(\frac{c_A^2}{c^2} - \frac{(k \cdot c_A)^2}{k^2 c^2}\right) S \right] ds.$$

(1.154)

anisotropy of the last term of equation (1.154) allows us to separate, at least partly, the magnetic effects from the variations of the sound speed and the acoustic cut-off frequency. The acoustic cut-off frequency, ω_c may be perturbed by surface magnetic fields and by temperature and density inhomogeneities. The effect of the cut-off frequency variation depends strongly on the wave frequency, and, therefore, it results in a frequency dependence in τ_{mean}.

In practice, the travel times are measured from the cross-covariance functions between selected central points on the solar surface and surrounding quadrants symmetrical relative to the North, South, East and West directions. In each quadrant, the travel times are averaged over narrow ranges of the travel distance, Δ. The travel times of the northward-directed waves are subtracted from the times of the south-directed waves to yield the time, $\tau_{\text{diff}}^{\text{NS}}$, which is predominantly sensitive to subsurface north–south flows. Similarly, the time differences, $\tau_{\text{diff}}^{\text{EW}}$, between westward- and eastward directed waves yields a measure of the east-ward flows. The time, $\tau_{\text{diff}}^{\text{oi}}$, between the outward- and inward-directed waves, averaged over the full annuli, is mainly sensitive to vertical flows and divergence of the horizontal flows. This represents a cross-talk effect between the vertical flows and horizontal flows, which is difficult to resolve when the vertical flows are weak [159].

Thus, the effects of flows and structural perturbations are separated from each other by taking the difference and the mean of the reciprocal travel times:

$$\delta\tau_{\text{diff}} \approx -2\int_\Gamma \frac{(nU)}{c^2} ds;$$

(1.155)

Fig. 1.38 Travel-time sensitivity kernels in the first Born approximation for sound–speed variations as a function of the horizontal, x, and vertical, y, coordinates for: **a** the first-bounce signal for distance $\Delta = 6°$, **b** the second-bounce signal for $\Delta = 60°$. The *solid curves* show the corresponding ray paths at frequency $\nu = 3$ mHz [162]

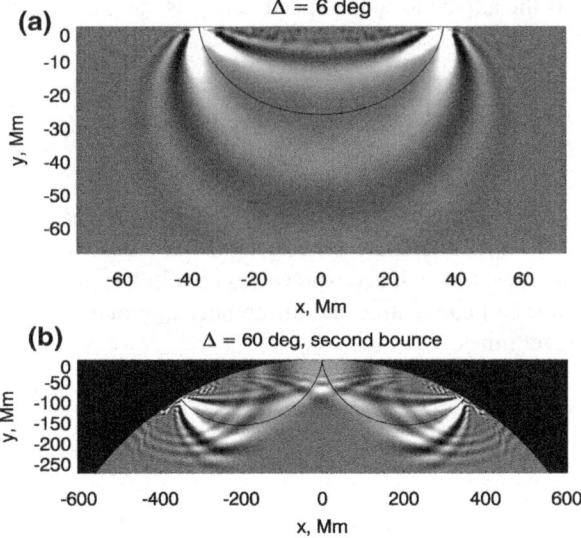

$$\delta\tau_{\text{mean}} \approx -\int_{\Gamma} \frac{\delta w}{c} S ds, \qquad (1.156)$$

where c is the adiabatic sound speed, $\boldsymbol{n}$ is a unit vector tangent to the ray, $S = k/\omega$ is the phase slowness, δw is the local wave speed perturbation:

$$\frac{\delta w}{c} = \frac{\delta c}{c} + \frac{1}{2}\left(\frac{c_A^2}{c^2} - \frac{(k c_A)^2}{k^2 c^2}\right). \qquad (1.157)$$

Magnetic field causes anisotropy of the mean travel times, which allows us to separate, in principle, the magnetic effects from the variations of the sound speed (or temperature). So far, only a combined effect of the magnetic fields and temperature variations has been measured reliably.

1.7.5 Born Approximation

The development of a more accurate theory for the travel times, based on the Born approximation is currently under way [156–158, 160, 161].

One unexpected feature of the single-source travel-time kernels calculated in the Born approximation is that these kernels have zero value along the ray path (called 'banana–doughnut kernels'). Examples of the Born kernels for the first and the second bounces are shown in Fig. 1.38. The kernels are mostly sensitive to perturbations within the first Fresnel zone.

Fig. 1.39 Tests of the ray and Born approximations: travel times for smooth spheres as functions of sphere radius at half maximum. The *solid lines* are the numerical results. The *dashed curves* are the Born approximation travel times and the *dotted lines* are the first order ray approximation. The *left panel* shows the two perturbations of the relative amplitude, $A = \pm 0.05$. The *right panel* is for the cases $A = \pm 0.1$ [160]

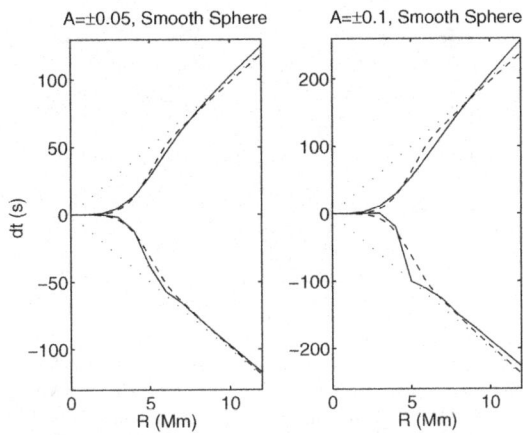

Figure 1.39 shows the test results for both the ray and Born approximations for a simple model of a smooth sphere in an uniform medium by comparing with precise numerical results [160]. These results show that for typical perturbations in the solar interior the Born approximation is sufficiently accurate, while the ray approximation significantly overestimates the travel times for perturbations smaller than the size of the first Fresnel zone. That means that the inversion results based on the ray theory may underestimate the strength of the small-scale perturbations. The comparison of the inversion results for sub-surface sound–speed structures beneath sunspots have showed a very good agreement between the ray-paths and Born theories [158].

1.8 Inversion Results of Solar Acoustic Tomography

The results of test inversions (e.g. [50, 55, 159]) demonstrate an accurate reconstruction of sound–speed variations and the horizontal components of subsurface flows. However, vertical flows in deep layers are not resolved because of the predominantly horizontal propagation of the rays in these layers. The vertical velocities are also systematically underestimated in the upper layers. When the vertical flow is weak, e.g. such as in supergranulation, the vertical velocity is not estimated correctly, because the trave-time signal is dominated by the horizontal flow divergence. In such situation, it is difficult to determine even the direction of the vertical flow [55]. Similarly, the sound–speed variations are underestimated in the deep layers and close to the surface. These limitations of the solar tomography should be taken into account in interpretation of the inversion results.

Here, I briefly present some examples of the local helioseismology inferences obtained by inversion of acoustic travel times.

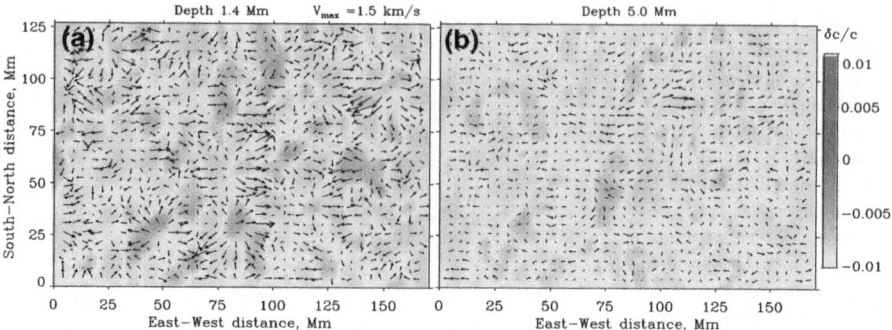

Fig. 1.40 The supergranulation horizontal flow velocity field (*arrows*) and the sound–speed perturbation (*color background*) at the depths of 1.4 Mm (**a**) and 5.0 Mm (**b**), as inferred from the SOHO/MDI high-resolution data of 27 January 1996 [50]

1.8.1 Diagnostics of Supergranulation

The data used were for 8.5 h on 27 January, 1996 from the high resolution mode of the MDI instrument. The results of inversion of these data are shown in Fig. 1.40 [50]. It has been found that, in the upper layers, 2–3 Mm deep, the horizontal flow is organized in supergranular cells, with outflows from the center of the supergranules. The characteristic size of the cells is 20–30 Mm. Comparing with MDI magnetograms, it was found that the cell boundaries coincide with the areas of enhanced magnetic field. These results are consistent with the observations of supergranulation on the solar surface. However, in the layers deeper than ~5 Mm, the supergranulation pattern disappears. The inversions show an evidence of reverse converging flows at the depth of ~10 Mm [159]. This means that supergranulation is a relatively shallow phenomenon.

1.8.2 Structure and Dynamics of Sunspot

The high-resolution data from the SOHO and HINODE space missions have allowed us to investigate the structure and dynamics beneath sunspots. Figure 1.41 shows an example of the internal structure of a large sunspot observed on June 17, 1998 [163]. An image of the spot taken in the continuum is shown at the top. The wave-speed perturbations under the sunspot are much stronger than those of the emerging flux, and can reach ~3 km/s. It is interesting that beneath the spot the perturbation is negative in the subsurface layers and becomes positive in the deeper interior. One can suggest that the negative perturbations beneath the spot are, probably, due to the lower temperature. It follows that magnetic inhibition of convection that makes sunspots cooler is most effective within the top 2–3 Mm of the convection zone. The strong positive perturbation below suggests that the deep sunspot structure is hotter

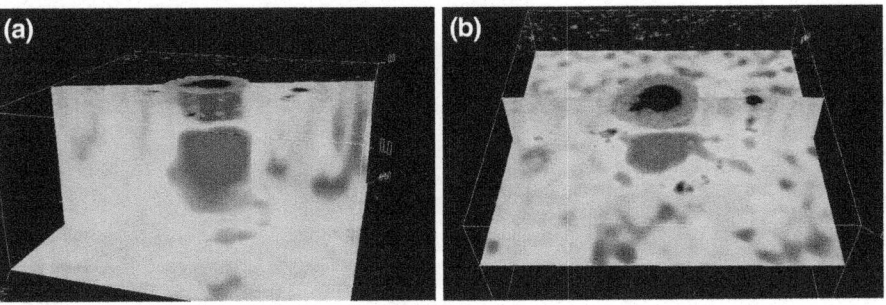

Fig. 1.41 The sound–speed perturbation in a large sunspot observed on June 20, 1998, are shown as *vertical* and *horizontal cuts*. The horizontal size of the box is 13° (158 Mm), the depth is 24 Mm. The positive variations of the sound speed are shown in *red*, and the negative variations (just beneath the sunspot) are in *blue*. The upper semitransparent panel is the surface intensity image (*dark color* shows umbra, and *light color* shows penumbra). In panel **b** the horizontal sound–speed plane is located at the depth of 4 Mm, and shows long narrow structures ('fingers') connecting the main sunspot structure with surrounding pores the same magnetic polarity as the spot [163]

than the surrounding plasma. However, the effects of temperature and magnetic field have not been separated in these inversions. Separating these effects is an important problem of solar tomography. These data also show at a depth of ∼4 Mm connections to the spot of small pores, which have the same magnetic polarity as the main spot. The pores of the opposite polarity are not connected to the main sunspot. This suggests that sunspots represent a tree-like structure in the upper convection zone.

Figure 1.42 shows the subsurface structures and flows beneath a sunspot obtained from HINODE [164]. A vertical cut along the East–West direction approximately in the middle of a large sunspot observed in AR 10953, May 2, 2007, (Fig. 1.42a), shows that the wave speed anomalies extend about half of the sunspot size beyond the sunspot penumbra into the plage area. In the vertical direction, the negative wave speed perturbation extends to a depth of 3–4 Mm. The positive perturbation is about 9 Mm deep, but it is not clear whether it extends further, because our inversion cannot reach deeper layers because of the small field of view. Similar two-layer sunspot structures were observed before from SOHO/MDI [163] (Fig. 1.41). But, it is striking that the new images strongly indicate on the cluster structure of the sunspot [165]. This was not previously seen in the tomographic images of sunspots obtained with lower resolution.

The high-resolution flow field below the sunspot is also significantly more complicated than the previously inferred from SOHO/MDI [166], but reveals the same general converging downdraft pattern. A vertical view of an averaged flow field (Fig. 1.42b) shows nicely the flow structure beneath the active region. Strong downdrafts are seen immediately below the sunspot's surface, and extend up to 6 Mm in depth. A little beyond the sunspot's boundary, one can find both upward and inward flows. Clearly, large-scale mass circulations form outside the sunspot, bringing plasma down along the sunspot's boundary, and back to the photosphere within about twice of the sunspot's radius. It is remarkable that such an apparent mass

Fig. 1.42 Wave speed perturbation and flow velocities beneath sunspots from HINODE data [164]

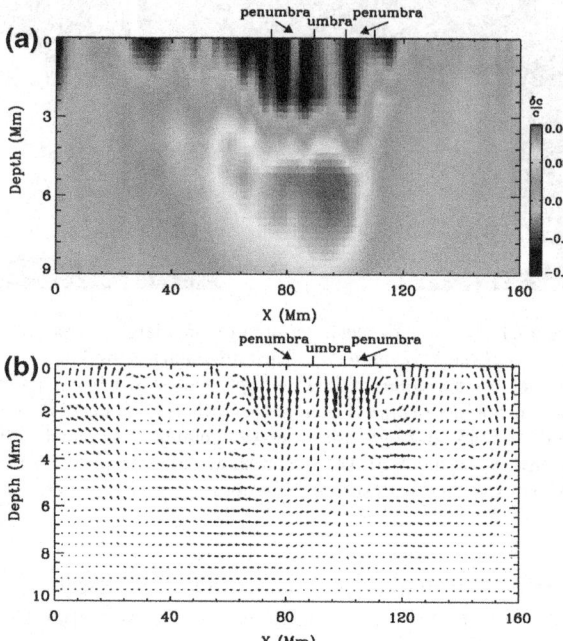

circulation is obtained directly from the helioseismic inversions without using any additional constraints, such as forced mass conservation. Previously, the circulation pattern was not that clear.

1.8.3 Large-Scale and Meridional Flows

Time–distance helioseismology [167] and also local measurements of the p-mode frequency shifts by the 'ring-diagram' analysis [134, 137, 138], have provided synoptic maps of subsurface flows over the whole surface of the Sun. Figure 1.43 shows a portion of a high-resolution synoptic flow map at the depth of 2 Mm below the surface. In addition, to the supergranulation pattern these maps reveal large-scale converging plasma flow around the active regions where magnetic field is concentrated. These flows are particularly well visible in low-resolution synoptic flow maps (Fig. 1.44). The characteristic speed of these flows is about 50 m/s.

These stable long-living flow patterns affect the global circulation in the Sun. It is particularly important that these flows change the mean meridional flow from the equator to the poles, slowing it down during the solar maximum years (Fig. 1.45). This may have important consequences for the solar dynamo theories which invoke the meridional flow to explain the magnetic flux transport into the polar regions and

Fig. 1.43 A portion of a synoptic subsurface flow map at depth of 2 Mm. The *color background* shows the distribution of magnetic field on the surface [167]

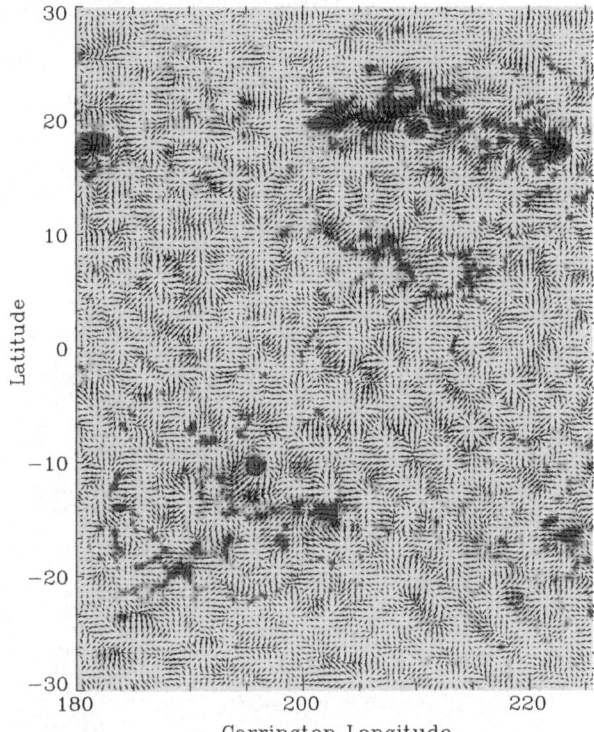

the polar magnetic field polarity reversals usually happening during the period of maximum of solar activity.

1.9 Conclusion and Outlook

During the past decade thanks to the long-term continuous observations from the ground and space the physics of solar oscillations made a tremendous progress in understanding the mechanism of solar oscillations, and in developing new techniques for helioseismic diagnostics of the solar structure and dynamics. However, many problems are still unresolved. Most of them are related to phenomena in strong magnetic field regions and in the deep interior. The prime helioseismology tasks are to detect processes of magnetic field generation and transport in the solar interior, and formation of active regions and sunspots. This will help us to understand the physics of the solar dynamo and the cyclic behavior of solar activity.

For solving these tasks it is very important to continue developing realistic MHD simulations of solar convection and oscillations, and to obtain continuous high-resolution helioseismology data for the whole Sun. The recent observations from

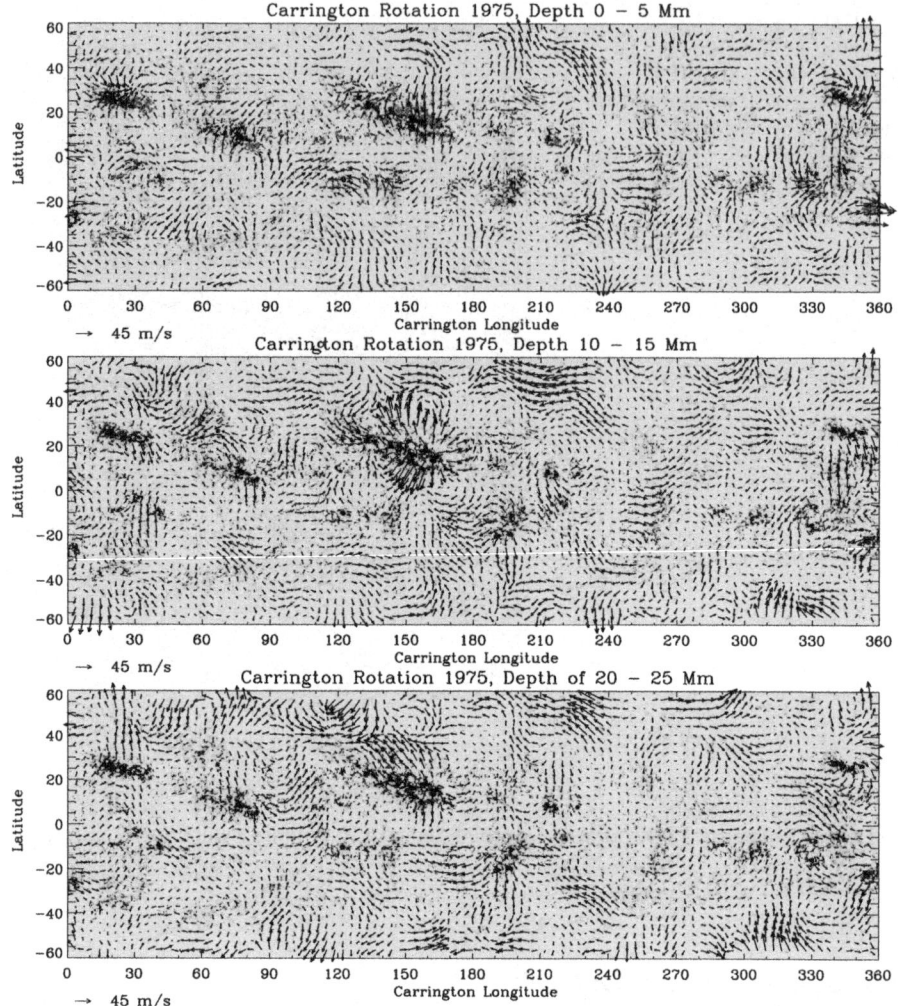

Fig. 1.44 Subsurface synoptic flow maps at three depths. The *color background* shows the distribution of magnetic field on the surface [167]

HINODE have demonstrated advantages of high-resolution helioseismology, but unfortunately such data are available only for small regions and for short periods of time. A new substantial progress in observations of solar oscillations is expected from the Solar Dynamics Observatory (SDO) space mission launched in February 2010.

The Helioseismic and Magnetic Imager (HMI) instrument on SDO provides uninterrupted Doppler shift measurements over the whole visible disk of the Sun with a spatial resolution of 0.5 arcsec per pixel (4096 × 4096 images) and 40–50 s time

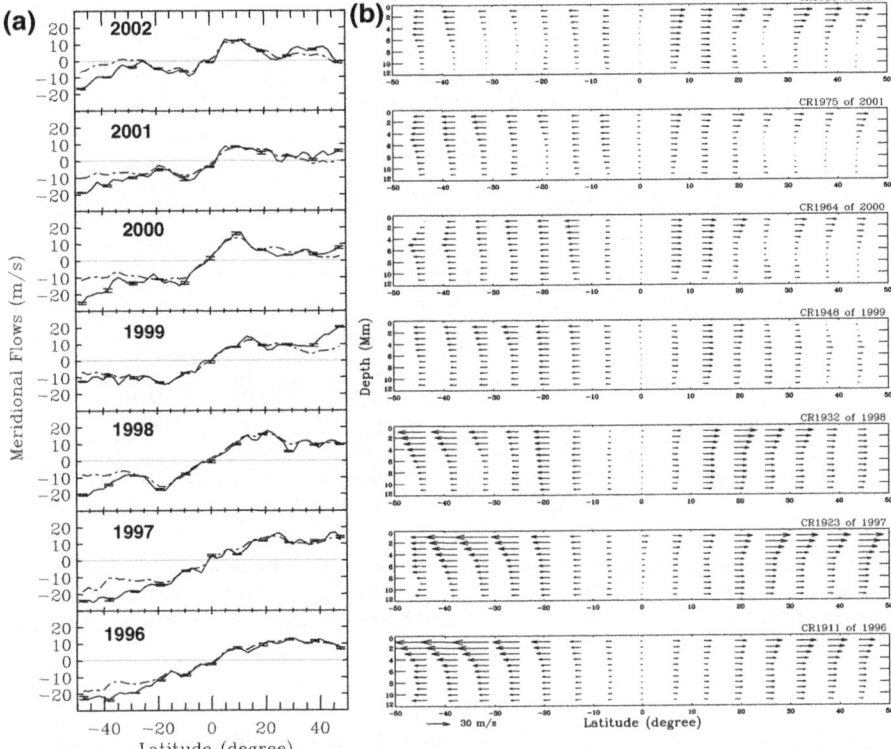

Fig. 1.45 Evolution of subsurface meridional flows during 1996–2002 for various Carrington rotations [167]

cadence. The total amount of data from this instrument will reach 2 Tb per day. This tremendous amount of data will be processed through a specially developed data analysis pipeline and will provide high-resolution maps of subsurface flows and sound–speed structures [53]. These data will enable investigations of the multi-scale dynamics and magnetism of the Sun and also contribute to our understanding of the Sun as a star.

The tools that will be used in the HMI program include: helioseismology to map and probe the solar convection zone where a magnetic dynamo likely generates this diverse range of activity; measurements of the photospheric magnetic field which results from the internal processes and drives the processes in the atmosphere; and brightness measurements which can reveal the relationship between magnetic and convective processes and solar irradiance variability.

Helioseismology, which uses solar oscillations to probe flows and structures in the solar interior, is providing remarkable new perspectives about the complex interactions between highly turbulent convection, rotation and magnetism. It has revealed a region of intense rotational shear at the base of the convection zone, called the

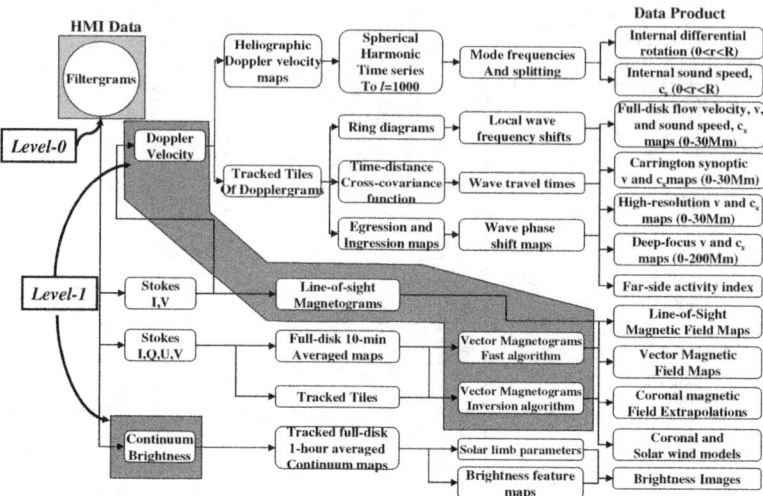

Fig. 1.46 A schematic illustration of the SDO–Solar Dynamics Observatory HMI data analysis pipeline and data products. The *dark shaded area* indicates Level-1 data products. The boxes to the right of this area represent intermediate and final Level-2 data products. The data products are described in detail in the HMI Science Plan [53]

tachocline, which is the likely seat of the global dynamo. Convective flows also have a crucial role in advecting and shearing the magnetic fields, twisting the emerging flux tubes and displacing the photospheric footpoints of magnetic structures present in the corona. Flows of all spatial scales influence the evolution of the magnetic fields, including how the fields generated near the base of the convection zone rise and emerge at the solar surface, and how the magnetic fields already present at the surface are advected and redistributed. Both of these mechanisms contribute to the establishment of magnetic field configurations that may become unstable and lead to eruptions that affect the near-Earth environment.

New methods of local-area helioseismology have begun to reveal the great complexity of rapidly evolving 3D magnetic structures and flows in the sub-surface shear layer in which the sunspots and active regions are embedded. Most of these new techniques were developed during analysis of MDI observations. As useful as they are, the limitations of MDI telemetry availability and the limited field of view at high resolution has prevented the full exploitation of the methods to answer the important questions about the origins of solar variability. By using these techniques for continuous full-disk high-resolution observations, HMI will enable detailed probing of dynamics and magnetism within the near-surface shear layer, and provide sensitive measures of variations in the tachocline.

The scientific operation modes and data products can be divided into four main areas: global helioseismology, local-area helioseismology, line-of-sight and vector magnetography and continuum intensity studies. The principal data flows and products are summarized in Fig. 1.46.

1.9.1 Global Helioseismology: Diagnostics of Global Changes Inside the Sun

The traditional normal-mode method described in Sects. 1.4 and 1.5, will provide large-scale axisymmetrical distributions of sound speed, density, adiabatic exponent and flow velocities through the whole solar interior from the energy-generating core to the near-surface convective boundary layer. These diagnostics will be based on frequencies and frequency splitting of modes of angular degree up to 1,000, obtained for several day intervals each month and up to $l = 300$ for each 2-month interval. These will be used to produce a regular sequence of internal rotation and sound–speed inversions to allow observation of the tachocline and average near surface shear.

1.9.2 Local-Area Helioseismology: 3D Imaging of the Solar Interior

The new methods of local-area helioseismology (Sects. 1.6–1.7), time–distance technique, ring-diagram analysis and acoustic holography represent powerful tools for investigating physical processes inside the Sun. These methods are based on measuring local properties of acoustic and surface gravity waves, such as travel times, frequency and phase shifts. They will provide images of internal structures and flows on various spatial and temporal scales and depth resolution. The targeted high-level regular data products include:

- Full-disk velocity and sound–speed maps of the upper convection zone (covering the top 30 Mm) obtained every 8 h with the time–distance methods on a Carrington grid;
- Synoptic maps of mass flows and sound–speed perturbations in the upper convection zone for each Carrington rotation with a 2-degree resolution, from averages of full disk time–distance maps;
- Synoptic maps of horizontal flows in upper convection zone for each Carrington rotation with a 5-degree resolution from ring-diagram analyses.
- Higher-resolution maps zoomed on particular active regions, sunspots and other targets, obtained with 4–8-h resolution for up to 9 days continuously, from the time–distance method;
- Deep-focus maps covering the whole convection zone depth, 0–200 Mm, with 10–15-degree resolution;
- Far-side images of the sound–speed perturbations associated with large active regions every 24 h.

The HMI science investigation addresses the fundamental problems of solar variability with studies in all interlinked time and space domains, including global scale, active regions, small scale, and coronal connections. One of the prime objectives of the Living With a Star program is to understand how well predictions of evolving space weather variability can be made. The HMI investigation will examine these questions in parallel with the fundamental science questions of how the Sun varies and how that variability drives global change and space weather.

Acknowledgment This work was supported by the CNRS, the International Space Science Institute (Bern), Nordita (Stockholm) and NASA.

References

1. Eddington, A.S.: The Internal Constitution of the Stars. Cambridge Univ. Press, Cambridge (1926)
2. Leighton, R.B., Noyes, R.W., Simon, G.W.: Astrophys. J. **135**, 474 (1962)
3. Mein, P.: Annales d'Astrophysique **29**, 153 (1966)
4. Frazier, E.N.: Astrophys. J. **152**, 557 (1968)
5. Ulrich, R.K.: Astrophys. J. **162**, 993 (1970)
6. Deubner, F.: Astron. Astrophys. **44**, 371 (1975)
7. Rhodes, E.J. Jr., Ulrich, R.K., Simon, G.W.: Astrophys. J. **218**, 901 (1975)
8. Ando, H., Osaki, Y.: Pub. Astron. Soc. Jpn **29**, 221 (1977)
9. Bahcall, J.N., Bahcall, N.A., Ulrich, R.K.: Astrophys. J. **156**, 559 (1969)
10. Hill, H.A., Stebbins, R.T., Brown, T.M.: In: Bulletin of the American Astronomical Society, vol. 7, p. 478 (1975)
11. Severny, A.B., Kotov, V.A., Tsap, T.T.: Nature **259**, 87 (1976)
12. Brookes, J.R., Isaak, G.R., van der Raay, H.B.: Nature **259**, 92 (1976)
13. Scherrer, P.H., Wilcox, J.M., Kotov, V.A., Severny, A.B., Tsap, T.T.: Nature **277**, 635 (1979)
14. Grec, G., Fossat, E., Pomerantz, M.: Nature **288**, 541 (1980)
15. Pallé, P.L., Roca Cortés, T., Gelly, B., the GOLF Team: In: Korzennik, S. (ed.) Structure and Dynamics of the Interior of the Sun and Sun-like Stars, vol. 418, p. 291. ESA Special Publication (1998)
16. Claverie, A., Isaak, G.R., McLeod, C.P., van der Raay, H.B., Cortes, T.R. Nature **282**, 591 (1979)
17. Vandakurov, Y.V.: Sov. Astron. **11**, 630 (1968)
18. Iben, I. Jr., Mahaffy, J.: Astrophys. J. Lett. **209**, L39 (1976)
19. Christensen-Dalsgaard, J., Gough, D.O., Morgan, J.G.: Astron. Astrophys. **73**, 121 (1979)
20. Christensen-Dalsgaard, J., Gough, D.O.: Astron. Astrophys. **104**, 173 (1981)
21. Duvall, T.L. Jr., Harvey, J.W.: Nature **302**, 24 (1983)
22. Ahmad, Q.R., et al.: Phys. Rev. Lett. **89**(1), 011301 (2002)
23. Pekeris, C.L.: Astrophys. J. **88**, 189 (1938)
24. Cowling, T.G.: Mon. Not. Roy. Astron. Soc. **101**, 367 (1941)
25. Ledoux, P., Walraven, T.: Handbuch der Physik **51**, 353 (1958)
26. Gough, D.O., Thompson, M.J.: Mon. Not. Roy. Astron. Soc. **242**, 25 (1990)
27. Dziembowski, W., Goode, P.R.: Memorie della Societa Astronomica Italiana **55**, 185 (1984)
28. Dziembowski, W.A., Goode, P.R.: Astrophys. J. **347**, 540 (1989)
29. Dziembowski, W.A., Goode, P.R.: Astrophys. J. **625**, 548 (2005)
30. Rhodes, E.J. Jr., Ulrich, R.K., Deubner, F.: Astrophys. J. **227**, 629 (1979)
31. Ulrich, R.K., Rhodes, E.J. Jr., Deubner, F.: Astrophys. J. **227**, 638 (1979)

32. Deubner, F., Ulrich, R.K., Rhodes, E.J. Jr.: Astron. Astrophys. **72**, 177 (1979)
33. Duvall, T.L. Jr., Harvey, J.W.: Nature **310**, 19 (1984)
34. Duvall, T.L. Jr., Dziembowski, W.A., Goode, P.R., Gough, D.O., Harvey, J.W., Leibacher, J.W.: Nature **310**, 22 (1984)
35. Brown, T.M., Morrow, C.A.: Astrophys. J. Lett. **314**, L21 (1987)
36. Kosovichev, A.G.: Sov. Astron. Lett. **14**, 145 (1988)
37. Brown, T.M., Christensen-Dalsgaard, J., Dziembowski, W.A., Goode, P., Gough, D.O., Morrow, C.A.: Astrophys. J. **343**, 526 (1989)
38. Rosner, R., Weiss, N.O.: Nature **317**, 790 (1985)
39. Parker, E.N.: Astrophys. J. **408**, 707 (1993)
40. Harvey, J.W., Abdel-Gawad, K., Ball, W., Boxum, B., Bull, F., Cole, J. Cole, L., Colley, S., Dowdney, K., Drake, R.: In: Rolfe, E.J.(ed.) Seismology of the Sun and Sun-Like Stars, vol. 286, pp. 203–208. ESA Special Publication (1988)
41. Brookes, J.R., Isaak, G.R., van der Raay, H.B.: Mon. Not. Roy. Astron. Soc. **185**, 1 (1978)
42. Isaak, G.R., McLeod, C.P., Palle, P.L., van der Raay, H.B., Roca Cortes, T.: Astron. Astrophys. **208**, 297 (1989)
43. Domingo, V., Fleck, B., Poland, A.I.: Sol. Phys. **162**, 1 (1995)
44. Scherrer, P.H., Bogart, R.S., Bush, R.I., Hoeksema, J.T., Kosovichev, A.G., Schou, J., Rosenberg, W., Springer, L., Tarbell, T.D., Title, A., Wolfson, C.J., Zayer, I., MDI Engineering Team: Sol. Phys. **162**, 129 (1995)
45. Leibacher, J.: In: IAU Joint Discussion, vol. 12, p. 46 (2003)
46. Gough, D.O., Toomre, J.: Sol. Phys. **82**, 401 (1983)
47. Hill, F.: Astrophys. J. **333**, 996 (1988)
48. Duvall, T.L. Jr., Jefferies, S.M., Harvey, J.W., Pomerantz, M.A.: Nature **362**, 430 (1993)
49. Kosovichev, A.G.: Astrophys. J. Lett. **469**, L61 (1996)
50. Kosovichev, A.G., Duvall, T.L. Jr.: In: Pijpers, F.P., Christensen-Dalsgaard, J., Rosenthal, C.S. (eds.) SCORe'96: Solar Convection and Oscillations and their Relationship, Astrophysics and Space Science Library, vol. 225, pp. 241–260 (1997)
51. Chang, H., Chou, D., Labonte, B., The TON Team: Nature **389**, 825 (1997)
52. Lindsey, C., Braun,D.C.: Sol. Phys. **192**, 261 (2000)
53. Kosovichev, A.G., HMI Science Team: Astronomische Nachrichten **328**, 339 (2007)
54. Stein, R.F., Nordlund, A.: Astrophys. J. Lett. **342**, L95 (1989)
55. Zhao, J., Georgobiani, D., Kosovichev, A.G., Benson, D., Stein, R.F., Nordlund, Å.: Astrophys. J. **659**, 848 (2007)
56. Jacoutot, L., Kosovichev, A.G., Wray, A.A., Mansour, N.N.: Astrophys. J. **682**, 1386 (2008)
57. Hanasoge, S.M., Duvall, T.L. Jr., Couvidat, S.: Astrophys. J. **664**, 1234 (2007)
58. Parchevsky, K.V., Zhao, J., Kosovichev, A.G.: Astrophys. J. **678**, 1498 (2008)
59. Hartlep, T., Zhao, J., Mansour, N.N., Kosovichev, A.G.: Astrophys. J. **689**, 1373 (2008)
60. Rimmele, T.R., Goode, P.R., Harold, E., Stebbins, R.T.: Astrophys. J. Lett. **444**, L119 (1995)
61. Skartlien, R., Rast, M.P.: Astrophys. J. **535**, 464 (2000)
62. Stein, R.F., Nordlund, Å..: Astrophys. J. **546**, 585 (2001)
63. Germano, M., Piomelli, U., Moin, P., Cabot, W.H.: Phys. Fluids **3**, 1760 (1991)
64. Moin, P., Squires, K., Cabot, W., Lee, S.: Phys. Fluids **3**, 2746 (1991)
65. Baudin, F., Samadi, R., Goupil, M., Appourchaux, T., Barban, C., Boumier, P., Chaplin, W.J., Gouttebroze, P.: Astron. Astrophys. **433**, 349 (2005)
66. Jacoutot, L., Kosovichev, A.G., Wray, A., Mansour, N.N.: Astrophys. J. Lett. **684**, L51 (2008)
67. Duvall, T.L. Jr., Jefferies, S.M., Harvey, J.W., Osaki, Y., Pomerantz, M.A.: Astrophys. J. **410**, 829 (1993)
68. Toutain, T., Appourchaux, T., Fröhlich, C., Kosovichev, A.G., Nigam, R., Scherrer, P.H.: Astrophys. J. Lett. **506**, L147 (1998)
69. Gabriel, M.: Astron. Astrophys. **265**, 771 (1992)
70. Nigam, R., Kosovichev, A.G., Scherrer, P.H., Schou, J.: Astrophys. J. Lett. **495**, L115 (1998)
71. Mitra-Kraev, U., Kosovichev, A.G., Sekii, T.: Astron. Astrophys. **481**, L1 (2008)

72. Roxburgh, I.W., Vorontsov, S.V.: Mon. Not. Roy. Astron. Soc. **292**, L33 (1997)
73. Severino, G., Magrì, M., Oliviero, M., Straus, T., Jefferies, S.M.: Astrophys. J. **561**, 444 (2001)
74. Wachter, R., Kosovichev, A.G.: Astrophys. J. **627**, 550 (2005)
75. Georgobiani, D., Stein, R.F., Nordlund, Å..: Astrophys. J. **596**, 698 (2003)
76. Kosugi, T., Matsuzaki, K., Sakao, T., Shimizu, T., Sone, Y., Tachikawa, S., Hashimoto, T., Minesugi, K., Ohnishi, A., Yamada, T., Tsuneta, S., Hara, H., Ichimoto, K., Suematsu, Y., Shimojo, M., Watanabe, T., Shimada, S., Davis, J.M., Hill, L.D., Owens, J.K., Title, A.M., Culhane, J.L., Harra, L.K., Doschek, G.A., Golub, L.: Sol. Phys. **243**, 3 (2007)
77. Tsuneta, S., Ichimoto, K., Katsukawa, Y., Nagata, S., Otsubo, M., Shimizu, T., Suematsu, Y., Nakagiri, M., Noguchi, M., Tarbell, T., Title, A., Shine, R., Rosenberg, W., Hoffmann, C., Jurcevich, B., Kushner, G., Levay, M., Lites, B., Elmore, D., Matsushita, T., Kawaguchi, N., Saito, H., Mikami, I., Hill, L.D., Owens, J.K.: Sol. Phys. **249**, 167 (2008)
78. Nigam, R., Kosovichev, A.G.: Astrophys. J. Lett. **505**, L51 (1998)
79. Duvall, T.L., Kosovichev, A.G. Jr., Murawski, K.: Astrophys. J. Lett. **505**, L55+ (1998)
80. Murawski, K., Duvall, T.L. Jr., Kosovichev, A.G.: In: Korzennik, S. (ed.) Structure and Dynamics of the Interior of the Sun and Sun-like Stars, vol. 418, p. 825. ESA Special Publication (1998)
81. Braun, D.C., Duvall, T.L. Jr., Labonte, B.J.: Astrophys. J. Lett. **319**, L27 (1987)
82. Cally, P.S.: Mon. Not. Roy. Astron. Soc. **395**, 1309 (2009)
83. Parchevsky, K.V., Kosovichev, A.G.: Astrophys. J. Lett. **666**, L53 (2007)
84. Parchevsky, K., Kosovichev, A., Khomenko, E., Olshevsky, V., Collados, M.: ArXiv e-prints: 1002.1117 (2010)
85. Nagashima, K., Sekii, T., Kosovichev, A.G., Shibahashi, H., Tsuneta, S., Ichimoto, K., Katsukawa, Y., Lites, B., Nagata, S., Shimizu, T., Shine, R.A., Suematsu, Y., Tarbell, T.D., Title, A.M.: Pub. Astron. Soc. Jpn **59**, 631 (2007)
86. Schüssler, M., Vögler, A.: Astrophys. J. Lett. **641**, L73 (2006)
87. Braun, D.C., Lindsey, C., Fan, Y., Jefferies, S.M.: Astrophys. J. **392**, 739 (1992)
88. Hanasoge, S.M.: Astrophys. J. **680**, 1457 (2008)
89. Khomenko, E., Collados, M.: Astron. Astrophys. **506**, L5 (2009)
90. Kosovichev, A.G., Zharkova, V.V.: Nature **393**, 317 (1998)
91. Kosovichev, A.G.: In: Leibacher, J., Stein. R.F., Uitenbroek, H. (eds.) Solar MHD Theory and Observations: A High Spatial Resolution Perspective, Astronomical Society of the Pacific Conference Series, San Francisco, vol. 354, p. 154 (2006)
92. Kosovichev, A.G.: In: Proceedings of SOHO 18/GONG 2006/HELAS I, Beyond the spherical Sun, vol. 624, ESA Special Publication (2006)
93. Kosovichev, A.G.: Sol. Phys. **238**, 1 (2006)
94. Donea, A., Braun, D.C., Lindsey, C.: Astrophys. J. Lett. **513**, L143 (1999)
95. Donea, A., Lindsey, C.: Astrophys. J. **630**, 1168 (2005)
96. Donea, A., Besliu-Ionescu, D., Cally, P.S., Lindsey, C., Zharkova, V.V.: Sol. Phys. **239**, 113 (2006)
97. Kosovichev, A.G.: Astrophys. J. Lett. **670**, L65 (2007)
98. Tassoul, M.: Astrophys. J. Suppl. **43**, 469 (1980)
99. Gough, D.O.: In: Astrophysical Fluid Dynamics—Les Houches 1987, pp. 399–560 (1993)
100. Duvall, T.L. Jr.: Nature **300**, 242 (1982)
101. Christensen-Dalsgaard, J., Dappen, W., Ajukov, S.V., Anderson, E.R., Antia, H.M., Basu, S., Baturin, V.A., Berthomieu, G., Chaboyer, B., Chitre, S.M., Cox, A.N., Demarque, P., Donatowicz, J., Dziembowski, W.A., Gabriel, M., Gough, D.O., Guenther, D.B., Guzik, J.A., Harvey, J.W., Hill, F., Houdek, G., Iglesias, C.A., Kosovichev, A.G., Leibacher, J.W., Morel, P., Proffitt, C.R., Provost, J., Reiter, J., Rhodes, E.J. Jr., Rogers, F.J., Roxburgh, I.W., Thompson, M.J., Ulrich, R.K.: Science **272**, 1286 (1996)
102. Christensen-Dalsgaard, J., Gough, D.O., Thompson, M.J.: In: Rolfe, E.J. (ed.) Seismology of the Sun and Sun-Like Stars, vol. 286, pp. 493–497. ESA Special Publication (1988)

103. Gough, D.O. (ed.): In: NATO ASIC Proc. 169: Seismology of the Sun and the Distant Stars, pp. 125–140 (1986)
104. Kosovichev, A.G., Parchevskii, K.V.: Sov. Astron. Lett. **14**, 201 (1988)
105. Duvall, T.L. Jr., Harvey, J.W., Libbrecht, K.G., Popp, B.D., Pomerantz, M.A.: Astrophys. J. **324**, 1158 (1988)
106. Christensen-Dalsgaard, J.: Mon. Not. Roy. Astron. Soc. **199**, 735 (1982)
107. Schou, J., Kosovichev, A.G., Goode, P.R., Dziembowski, W.A.: Astrophys. J. Lett. **489**, L197 (1997)
108. Rozelot, J.P., Lefebvre, S., Pireaux, S., Ajabshirizadeh, A.: Sol. Phys. **224**, 229 (2004)
109. Lefebvre, S., Kosovichev, A.G.: Astrophys. J. Lett. **633**, L149 (2005)
110. Lefebvre, S., Kosovichev, A.G., Rozelot, J.P.: Astrophys. J. Lett. **658**, L135 (2007)
111. Dziembowski, W.A., Pamyatnykh, A.A., Sienkiewicz, R.: Mon. Not. Roy. Astron. Soc. **244**, 542 (1990)
112. Gough, D.O., Kosovichev, A.G.: In: Rolfe, E.J. (ed.) Seismology of the Sun and Sun-Like Stars, vol. 286, pp. 195–201. ESA Special Publication (1988)
113. Gough, D.O., Kosovichev, A.G.: In: Berthomieu, G., Cribier, M. (eds.) IAU Colloq. 121: Inside the Sun, Astrophysics and Space Science Library, vol. 159, p. 327. Kluwer Acedemic Publichers, Docdrecht (1990)
114. Kosovichev, A.G.: J. Comput. Appl. Math. **109**, 1 (1999)
115. Backus, G.E., Gilbert, J.F.: Geophys. J. **16**, 169 (1968)
116. Tikhonov, V.Y., Arsenin, A.N.: Solution of Ill-Posed Problems. Winston & Sons, Washington (1977)
117. Asplund, M., Grevesse, N., Sauval, A.J., Scott, P.: Ann. Rev. Astron. Astrophys. **47**, 481 (2009)
118. Bahcall, J.N., Serenelli, A.M., Basu, S.: Astrophys. J. Lett. **621**, L85 (2005)
119. Elliott, J.R., Kosovichev, A.G.: Astrophys. J. Lett. **500**, L199+ (1998)
120. Däppen, W.: In: Demircan. O., Selam, S.O., Albayrak, B. (eds.) Solar and Stellar Physics Through Eclipses, Astronomical Society of the Pacific Conference Series, vol. 370, p. 3. San Francisco (2007)
121. Lynden-Bell, D., Ostriker, J.P.: Mon. Not. Roy. Astron. Soc. **136**, 293 (1967)
122. Schou, J., Antia, H.M., Basu, S., Bogart, R.S., Bush, R.I., Chitre, S.M., Christensen-Dalsgaard, J., di Mauro, M.P., Dziembowski, W.A., Eff-Darwich, A., Gough, D.O., Haber, D.A., Hoeksema, J.T., Howe, R., Korzennik, S.G., Kosovichev, A.G., Larsen, R.M., Pijpers, F.P., Scherrer, P.H., Sekii, T., Tarbell, T.D., Title, A.M., Thompson, M.J., Toomre, J.: Astrophys. J. **505**, 390 (1998)
123. Appourchaux, T., Liewer, P., Watt, M., Alexander, D., Andretta, V., Auchère, F., D'Arrigo, P., Ayon, J., Corbard, T., Fineschi, S., Finsterle, W., Floyd, L., Garbe, G., Gizon, L., Hassler, D., Harra, L., Kosovichev, A., Leibacher, J., Leipold, M., Murphy, N., Maksimovic, M., Martinez-Pillet, V., Matthews, B.S.A., Mewaldt, R., Moses, D., Newmark, J., Régnier, S., Schmutz, W., Socker, D., Spadaro, D., Stuttard, M., Trosseille, C., Ulrich, R., Velli, M., Vourlidas, A., Wimmer-Schweingruber, C.R., Zurbuchen, T.: Exp. Astron. **23**, 1079 (2009)
124. D'Silva, S., Howard, R.F.: Solar Phys. **148**, 1 (1993)
125. Brandenburg, A.: Astrophys. J. **625**, 539 (2005)
126. Benevolenskaya, E.E., Hoeksema, J.T., Kosovichev, A.G., Scherrer, P.H.: Astrophys. J. Lett. **517**, L163 (1999)
127. Howe, R.: Adv. Sp. Res. **41**, 846 (2008)
128. Vorontsov, S.V., Christensen-Dalsgaard, J., Schou, J., Strakhov, V.N., Thompson, M.J.: Science **296**, 101 (2002)
129. Howard, R., Labonte, B.J.: Astrophys. J. Lett. **239**, L33 (1980)
130. Kosovichev, A.G., Schou, J.: Astrophys. J. Lett. **482**, L207 (1997)
131. Howe, R., Christensen-Dalsgaard, J., Hill, F., Komm, R.W., Larsen, R.M., Schou, J., Thompson, M.J., Toomre, J.: Science **287**, 2456 (2000)
132. Rempel, M.: Astrophys. J. **655**, 651 (2007)

133. Spruit, H.C.: Sol. Phys. **213**, 1 (2003)
134. Haber, D.A., Hindman, B.W., Toomre, J., Bogart, R.S., Larsen, R.M., Hill, F.: Astrophys. J. **570**, 855 (2002)
135. Basu, S., Antia, H.M., Bogart, R.S.: Astrophys. J. **610**, 1157 (2004)
136. Gough, D.: Sol. Phys. **100**, 65 (1985)
137. Haber, D.A., Hindman, B.W., Toomre, J., Bogart, R.S., Thompson, M.J., Hill, F.: Sol. Phys. **192**, 335 (2000)
138. Haber, D.A., Hindman, B.W., Toomre, J., Thompson, M.J.: Sol. Phys. **220**, 371 (2004)
139. Komm, R., Morita, S., Howe, R., Hill, F.: Astrophys. J. **672**, 1254 (2008)
140. Dikpati, P., Giman, M.: Astrophys. J. **638**, 564 (2006)
141. Hindman, B.W., Haber, D.A., Toomre, J.: Astrophys. J. **653**, 725 (2006)
142. Claerbout, J.F.: Geophysics **33**, 264 (1968)
143. Rickett, J.E., Claerbout, J.F.: Sol. Phys. **192**, 203 (2000)
144. Lindsey, C., Braun, D.C.: Astrophys. J. **485**, 895 (1997)
145. Chou, D., Chang, H., Sun, M., Labonte, B., Chen, H., Yeh, S., The TON Team.: Astrophys. J. **514**, 979 (1999)
146. Chen, H., Chou, D., Chang, H., Sun, M., Yeh, S., Labonte, B., The TON Team.: Astrophys. J. Lett. **501**, L139 (1998)
147. Braun, D.C., Lindsey, C.: Sol. Phys. **192**, 307 (2000)
148. Chou, D., Duvall, T.L. Jr.: Astrophys. J. **533**, 568 (2000)
149. Chou, D.: In: Shibata, K., Nagata, S., Sakurai, T. (eds.) New Solar Physics with Solar-B Mission, Astronomical Society of the Pacific Conference Series, vol. 369, p. 313. San Francisco (2007)
150. Nigam, R., Kosovichev, A.G.: Astrophys. J. **708**, 1475 (2010)
151. Ben-Menahem, A.: Bull. Seismol. Soc. Am. **54**, 1351 (1964)
152. Tindle, C., Guthrie, K.: J. Sound Vib. **34**, 291 (1974)
153. Unno, W., Osaki, Y., Ando, H., Saio, H., Shibahashi, H.: Nonradial Oscillations of Stars. Univ. of Tokyo press, Tokyo (1989)
154. Bracewell, R.N.: The Fourier Transform and Its Applications. Mc Graw-Hill Inc, US (1986)
155. Duvall, T.L. Jr.: In: Ulrich, R,K., Rhodes, E.J. Jr., Dappen, W. (eds.) GONG 1994. Helio- and Astro-Seismology from the Earth and Space, Astronomical Society of the Pacific Conference Series, vol. 76, pp. 465–474. San Francisco (1995)
156. Birch, A.C., Kosovichev, A.G.: Sol. Phys. **192**, 193 (2000)
157. Birch, A.C., Kosovichev, A.G., Duvall, T.L. Jr.: Astrophys. J. **608**, 580 (2004)
158. Couvidat, S., Birch, A.C., Kosovichev, A.G.: Astrophys. J. **640**, 516 (2006)
159. Zhao, J., Kosovichev, A.G.: In: Sawaya-Lacoste, H. (ed.) GONG+ 2002. Local and Global Helioseismology: The Present and Future, vol. 517, pp. 417–420. ESA Special Publication (2003)
160. Birch, A.C., Kosovichev, A.G., Price, G.H., Schlottmann, R.B.: Astrophys. J. Lett. **561**, L229 (2001)
161. Gizon, L., Birch, A.C., Bush, R.I., Duvall, T.L. Jr., Kosovichev, A.G., Scherrer, P.H., Zhao, J.: In: Battrick, B., Sawaya-Lacoste, H., Marsch, E., Martinez Pillet, V., Fleck, B., Marsden, R. (eds.) Solar Encounter. Proceedings of the First Solar Orbiter Workshop, vol. 493, pp. 227–231. ESA Special Publication (2001)
162. Kosovichev, A.G., Duvall, T.L. Jr.: In: Keil, S.L., Avakyan, S.V. (eds.) Society of Photo-Optical Instrumentation Engineers (SPIE) Conference Series, vol. 4853, pp. 327–340. Presented at the Society of Photo-Optical Instrumentation Engineers (SPIE) Conference (2003)
163. Kosovichev, A.G., Duvall, T.L.J., Scherrer, P.H.: Sol. Phys. **192**, 159 (2000)
164. Zhao, J., Kosovichev, A.G., Sekii, T.: Astrophys. J. **708**, 304 (2010)
165. Parker, E.N.: Astrophys. J. **230**, 905 (1979)
166. Zhao, J., Kosovichev, A.G., Duvall, T.L. Jr.: Astrophys. J. **557**, 384 (2001)
167. Zhao, J., Kosovichev, A.G.: Astrophys. J. **603**, 776 (2004)

Part II
Section 1: The Sun as a star

R. Muller (Latt-Toulouse University, France)
I. Kitiashlivi (Stanford University, USA)
K. Belkacem (Université de Liège)
J.P. Rozelot (OCA-Nice University, France)

Chapter 2
The Quiet Solar Photosphere: Dynamics and Magnetism

Richard Muller

Abstract The Sun is the only solar-type star where the dynamics and the magnetism can be studied in detail and the physical process involved understood, in particular those which occur at very small scales. This lecture is restricted to the quiet solar photosphere. The properties of the three cellular scales of motions observed at the solar surface (granulation, mesogranulation, supergranulation) are presented, as well as the numerical simulations which reproduce most granulation properties very satisfactorily. The granular convection is driven by radiative cooling through the surface. In these simulations, the mesogranulation appears to be an extension to deeper layers of the surface granular convection. The mesogranulation also appears as convective, indirectly driven by the surface radiative cooling. However, several alternative origins for both the mesogranulation and the supergranulation, have been proposed too. On the same way, the magnetic structure of the quiet photosphere is described, including the network and the fields. Their origin is discussed on the basis of the properties of the magnetic elements and of the results of numerical simulations of magneto-convection and of local dynamo. In the network, that means at the supergranular boundaries, the field is concentrated in the form of vertical flux tubes of sizes smaller than a few hundred kilometers and of magnetic field strength 1–2 kG. They are visible as bright points located in the intergranular lanes. The presence of magnetic field is also ubiquitous inside the supergranules, where it is known as IntraNetwork magnetic field. This field is much different from the network field, consisting of small loops of size $1''-2''$, closely related to granules. The field strength is much lower than in the network, not exceeding a few hundreds Gauss, and of flux lower by one or two orders of magnitude. The time scale of both kinds of field is short, less than 10 min, determined by the evolution of the neighbouring granules. The origin of the intranetwork field is not yet clarified: are they fragments of

Richard Muller (✉)
Laboratoire d'Astrophysique de Toulouse-Tarbes, Université de Toulouse, CNRS Observatoire du Pic du Midi, 57 Avenue d'Azereix 65000 Tarbes, France
e-mail: muller@ast.obs-mip.fr

J.-P. Rozelot and C. Neiner (eds.), *The Pulsations of the Sun and the Stars*,
Lecture Notes in Physics 832, DOI: 10.1007/978-3-642-19928-8_2,
© Springer-Verlag Berlin Heidelberg 2011

magnetic flux rising from deep layers and reprocessed close to the surface by convective motions, or generated near the surface by a fast local dynamo?

2.1 Introduction

The waves which propagate below the surface of the Sun and of solar type stars, are affected by the strongly turbulent and highly magnetized upper convection zone. It is thus important to know the dynamic and magnetic properties of these layers on the Sun, which are observed in much detail at its surface, in order to derive informations from the propagating acoustic waves. This review is restricted to the quiet Sun, excluding active regions. The first section is devoted to the various dynamical patterns observed in the photosphere, the second one to the magnetic structure. In both sections, the observed properties are presented first, followed by the results of numerical simulations. The numerical simulations of convection reproduce remarkably well the granulation at the surface, and reveal properties on deeper layers not accessible by direct observations. Magneto-convection simulations allow us to better understand the origin and the properties of the small magnetic elements which are observed in the quiet photosphere, and which expands in higher layers.

2.2 Solar Surface Flows

Three scales of motions, more or less distinct, are observed at the surface of the Sun: the granulation, which is the directly visible manifestation of surface convection; the mesogranulation, which is detected by the associated vertical or horizontal velocities; the supergranulation, detected by the associated horizontal velocities and by magnetic flux concentrations at the cell boundaries.

They have long been believed to be due to the recombination of Hydrogen just below the surface (granulation) and Helium (HeI, 8,000 km below the surface (mesogranulation), HeII, 20,000 km below supergranulation). There are now some doubts with this interpretation. Thus, the origin of the granulation appears well established, which is not the case yet for the two larger scales, although a convective origin is prevailing. www://svmlight.joachims.org/

2.2.1 The Solar Granulation

The solar surface is now observed with a spatial resolution reaching the diffraction limit of the telescopes, thanks to the quality of the sites were they are located and of sophisticated and reliable restoration techniques (speckle and phase diversity). It is as small as $0''1 = 70$ km at the largest one, the 1 m Swedish Solar Telescope

Fig. 2.1 The solar granulation observed with HINODE with a spatial resolution of $0''2$; field of view: $110'' \times 55''$ $(80,000 \times 40,000\,\mathrm{km})$

in La Palma (Canary Islands) [1]. In images obtained with such a resolution, the photosphere appears highly structured, because large variations of temperature and density take place just below the surface. Large variations of the emergent intensities are implied, visible in what is known as the granulation pattern (Fig. 2.1).

The size histogram of granules is decreasing continuously from the resolution limit of the best observations ($0''1 = 70\,\mathrm{km}$) to about $4''$ ($3,000\,\mathrm{km}$) for the largest granules. That means that there are much more small than large granules, implying that one cannot define a characteristic scale, strictly speaking. However, it is commonly admitted that the typical scale of granules is in the range $1''$–$2''$ (700–1,400 km).

Their lifetime is typically 5–10 min, the largest granules living longer; this corresponds to the returning time of convective cells. The common evolution of a granule can be described as follows: a new granule, born either by spontaneous appearance in an intergranular space, either from the merging of two granules or, more frequently, as a fragment of a large splitting granule, first expands, then splits into several fragments when it reaches a size larger than about $2''$, or merge with another granule, or simply fade away [2]. The most spectacular and vigorous example of granule fragmentation is represented by exploding granules, first observed at the Pic du Midi Observatory in 1969 [3]. In this case, as the granule expands, a dark area appears in the center and the granule splits up to six or seven fragments (Fig. 2.2). Granules can also vanish or merge with another granule. Repeating fragmenting and exploding granules form a family of granules [4–6]. A family originates from a single granule at its beginning. Its typical lifetime is 1–2 h (but lifetimes as long as 8 h were reported) and its typical spatial extension is $5''$–$10''$.

Bright granules are associated with upflows, while the surrounding dark intergranular lanes are associated with downflows. Such a correlation is characteristic of convective cells. The convective nature of the solar granulation is confirmed by numerical simulations (see Sect. 2.2.5). The typical vertical velocity is 1–$2\,\mathrm{km\,s}^{-1}$. Inside granules, the speed of the divergent horizontal flow is also about 1–$2\,\mathrm{km\,s}^{-1}$. The granular convective pattern penetrates into the photosphere, with a speed decreasing with height; velocities of several hundreds $\mathrm{m\,s}^{-1}$ are still measured in the outer photosphere, 500 km above the surface. But the intensity (which means the temperature) fluctuations vanish at a much lower height, about 100 km above the surface [7].

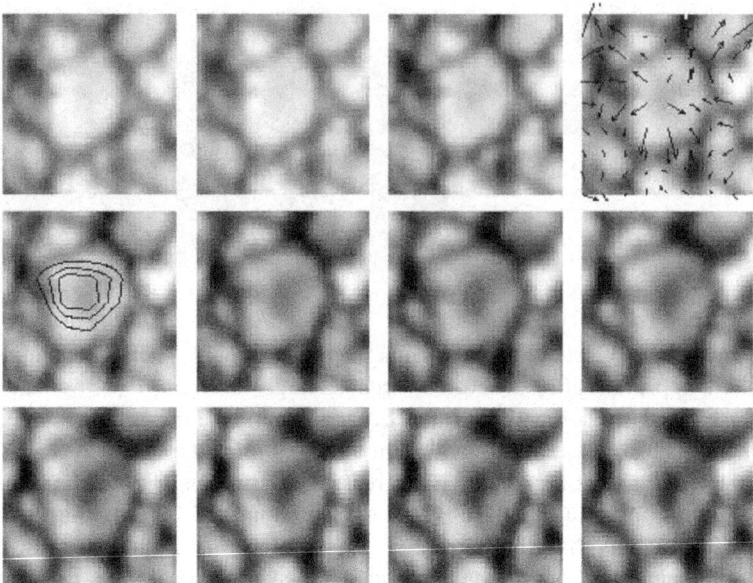

Fig. 2.2 Exploding granule; sizebox: $4'' \times 4''$; time interval: 30 s (from the Pic du Midi Observatory)

Observations and numerical simulations clearly show that the origin of the granulation is convective. But some turbulent properties have been inferred from observations, like the increasing number of small granules, the presence of fine structure inside granules, the slope of the power spectrum of the emergent intensity fluctuations, which is close to $-5/3$, the fractal dimension, the large width of spectral lines, indicating turbulence, in the interface between granules and intergranules, which are regions subject to shear between ascending and descending plasmas, as well as from numerical simulations (the downflowing plasma is more turbulent than the upflowing plasma, which rises gently). It is not surprising to find some turbulent properties for the solar granulation, owing to the very low gaz viscosity and extremely high resulting Rayleigh number.

2.2.2 The Solar Mesogranulation

The mesogranulation, of scale intermediate between the granulation and the supergranulation, was detected for the first time in 1981 [8] at Sacramento Peak Observatory (USA), as a pattern of vertical velocities of about $100 \, \text{m s}^{-1}$ (Fig. 2.3). It was revealed on Dopplergrams averaged over about 1 h, so that the granulation velocity pattern is sufficiently smoothed. It also appears as a pattern of diverging cells visible in maps of horizontal flows obtained by tracking the motions of granules at the solar surface [9, 10]. An example is shown in Fig. 2.4. The amplitude of the

Fig. 2.3 Supergranulation (*upper panel*) and Mesogranulation (*lower panel*) as determined from granulation Dopplergrams time averaged over a 60 min period. The field of view is $160'' \times 60''$. The Supergranulation is revealed by a spatial averaging with a $9'' \times 9''$ running window. The mesogranulation, of much smaller scale, is revealed with $3'' \times 3''$ averaging (from November et al. [8])

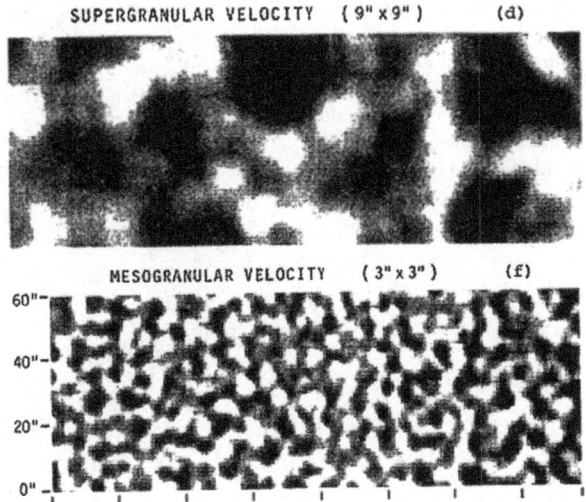

vertical and horizontal velocities of mesogranules are typically $50–100 \, \mathrm{m \, s^{-1}}$ and $300–500 \, \mathrm{m \, s^{-1}}$, respectively; the time scale is of the order of 1 h. Mesogranules are advected outward from the center to the boundary of supergranules by the supergranular flow with a speed of $300–500 \, \mathrm{m \, s^{-1}}$ [11]. Inside mesogranules, granules are more vigorous than outside: they are larger, brighter and exploding with more energy. Mesogranules coincide with families of granules [5, 6]; their center is brighter than their boundary. The correlation between upflows and excess of brightness at cell centres and between downflows and deficit of brightness at the external boundaries, together with the presence of a diverging cellular flow, are in favor of a convective origin. But, because of the absence of two peaks in the power spectra (see Sect. 2.2.4), it is not clear whether it is a distinct scale from the granulation.

2.2.3 The Solar Supergranulation

The supergranulation, discovered in 1954 [12] is a pattern of horizontal flows detected by Doppler measurements away from the disk centre [13], where the line of sight components can be detected (Fig. 2.5). This cellular horizontal flow can also be revealed near disk centre by tracking the motion of granules or mesogranules on the solar surface ([9, 11, 14–16], Fig. 2.6). The size and lifetime of a typical supergranule are $50''$ (30,000 km) and 24 h, respectively. Helioseismological soundings show that supergranular cells do not extend deeper than 5,000 km below the surface. The horizontal flow diverging outward from cell centers to boundaries, is $300–400 \, \mathrm{m \, s^{-1}}$. The vertical flow is very weak and hard to be detected, because of the high level of the granulation noise, and of the presence of magnetic fields in the boundaries which complicates the measurements. Downflows of the order of $100 \, \mathrm{m \, s^{-1}}$ are reported

Fig. 2.4 The velocity pattern at the solar surface, derived with the help of granules used as tracers; the cells of diverging vectors correspond to mesogranules (from Roudier et al. [10])

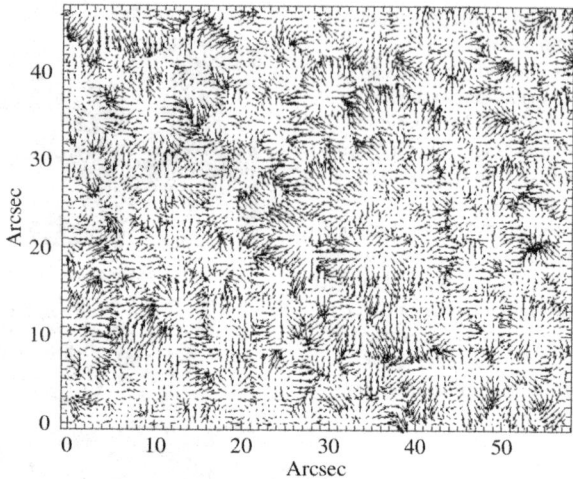

Fig. 2.5 The cellular flows in a full-disk SOHO/MDI Dopplergram. The flow field is dominated by the largely horizontal flows in the supergranules with typical velocities of 300–400 m s^{-1}

(see [17] for the references). A weak excess of temperature of 1–3 K has been recently reported [18]. A diverging flow outward from cell centers, a weaker temperature and the presence of downflows at the boundaries, are indications of convection in supergranules. But, on the other hand, the probability distribution function of the divergence field computed by [16] would show, according to these authors, the signature of intermittency of the supergranulation and thus its turbulent nature. Consequently, the origin of the supergranulation is still an open question (see Sect. 2.2.7).

Fig. 2.6 The supergranulation velocity field as derived from granulation tracking, with divergence contours superimposed (scales shorter than 8 Mm have been filtered out and a time window of 150 min has been used) (from Rieutord et al. [16])

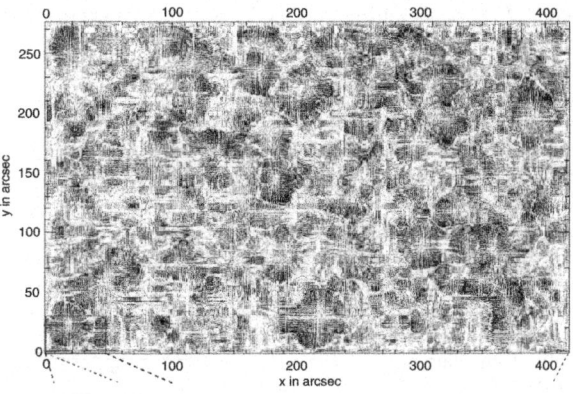

2.2.4 The Power Spectrum of Photospheric Flows

Power spectra of horizontal photospheric flows, which are convenient tools to examine the properties of velocity patterns, have been derived either from full disk dopplergrams [19] or from coherent structure tracking in limited areas near the disk centre [16] (in this case, the coherent structures are granules). The spectrum derived from dopplergrams acquired by the MDI instrument on the SOHO spacecraft has two distinct peaks (Fig. 2.7), corresponding to granules and supergranules, but there is no distinct feature visible at wavenumbers corresponding to mesogranules. The power spectra of the horizontal flows derived with agranulee tracking technique, as shown in Fig. 2.4, but in a much larger field of view, also show a conspicuous peak at the supergranular characteristic scales [16]. In this analysis, the supergranular peak appears in the power spectra obtained with two different data sets: one obtained with TRACE in the space (field of view: $1000'' \times 1000''$, pixel size: $0''5$), the other obtained with the wide-field camera CALAS at the Pic du Midi Observatory ($400'' \times 400''$, pixel size: $0''1$). Regarding these converging results, derived from very different kinds of data, one may be quite confident that the power spectrum of photospheric flows has a real peak at the typical scale of the supergranulation, which thus appears as distinct from granulation and mesogranulation. On the other hand, mesogranulation seems to be a simple extension of granulation to larger scales, probably associated to families of granules. In the numerical models presented in the next section, mesogranulation also appears as an extension of granulation seated in deeper layers.

2.2.5 Numerical Simulation of the Solar Convection; Origin of the Granulation

3D as well as 2D numerical simulations have been computed to be compared with the solar granulation, in order to understand the solar convection and get informations

Fig. 2.7 Doppler velocity power spectrum for SOHO/MDI data. The supergranule peak at $1 \approx 120$ (corresponding to a size of 35 Mm) is prominent as is the broad peak at $1 > 1,000$ (or at size <4.5 Mm) corresponding to the range of mesogranules and granules. There is no spectral feature suggesting a component for mesogranules cells (from Hathaway et al. [19])

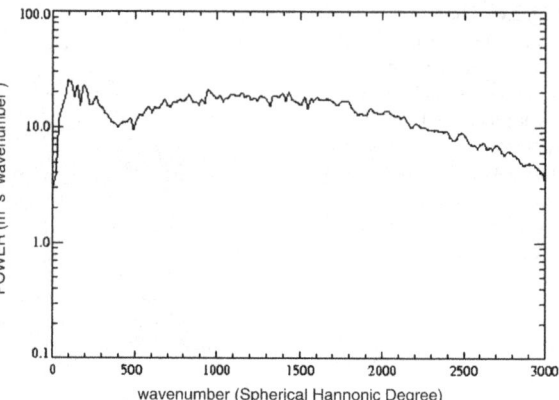

not available by direct observations. 2D models are less realistic, but are useful as they allow us to have an easier control of the effect of various physical parameters on the results. 3D simulations are more realistic, of course, but it is more difficult to interpret the results. Because the appearance of the granulation at the surface is well reproduced by the 3D simulations, as well as many different diagnostics, one may be confident that the physical origin of the solar granulation is well understood. Consequently, I will only present here the results from the 3D simulations, which are all in remarkable agreement, despite different assumptions concerning, in particular, fluid compressibility and fluid viscosity and the treatment of the radiative transfer (Stein and Nordlund [20–23], who were the precursors in convection simulation), and [17, 24–28] among others). These models are based on the solution of the basic equations of fluid dynamics, including radiative transfer (see [29] for the details). It is crucial that the radiative transfer is taken into account, because near the surface of the Sun, the energy flux changes from almost exclusively convective below the surface, to radiative above the surface. Because a large fraction of the internal energy is in the form of ionization energy near the surface, the equation of state includes the effect of ionization and excitation of hydrogen and other abundant elements, and the formation of H_2 molecules. The codes are stabilized by a numerical viscosity which is several orders of magnitudes larger than the viscosity on top of the convection zone of the Sun. This is their main weakness. In some models [17, 24–26], the fluid is incompressible, allowing an exceptionally large Rayleigh number and a vigorous convection, while in the more realistic models, the fluid is compressible, but at the expense of the vigor of the convection. The box size is limited by the computer capacities, in general to about $250 \times 250 \times 160$ grid points, covering a spatial domain 6×6 Mm horizontally and extending from the temperature minimum, 500 km above the surface, to 2,000 km below.

The numerical simulation show that the solar convection is driven by radiative cooling at the surface, producing the familiar granulation pattern. Stein and Nordlund [22], for example, describe the convective flow as follow: beneath the surface, it is asymmetric, consisting of a gentle, expanding, structureless and warm upflow,

Fig. 2.8 Snapshot of entropy fluctuations. Low entropy gas forms the cores of downdrafts that penetrate through the entire computational domain (from Stein and Nordlund [23])

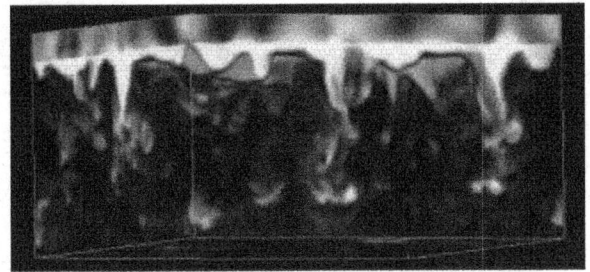

in which strong, isolated, converging, filamentary cool downflows are embedded (Figs. 2.8, 2.9). In the interior of granules, warm plasma ascends. As it approaches and passes through the optical surface, the plasma cools, recombines and loses entropy. It then turns over and converges into dark intergranular lanes and further into the vertices between granular cells. The vertices feed turbulent filamentary downdrafts below the surface, which merge into deeper, more widely spaced filaments. These downdrafts drive both larger scale cellular upflows and smaller scale turbulent motions. The horizontal flow of hot fluid, which is advected toward the cool filaments, has a hierarchical cellular appearance: very small cells at the surface and successively larger cells at larger depths are driven by the merging of the filamentary downdrafts. The surface appears highly structured (granulation), because large variations of temperature ($\approx 5,000$ K at the same geometrical depth) and density take place just below the surface. These large variations are a consequence of the rapid drop of the temperature in a thin layer (≈ 100 km), where the convective energy is transformed into radiative energy, and of the enormous temperature sensitivity of the opacity of the photospheric plasma. Large variations of emergent intensity are implied, visible as the granulation pattern. Large deep seated cells are only visible through the advection of smaller cells by their horizontal velocity fields. Most ascending particles never reach the surface; they return to the downflows beneath the surface. Both observed and simulated granulation look very similar (Fig. 2.10) and several granulation properties are well reproduced by the numerical simulations: the intensity contrast: 20–30% from the models, 12% for images restored for blurring; granule fragmentation; C-shape of photospheric lines (the combination of the blue-shifted line profile of hot upflowing plasma in granules with the weaker red-shifted profile of cold downflowing plasma in intergranules, results in an asymmetric profile, whose bisector has a shape resembling the letter C, Fig. 2.11). It must be noted, however, that in the real Sun, the upper convection zone and the photosphere are pervaded by a non negligible amount of magnetic flux, which can disturb the dynamical processes, as we shall see in Sect. 2.3.8.

Fig. 2.9 A schematic representation of convective flows below the surface: *flow lines* show merging of the downdrafts on successively larger scales; the warm ascending plasma turns out over and converges into cool downdrafts (from Spruit et al. [21])

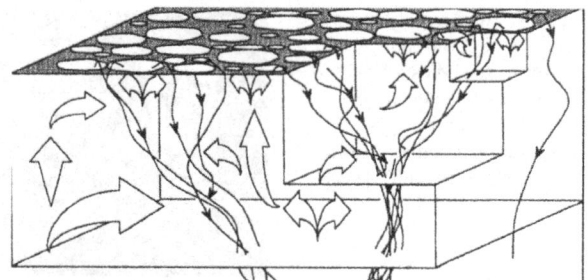

Fig. 2.10 Comparison of granulation as seen in the emergent intensity from a simulation, and as observed by the Swedish Vacuum Solar telescope (50 cm). The *top row* shows a simulation; the *middle row* shows this image smoothed by an Airy plus exponential point-spread function; the *bottom row* shows a white light images from La Palma. Note the similar appearance of the smoothed simulation and the observed granulation (from Stein and Nordlund [22])

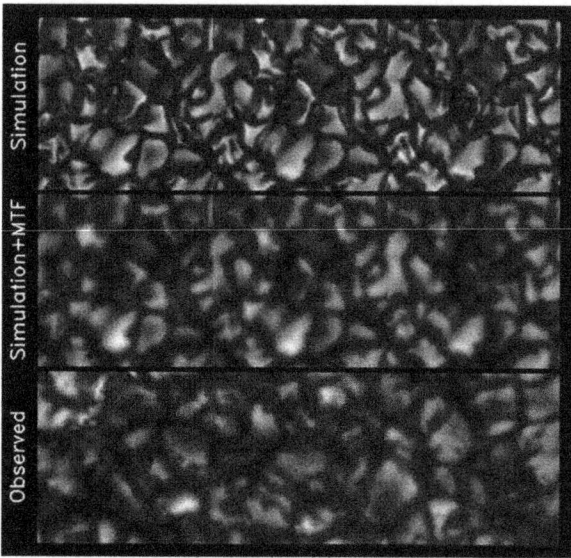

2.2.6 Origin of the Solar Mesogranulation

Two main origins have been suggested for the mesogranulation: convection or interaction between granular flows:

Convection. A pattern of convective cells of mesogranular scale appears at the bottom of the computational box (500 km below the surface), as a result of the numerical simulations of convection presented by Stein and Nordlund [20, 22, 23]. These convective cells are driven by merging of filamentary downdrafts in deeper layers; they advect the granules above. granulation movies also show that granules are advected by mesogranular flows.

In the simulation made by Rincon et al. [30] of fully compressible turbulent convection in a polytropic atmosphere, the turbulent energy power spectrum is dominated by a mesoscale pattern. The authors show that this pattern has a genuine convective origin, and suggest that mesogranulation is the dominant convective mode below the

Fig. 2.11 *Top*: synthetic (*dash-dotted*) and observed (*full drawn*) profiles of FeI 5507. *Bottom*: bisectors of the same line: synthetic (*dash-dotted*) and observed (*full drawn*) (from Nordlund [29])

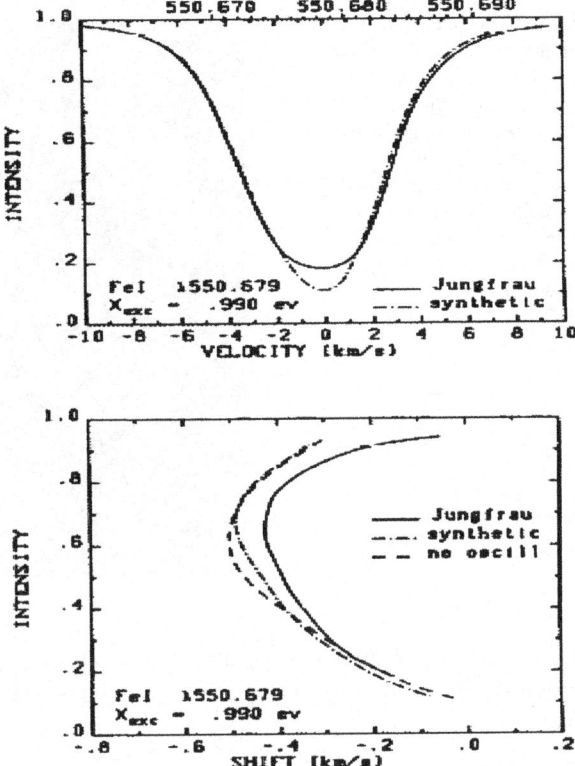

photosphere. However, the results have to be considered with caution, because of the absence of radiative transfer in the simulation.

Interactions between granules or intergranular flows. In Cattaneo et al.'s model [25] of compressible, non radiative atmosphere, the temperature fluctuations at the surface exhibit a mesoscale pattern superimposed to the more conspicuous granular pattern (Fig. 2.12). The authors suggest that their mesoscale pattern results from a collective non linear interaction between granules. But, such a mesogranular pattern doesn't appear so clearly in more realistic compressible and radiatiave atmospheres, neither in solar granulation images.

On the other hand, Rast [17] finds that spatial and temporal scales naturally arise through the collective advective interaction of many small-scale and short-life downflow plumes; in this simplified model the surface flows advect the plumes, which can merge together, eventually producing strong downflows spatially distributed with a characteristic length-scale; these long-lived downflows then define the vertices of the mesogranular flowss in a first step, then the supergranular flows as the downflow plumes merge further in deeper layers.

As the Stein and Nordlund's model [20, 22, 23] is computed in realistic physical conditions (except for the viscosity), one may believe that mesogranulation is of

Fig. 2.12 Temperature
fluctuations in Cattaneo
et al.'s simulation [25]; two
distinct cellular patterns
appear: convection cells
corresponding to granules
and larger cells of
mesogranular scale bounded
by darker (lower
temperature) lanes which
coincide with stronger
downflows

temperature

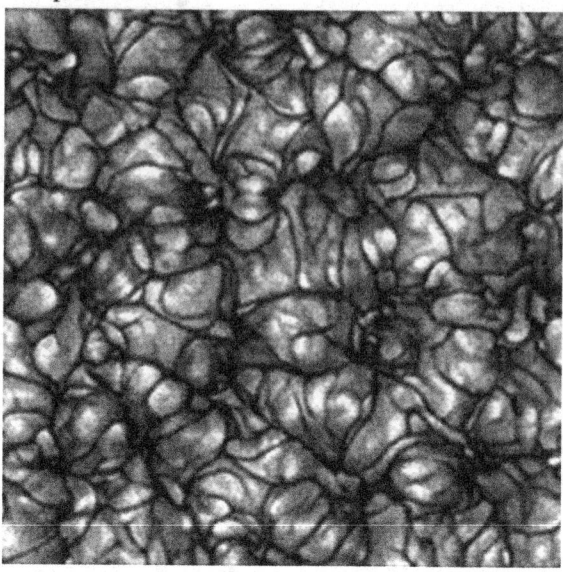

convective origin rather than resulting from interactions, the driving force being the
radiative cooling through the solar surface which primarily produces the granulation
pattern.

2.2.7 Origin of the Solar Supergranulation

Like for mesogranulation, convection and interactions between granules or between
downflow plumes have been suggested for the origin of the supergranulation. In
addition, an unconventional magnetically driven origin has also been proposed.

Convection: Recently, Stein and co-workers [31, 32] have adapted to mesogranular
and supergranular scales, the numerical simulation they have developed previously
to simulate granulation and mesogranulation. To that end, they increased the size of
the computational box from $6 \times 6 \times 2.5$ Mm to $96 \times 96 \times 20$ Mm, at the expense
of the spatial resolution, which was decreased from 25 to 150 km. With such a low
resolution, granulation cannot be simulated properly. The horizontal dimension of
the box is twice the typical size of supergranules; in the vertical dimension, the box
includes the hydrogen, first and most of the second helium ionization zones. convection is driven by buoyancy. Close to the surface buoyancy driving is balanced by
divergence of the kinetic energy, but deeper down it is balanced by dissipation. Granules are not fully resolved; mesogranular scale cells appear just below the surface; the
size of the cells increases with depth, as filamentary downflows merge and reach the
supergranular scale at the bottom of the box. The snapshots of streamlines in vertical

slices look very similar to the snapshot shown in Fig. 2.8; the only difference is the much larger real spacing between the downdrafts. The power spectrum is continuous from mesogranular to supergranular scales, in disagreement with the observed spectra, which show two distinct peaks (Sect. 2.2.4, Fig. 2.7). Unfortunately, it is not possible yet to simulate granulation, meso and supergranulation simultaneously in the same computational box.

Interactions: The same kinds of interactions have been proposed for the origin of supergranules as for the origin of mesogranules: collective interaction of exploding mesogranules [33]; advective interaction of downflow plumes [17]: same model as for the origin of mesogranules, but extended to the plumes spatially distributed at the mesogranular scale which merge deeper down to form a pattern of stronger downflows at the supergranular scale.

Magnetic driving: A novel and original explanatory model has been proposed by Crouch et al. [34], who suggest that the supergranular magnetic network is a spatial pattern that emerges autonomously from the advection of small-scale magnetic elements at the granular scale and their subsequent interaction, characterized by aggregations and cancellations. Once build up, the emergent network serves as a template to seed supergranular downflows, possibly by interfering with convection, localized cooling. That means that the merging magnetic network builds up before the supergranular horizontal flow develops.

It is interesting to note that the same various kinds of origin are proposed for supergranules and mesogranules. In the simulation of convection, where the scale of the developed cells depends, among others parameters, of the size and of the spatial resolution of the computational domain, the size of the cells increases smoothly from granulation scale at the optical depth unity (which is the simulated solar surface), to supergranulation scale at the bottom. Supergranulation appears indirectly driven by surface radiative cooling, [31, 32]. The corresponding power spectra should be single-peaked, extending from granular to supergranular sizes, since the transition between these scales is very smooth in the simulations. But this is not supported by the observations, since the observed power spectra have two distinct peaks, a wide one, centered at about 4″, which overlays both the granular and mesogranular ranges of sizes, and another one centered at about 50″, which corresponds to the mean size of supergranules.

2.3 Magnetic Field in the Quiet Photosphere

2.3.1 Introduction

Late type stars with deep convective envelopes are magnetically active. But it is only on the Sun that this kind of magnetism can be studied in details.

The solar magnetism extends over very wide ranges of absolute Magnetic flux and of lifetimes (Fig. 2.13), including: *large bipolar active regions*, containing sunspots,

Fig. 2.13 Full-disk
magnetogram from
SOHO/MDI. The
photospheric network is
clearly visible away from the
active regions.

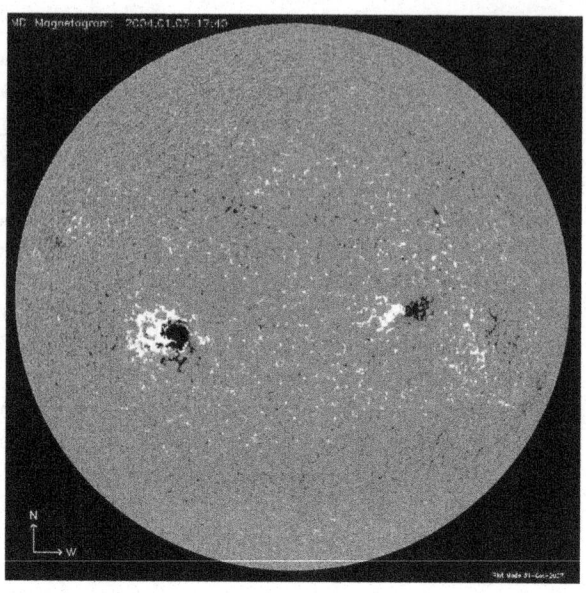

with flux larger than 5×10^{21} Mx (1 Mx is the magnetic flux of a 1 G field strength
over a 1 cm^2 area), living several weeks or more; *small bipolar active regions* containing pores, with flux in the range 1×10^{20}–5×10^{21} Mx, lasting several days;
ephemeral bipolar active regions (ERs), without neither sunspots nor pores, with
flux in the range 3×10^{18}–1×10^{20} Mx, lasting several hours; *individual elements
of the photospheric network*, at the boundaries of supergranules, whith flux in the
range 10^{17}–10^{18} Mx, lasting a few minutes, *individual elements inside the network*,
known as the Intranetwork (IN) magnetic field, with flux in the range 10^{16}–10^{17} Mx,
also lasting no more than a few minutes. The magnetic flux emerges onto the phos-
tosphere in the form of bipolar regions; while active regions containing sunspots
appear in the sunspot belts, on each side of the solar equator, ERs can emerge everywhere on the surface of the Sun, from equator to the poles. The magnetic flux also
emerges at the very small granulation scale (1,000 km), inside the network, which
means inside supergranules. The network is believed to be fed from active regions
and from ephemeral regions, by large scale circulation and diffusion by supergranules; it may also be fed by intranetwork flux, advected to supergranular boundaries
by supergranular flows. This point will be discussed in more details in Sect. 2.3.10.

I restrict this chapter to the quiet Sun magnetic field, not including active regions,
where it appears concentrated at the boundaries of supergranules, forming the photospheric network.

When the magnetic sensitivity is increased, the IN magnetic field, one order of
magnitude weaker in term of flux, is detected inside supergranules, i.e. inside the
network (Fig. 2.16). As we will see below, the properties of the magnetic field in the
network and inside the network are much different. While it is well established that

Fig. 2.14 Full-disk CaII
3,933 K-line filtergram,
taken with CLIMSO at the
Pic du Midi Observatory.
The photospheric network
forms bright cells away from
the active region

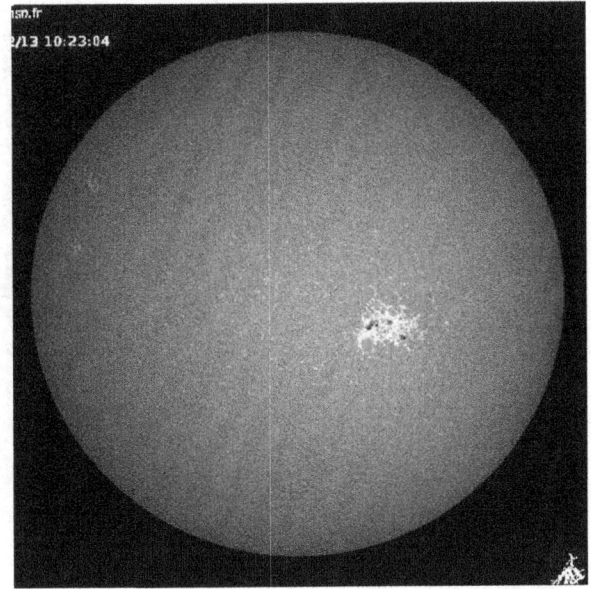

the magnetic field in active regions originates at the base of convection zone, the origin of the network and intranetwork fields is not well understood yet.

Recent observations of high spatial resolution and high sensitivity, performed with the spectro-magnetograph attached to the Solar Optical Telescope (SOT) on board the satellite HINODE as well as with several ground-based telescopes (the German Vacuum Tower Telescope (VTT) and the French THEMIS in Tenerife, Canary Islands, the Swedish SST at La Palma, Canary Islands, the Dunn telescope at Sac Peak, New Mexico), have revealed the ubiquitous presence of magnetic flux everywhere on the Sun, at the scale of the granulation, and provide magnetic flux measurements and magnetic field determinations with unprecedented precision, changing our view of the quiet Sun magnetic field.

2.3.2 The Photospheric Network

The photospheric network underlines the supergranule boundaries, and is formed by elements of small sizes (less than 200 km) and of strong field strengths (1–2 kG). It is cospatial with the bright chromospheric network which is visible in CaII 3,933 filtergrams (Fig. 2.14): a Calcium bright point corresponds to each magnetic element, indicating that they are hotter than the surrounding atmosphere (Fig. 2.15). The network is made of magnetic patches where one polarity is dominant in longitudinal field magnetograms (Fig. 2.16).

Fig. 2.15 Magnetogram of
the remnant of an active
region (*bottom figure*) and
the corresponding CaII 3933
filtergram (*top figure*) taken
with the 1 m Swedish Solar
Telescope in La Palma. It is
not a real quiet Sun
magnetogram, but it nicely
shows how the strong and
vertical magnetic field
concentrates at the
boundaries of supergranules,
as clearly visible in the upper
right of the field of view
(from Berger et al. [1])

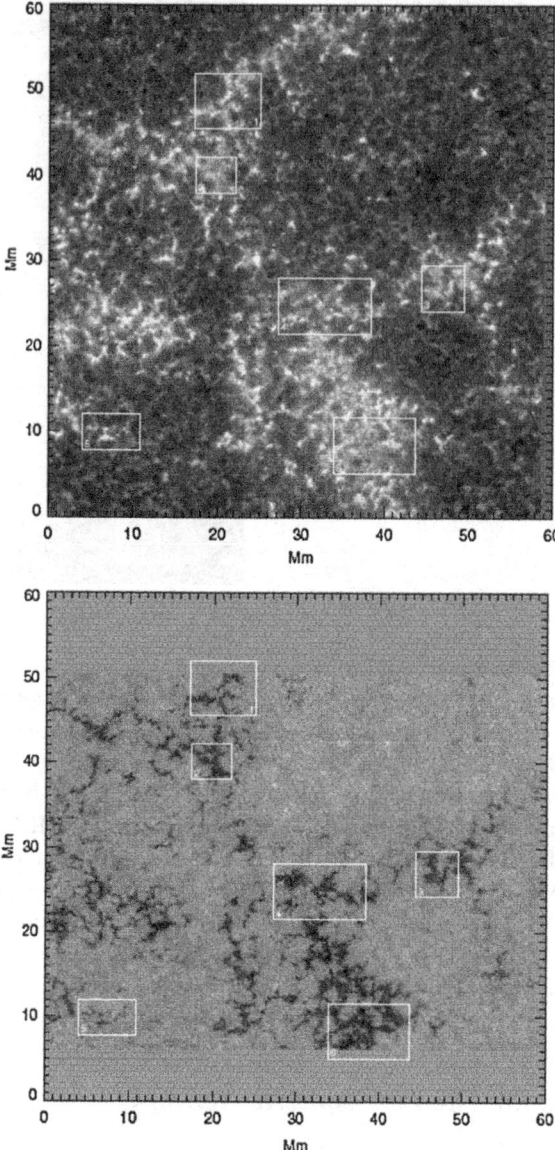

The high field strength and the small size were first inferred from low resolution
observations with an indirect method, called 'magnetic signal ratio' [35]. In this
method, the Stokes-V polarization signal is measured in two different lines belonging
to the same multiplet, like FeI 5250 and FeI 5247, which mainly differ in the Landé
factor, $g = 2$ and 3, respectively for these two lines. For weak magnetic fields, i.e.
smaller than about 1,000 G, the circular polarization signal measured in the two

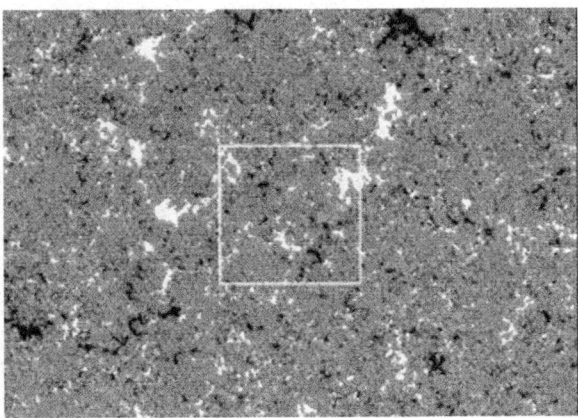

Fig. 2.16 High sensitivity and high resolution magnetogram of the vertical component of the magnetic field in the quiet Sun, obtained with the Solar Optical Telescope/Spectro-Polarimeter (SOT/SP) aboard the HINODE spacecraft. Field of view: $165'' \times 115''$ ($120\,Mm \times 84\,Mm$). It shows th photospheric network (formed by the large patches) and the intranetwork (IN) field (smaller patches inside the network) (from Lites et al. [62])

lines should be in the ratio 3/2. For stronger fields, it should be lower, because of a saturation effect in the line with the stronger Landé factor. In the network, Stenflo [35] found a ratio smaller than 3/2 and concluded that the field strength should be of the order of 1–2 kG. For such a high field strength, the measured flux implies that the size of the magnetic elements is small, in the range 100–300 km. Soon after, a pattern of very small bright points (called the 'Solar Filigree'), corresponding to the photospheric network observed with a high spatial resolution, was discovered by Dunn and Zirker in 1973 [36] in high resolution filtergrams, obtained in the wings of the line Hα, supporting Stenflo's finding (see Fig. 2.17, where the filigree is observed in the G-Band at 4,305 Å instead of the Hα line wings).

Small size and kG field strength, led Spruit in 1976 [37], to the concept of a vertical, evacuated, concentrated flux tube (Fig. 2.18); it explains successfully the excess of brightness in the network and facular elements and its centre to limb variation. The presence of kG magnetic field inside the flux tube, implies that the gas pressure is lower than outside; consequently the tube is partly evacuated and is oriented vertically by buoyancy. Because of this partial evacuation, the gas inside the tube is radiatively heated through the walls by the hotter gas outside, and one sees deeper and hotter layers inside the tube. This explains the observed excess of brightness in small size magnetic elements relatively to the surrounding atmosphere. The brightness excess increases in higher layers, because the evacuated, thin flux tube, is increasingly heated by the hotter tube walls. The concept of concentrated flux tubes has been confirmed by recent high resolution and high magnetic sensitivity observations made with HINODE/SOT, and with several earth based telescopes (the 1 m Swedish SVST, the German VTT, the French THEMIS, all located in the Canary Islands).

Fig. 2.17 The photospheric network visible in the G-Band lines of the CH molecule at 4,305 Å, as observed with HINODE/SOT. It surrounds a supergranule

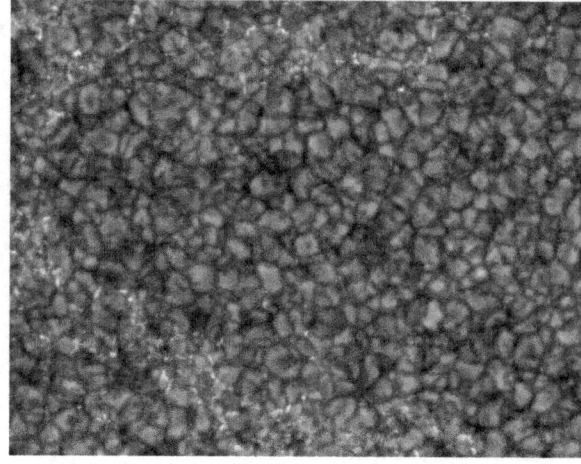

Fig. 2.18 A schematic representation of an evacuated magnetic flux tube (see the text). The diameter of the tube is 100 km below the surface. The gas inside is heated by the hotter external plasma. In the evacuated tube one sees deeper, hotter layers, explaining the excess of brightness in magnetic elements (from Spruit [37])

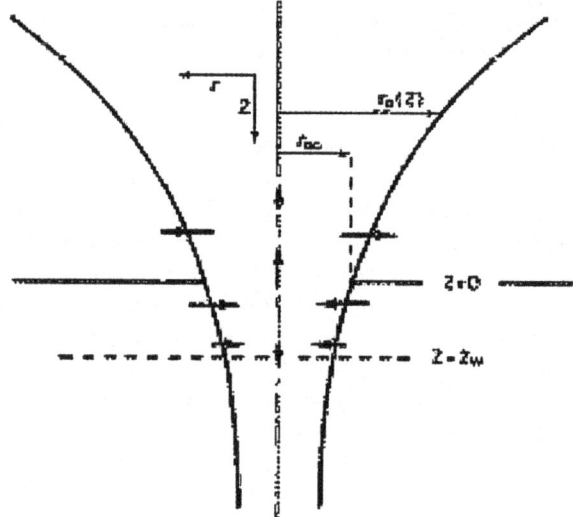

One can summarize the properties of the magnetic elements in the photospheric network as follow:

Magnetic field strength: 1–2 kG, first deduced from the indirect line ratio method described above; high field strengths have been confirmed by direct measurements in the IR, thanks to the wide separation of the circular Zeeman components in this wavelength range; they have also been confirmed by several other indirect determinations: by the line ratio method, but with higher spatial resolution than in the original Stenflo's observations from 1973; by LTE inversion of the Stokes I, V, Q, U profiles of photospheric lines, based on Milne-Eddington atmospheres, in which the temperature is assumed to vary linearly with the geometrical depth. The inversion returns the values of about ten parameters, including the three components of the magnetic field (strength, inclination, azimuth), the spatial filling factor, directly related to the size of the magnetic elements, the variation of the temperature with height, etc.

- *Magnetic field orientation*: essentially vertical, as shown by the quasi absence of linear polarization in observations made at the disk centre, and inferred from Stokes profiles.

- *Magnetic flux*: $1–5 \times 10^{17}$ Mx.

- *Flows in the tubes:* flow velocity is usually determined by the shift of the zero-crossing position of the Stokes-V profile (it is the wavelength position where the V-polarization is zero) in the magnetic element, compared to the position of the maximum of the Stokes-I profile averaged in the surrounding non magnetic area. When the spatial resolution is not better than $1''$, several magnetic elements must be contained in the resolution element to produce a detectable signal and the measured velocity is found to be ascending; but when the resolution is sufficient to isolate individual elements, the measured velocity can be ascending or descending. In recent high spatial resolution ($0''3$) made with SOT/HINODE, transient downflows as high as $7 \, \text{km s}^{-1}$ have been reported [38] during the cooling/concentration phase of a flux tube. The downward flow bounces back at the dense bottom layer, and upward motion appears, which may produce a shock wave in the upper photosphere. Such 'rebounds' have been observed by [39] and [40]. Upflows also appear in the 2D dynamical fluxtube model of Steiner et al. [41].

2.3.3 Observation of the Photospheric Network in the G-Band at 4,305 Å

The magnetic elements of the photospheric network are easily observed in filtergrams made in the bandhead of the CH molecule at 4,305 Å, known as the G-Band, with a filter 10 Å wide (Fig. 2.17). Because of the presence of many strong absorption lines, the equivalent height of formation is 100–200 km above the surface, where the contrast of the magnetic elements relatively to the quiet photosphere increases. The contrast is enhanced in the G-Band, because most lines being of molecular origin, the opacity inside the relatively cool magnetic flux tube is decreased, allowing us to see deeper, hotter layers than outside. Thanks to the very short exposure times allowed by the relatively wide bandpass of G-Band filters which freezes the atmospheric seeing, the magnetic elements can be observed with a spatial resolution close to the diffraction limit of ground-based telescopes ($0''1$, in the best case, at the Swedish 1m telescope in La Palma), which cannot be reached in magnetograms yet. They were first observed by Muller at the Pic du Midi Observatory in 1980 [42]. In G-Band images, magnetic elements are visible as tiny bright features, more or less elongated, smaller than $0''5$, located in intergranular lanes. Many of them are detected at the diffraction limit of the telescopes, wich means that they should be smaller than 100 km or so [43–46]. For simplicity we call the G-Band bright features 'G-Band Bright

Points', or 'GBPs', even if their shape is not strictly roundish, but elongated. GBP$
are commonly used as proxies of magnetic elements and are useful for improving
our understanding of the small scale solar magnetism. For example,time series o$
G-Band images allow us to investigate their dynamics and their interaction with granules, which is not so easy directly with magnetograms. They have shown that the
GBPs, and consequently the associated magnetic elements, evolve very dynamically
with time scales of a few minutes (which is the evolution time scale of granules), moving in the intergranular lanes, primarily driven by the evolution of the surrounding
granules [47–51]. Continuous fragmentations and merging of GBPs occur continuously, indicating that this should also be the fundamental mode of evolution of the
associated magnetic structures in the photospheric network. By the way, this should
also be the common behavior of the magnetic elements in active region plages.

All GBPs are spatially related to magnetic flux. But, while all isolated GBPs correspond to magnetic elements of nearly the same spatial extension, when several o$
them are close together, forming clusters or ribbons, the situation is more complicated: the magnetic flux extends over the whole cluster or ribbon area; the bright
elements coincide with flux maxima, but the space in between them is magnetized
too [1, 52], (Fig. 2.19).

GBPs are buffeted by the evolving granules; their proper motion is chaotic with a
mean speed of $1.4 \, \mathrm{km \, s^{-1}}$, and a maximum velocity of up to $5 \, \mathrm{km \, s^{-1}}$. These transversal motions can propagate along the flux tubes as transverse magneto-acoustic waves
up to the corona [53, 54]. The energy transported by these waves from the network
(i.e. from the quiet photosphere), is one order of magnitude larger than required
to heat the corona [55]. Thus, the buffeting of GBPs by granules can contribute
significantly to the heating of the solar corona.

2.3.4 Intranetwork Magnetic Field

Livingston and Harvey [56] detected for the first time in 1971, at the Kitt Peak Observatory (USA), magnetic flux inside the network (i.e. inside supergranules), in high
sensitivity line-of-sight circular polarization magnetograms made near the disk centre. That kind of magnetograms reveals the vertical component of the magnetic field,
relatively to the solar surface. This field is known as the Intranetwork (IN) Magnetic
Field and is made of mixed polarity elements, often forming bipoles (Figs. 2.16 and
2.20). The bipole elements have a size of $2''-3''$ and are separated by about $10''$ [57].
Both their size and flux change very rapidly, with a typical time scale of a few tens
of minutes [58]. The flux per element is of the order of $5 \times 10^{16} \, \mathrm{Mx}$, one order of
magnitude smaller than in the network.

The IN field tends to appear inside supergranules, $2''-5''$ in extension; the two
components move appart and reach a separation of about $10''$. They are advected
towards the supergranule boundaries by supergranular flows, where they can merge
with a network element of the same polarity, or cancel when they meet a network
element of opposite polarity.

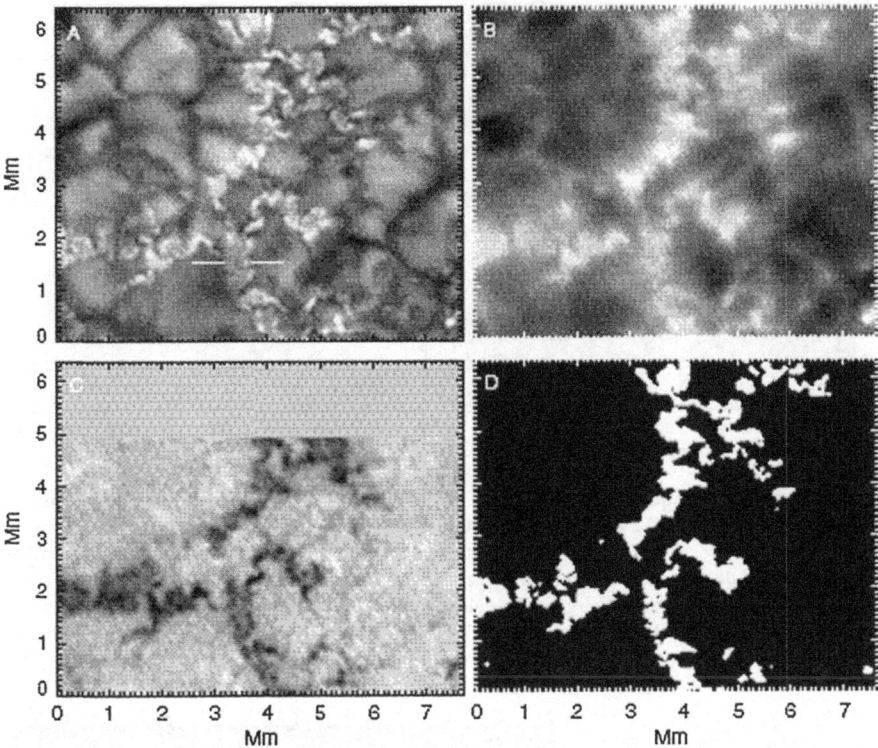

Fig. 2.19 Relation between G-Band Bright features (GBPs) and magnetic elements observed with a $0''1$ spatial resolution with the Swedish Telescope at La Palma. **a** G-Band 4,305 Å filtergram, **b** CaII 3,968 H-line filtergram, **c** FeI 6302 magnetogram, **d** Binary mask of the G-Band emission in panel (**a**) (from Berger et al. [1])

The IN magnetic flux is spatially displayed on a mesogranular scale ($5''$–$10''$), being stronger at the mesogranular cells boundaries than inside (Figs. 2.20 and 2.21), but remaining weaker than in the network at the supergranule boundaries. In higher resolution line-of-sight magnetograms approaching $0.5''$, like those obtained at the Sac Peak Observatory (USA) [59] with the Dunn Tower Telescope, or at Tenerife [60], with the German VTT (Fig. 2.22), the vertical component of the magnetic field appears to be closely related with the granulation pattern, located preferentially in intergranular lanes and associated with downflows. However, some magnetic flux is also found associated with granules. Magnetic flux in the range 5×10^{15}–5×10^{16} Mx are measured, and field strength in the range 200–1,000 G estimated from the Stokes-V profiles of the IR line FeI 15658, definitely weaker than in the network. These weak magnetic features evolve in close connection with the solar granulation.

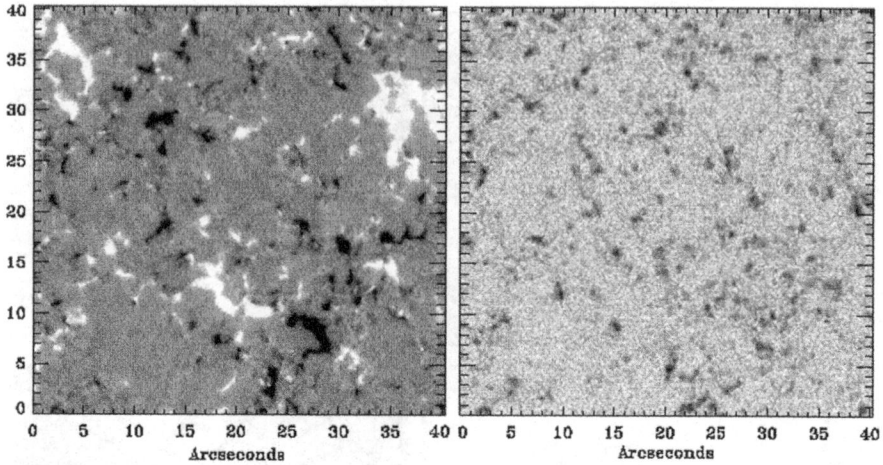

Fig. 2.20 Central $40''$ square highlighted in Fig. 2.14, shown in expanded view. Vertical flux F_V (*left*) and Transversal flux F_T (*right*) for the Quiet-Sun map of Fig. 2.14. The *gray scale* for F_V saturates at $\pm 50\,\mathrm{Mx\,cm^{-2}}$ ($1\,\mathrm{Mx\,cm^{-2}} = 1\,\mathrm{G}$), but it saturates at $200\,\mathrm{Mx\,cm^{-2}}$ for F_T. *White* in the *left panel* denotes positive flux, and *dark* in the *right panel* corresponds to high values of the transverse flux. Note the supergranular cell surrounded by strong flux in the *left panel* located between the coordinates 15:35 and 10:38, and the mesogranular size areas in both images having smaller flux (from Lites et al. [35])

Fig. 2.21 Line-of-sight magnetogram of a $25'' \times 15''$ area obtained with the 1 m Swedish Telescope at La Palma, showing only high flux signals above $50\,\mathrm{Mx\,cm^{-2}}$. Note the regular pattern with a size similar to the $5''$–$10''$ scale of the mesogranulation. Tick-marks correspond to 1 arcsec (from Cerdeña et al. [60])

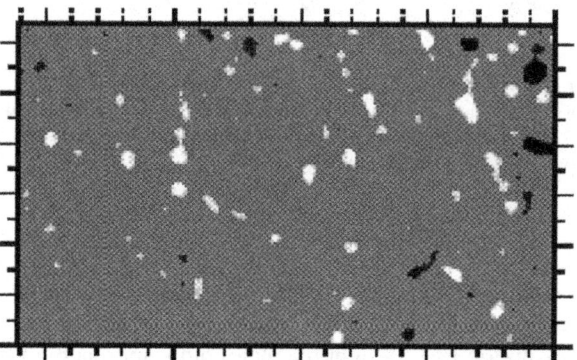

2.3.5 Horizontal Component of the Intranetwork Magnetic Field and Granular Magnetic Field

Horizontal magnetic flux structures were revealed with the Advanced Stokes Polarimeter (ASP) at the Sac Peak Observatory (USA), by Lites et al. [61], in 1996. The horizontal component was discovered much later than the vertical component, because the polarization signals in the Q and U Stokes parameters are as small as 0.1–0.2% of the continuum intensity, i.e. one order of magnitude weaker than

Fig. 2.22 Speckle reconstructed broad-band image overlead with a line-of-sight magnetogram of FeI 6302 Å, with contours at flux densities $F_V = \pm 30, \pm 50, \pm 70, \pm 90\,\mathrm{Mx}^{-2}$. The *solid* and *dotted contours* indicate opposite polarities. The distance between tick-marks is 1 arcsec (from Cerdeña et al. [60])

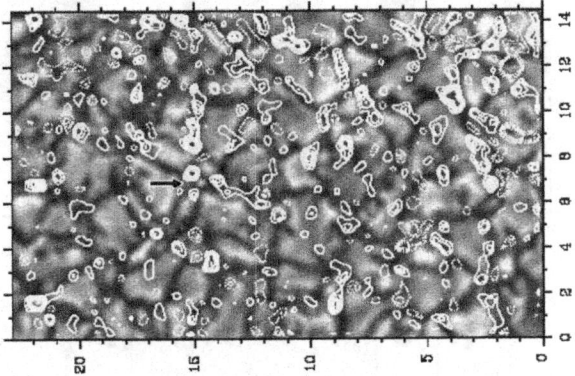

the Stokes-V signal. Since 2007, the horizontal component has been observed with higher spatial resolution ($0''3$) and magnetic sensitivity (0.05% of the continuum intensity), with the HINODE/SOT Spectro-Polarimeter, designed by Lites on the basis of the ASP. This spectro-polarimeter provides quasi-simultanous maps of the horizontal and vertical components of the solar magnetic field, where the IN field appears to be ubiquitous in the Quiet Sun. There is no direct spatial correspondence between the two components (Fig. 2.20). The vertical fields are concentrated in the intergranular lanes (Fig. 2.22); the weaker horizontal fields do not coincide with the locations of the vertical fields and occur preferentially at the edges of bright granules or inside them [61, 62]. This kind of spatial relation between the horizontal and the vertical components, suggests that the IN field consists of small Ω loops, a few arcseconds of dimension, vertically rooted in the intergranular lanes, connected by a mainly horizontal field above granules. The presence of Ω loops closely associated with granules has been recently confirmed by high resolution time series of vector magnetograms obtained with the German VTT at Tenerife [63] and with the SOT/Spectro-Magnetograph [64–66]. The emergence of small magnetic Ω loops is described as follow by Centeno et al. [64]: the horizontal magnetic field appears prior to any significant amount of vertical field (Fig. 2.23); as time goes on, the traces of the horizontal field disappear, while the vertical dipoles drift -carried by the plasma motions- toward the surrounding intergranular lanes. These events take place within typical granulation timescales.

2.3.6 Physical Properties of the Intranetwork Magnetic Field and Comparison with the Properties of the Network Field

Recently, our knowledge of the properties of the IN magnetic field has been very much improved, thanks to the high spatial resolution and high magnetic sensitivity observations made with the HINODE/SOT Spectro-Polarimeter mentioned above.

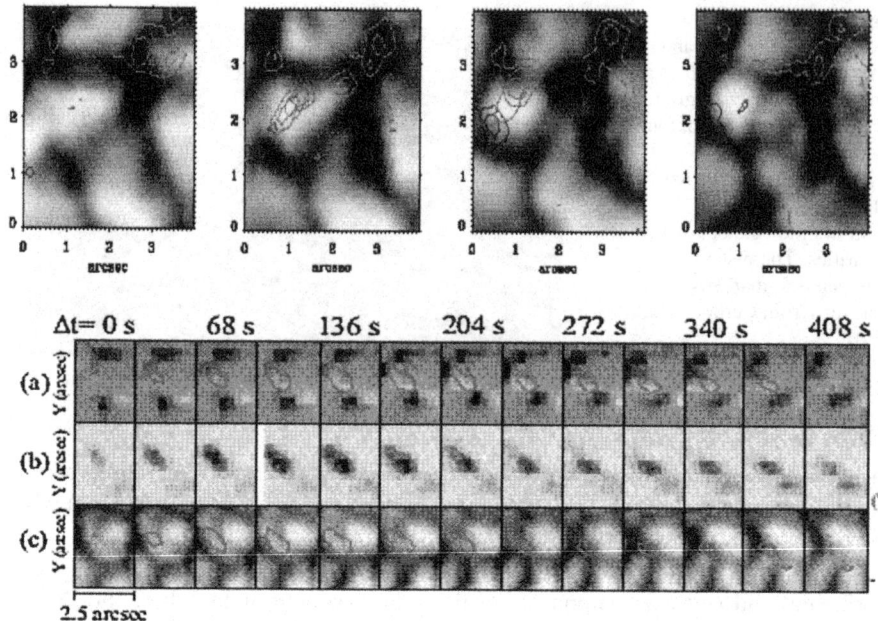

Fig. 2.23 Two examples of emergence of magnetic flux in the form of an Ω loop observed with the HINODE/SOT instrument. *Top*: four granulation images separated by 125 s, with the contours of magnetic signals superimposed, where the thick contours represent positive and negative circular polarisation signals, and the thin contours in between represent linear polarisation signal (from Centeno et al. [64]). *Bottom*: the rows show the time evolution of three different physical quantities. *Top row*: circular polarisation CP (vertical magnetic fiel); *middle row*: linear polarisation LP (horizontal magnetic field); *bottom row*: continuum intensity. The region where LP is larger than 0.3% is enclosed by *Black lines*. The emergence of the horizontal magnetic flux starts at $\Delta t = 0$ s. Note the clear correlation between the emerging magnetic flux and the granule location (from Ishikawa et al. [65])

Spectro-magnetograms can be obtained at a cadence of 2 min on a small area (4″ wide and 82″ long), allowing us to investigate the temporal evolution of IN magnetic elements. The four Stokes profiles I, V, Q, U are obtained in each position along the slit. Maps of about ten physical parameters, including field strength, inclination to the vertical, azimuth angle, velocity, temperature, filling factor, etc, are derived by inverting the four Stokes profiles in each pixel; a Milne-Eddington atmosphere is assumed [64, 66]. The main results obtained so far on the IN magnetic field are summarized below and compared with the properties of the network field.

Magnetic flux: in the range 5×10^{15}–5×10^{16} Hx, for individual elements, one order of magnitude smaller than in the network.

Field strength: less than 500 G, much weaker than the 1–2 kG network field.

Shape: horizontal as well as vertical magnetic fields are detected inside the network, whereas only vertical fields are observed in the network. The vertical fields are located in intergranular lanes, while the horizontal fields are mainly found inside granules. This suggests that the IN field is made with small Ω loops, closely related to granules, which thus appears much different from the network field, which is made of vertical, strong flux tubes, located in intergranular lanes.

The cross-section of the IN elements is less than 200 km, like for the network elements.

Timescale: the timescale of both the IN and network fields is of the order of the granulation lifetime: a few minutes.

Association with G- Band and CaII brightenings: contrary to the Network elements, there are no CaII brightenings associated with IN elements. Concerning G-Band brightenings, only the brightest IN elements located at the boundaries of mesogranules, can be associated with detectable magnetic flux.

2.3.7 Hidden Turbulent Flux

Close to the solar limb, the resonant photospheric lines are linearly polarized due to scattering of photons. The presence of a weak turbulent magnetic field reduces the degree of linear polarization [67]. This effect is known as the Hanlé effect. The strength of the depolarizing turbulent magnetic field of mixed polarity was estimated to 30 G [68, 69]. Later, comparing the depolarization in the weak C_2 lines of the Zwan system and in the resonant line SrI 4,607, Trujillo Bueno et al. [70] concluded that hidden (hidden because not spatially resolved) fields of 10 G are associated with the plasma of the center of granules and stronger fields, of the order of 200 G, in intergranular lanes. This picture fits relatively well with the IN granular field, and may correspond to the Stenflo's and Faurobert-Scholl's turbulent field.

2.3.8 Numerical Simulations of Small-Scale Magneto-Convection

The magneto-convection we are interested here works at the scale of the granulation and has to be distinguished, of course, from the large-scale magneto-convection which drives the solar cycle.

Initially, the simulation is set up as a purely hydrodynamical convection, like one of those mentioned in Sect. 2.2.5, starting from a plane-parallel model, extending, for example 800 km below to 600 km above the level of continuum optical depth unity. The horizontal size of the box is typically 6,000 km in each direction, with a resolution grid of $100 \times 288 \times 288$ pixels. Radiative transfer, which is the main driver of convection and has an important influence on the temperature structure and brightness of the magnetic field concentrations, is taken into account, as well as partial ionization, which strongly affects the efficiency of convective transport. After convection is fully developed, a magnetic field is inserted in various forms: a uniform horizontal field of 30 G [71]; a uniform vertical field of strengths varying from 0 to 200 G [26]; a uniform vertical field of 200 G [28]; a twisted flux tube

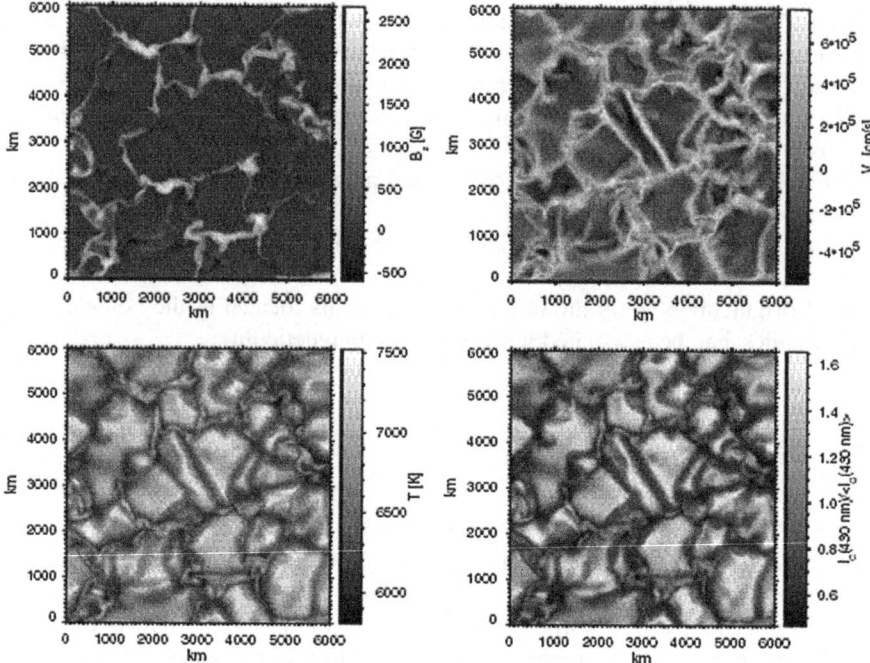

Fig. 2.24 Maps of physical quantities for a snapshot from the run with 200 G average vertical magnetic field obtained by Shelyag et al. [27]. *Upper left*: vertical component of the magnetic field at the level $\tau = 1$ at 5, 000 Å, corresponding to the visible solar surface. The image shows magnetic flux concentrations in the intergranular lanes. *Upper right*: vertical component of the velocity at the level $\tau = 1$ at 5, 000 Å. Positive (negative) values correspond to downflows (upflows). *Lower left*: gas temperature on the level $\tau = 1$ at 5, 000 Å. There are local enhancements of the temperature in the regions of strong magnetic field in intergranular lanes.*Lower right*: normalized continuum intensity at 4,300 Å. The image shows brightenings in the magnetic flux concentrations, which closely correspond to the local temperature enhancements and magnetic flux concentrations

of various field strengths and various twist wavelengths, emerging from below the photosphere [72]. In fact,the results from these models are all very similar. convective motions produce highly intermittent magnetic fields in the intergranular lanes that collect over the boundaries of the underlying mesogranular scale cells. The process of flux expulsion from granules and convective field amplification lead to a dichotomy of strong, mainly vertical fields embedded in the granular downflow network, and weak, randomly oriented fields filling the hot granular upflows. The strong fields form a magnetic network with thin, sheet-like structures extending along downflow lanes (Figs. 2.24, 2.25), which somewhat differs from the observed pattern, which is made not only with sheets, but also with many roundish aligned features (Fig. 2.26), especially in the quiet Sun. At the visible surface around optical depth unity, the strong field concentrations are in pressure balance with their weakly magnetized surroundings and reach field strengths of up to 2 kG, strongly exceeding the values

(a)

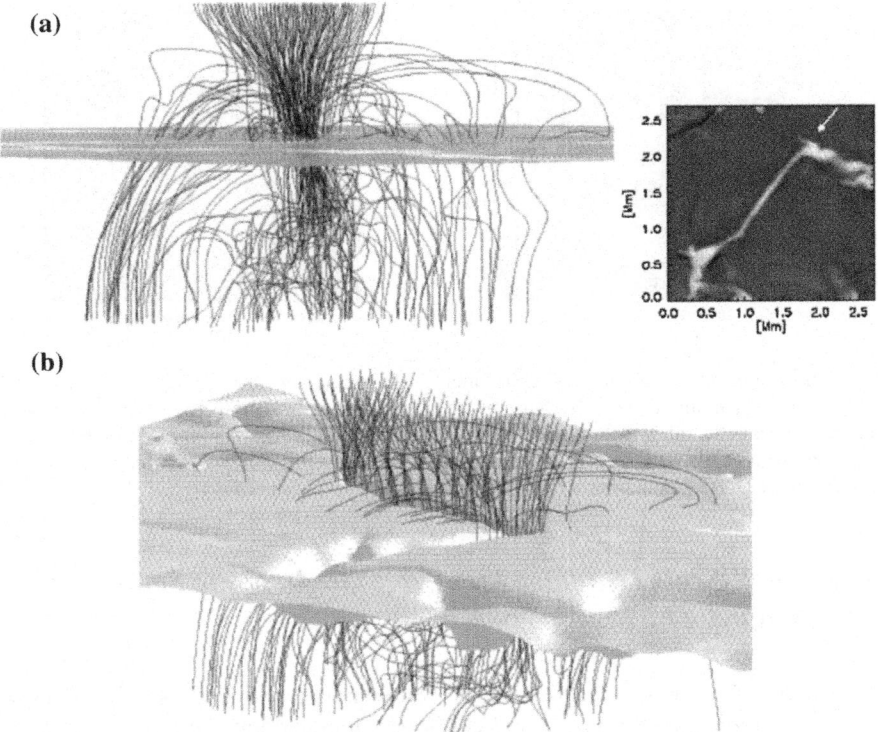

(b)

Fig. 2.25 Results from a numerical simulation by Vögler et al. [28], showing the distribution of the magnetic field lines. The simulation was started with a 200 G uniform vertical magnetic field

corresponding to equipartition with the kinetic energy density of the convective motions (which is 600 G only).

Simulated images of the emergent intensity smoothed by the diffraction of the telescope and by the degradation by the Earth's atmosphere resembles the observed images taken with the same resolution (Fig. 2.26). The observed magnetic elements are well reproduced by the simulations, but not their spatial distribution: in the real quiet Sun, strong fields are found at the boundaries of supergranules; but in the simulations, they form a pattern of mesogranular scale (where, in fact, the fields are observed to have only moderate strengths). The simulations thus appear to be representative of active regions rather than of the quiet Sun. Moreover, the appearance of a strong flux pattern at mesogranular scales may be artificially due to the size of the computational box, which is just a little larger than a mesogranule.

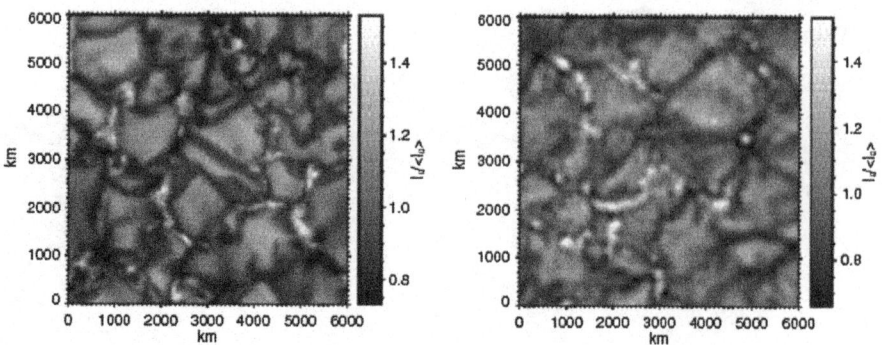

Fig. 2.26 Simulated and observed G-Band images (from Shelyag et al. [27]). *Left*: synthetic G-Band image of the simulated area after smoothing by the function mimicking the diffraction by the telescope and the image degradation by the Earth's atmosphere. *Right*: observed G-Band image with a similar area fraction of G-Band bright points as in the 200-G simulation (subfield of an image taken with the Dutch Open Telescope on La Palma, courtesy: P. Sütterlin)

2.3.9 Small-Scale Dynamo: A Possible Origin of the Small-Scale Magnetic Field

The small-scale magneto-convection we were dealing with in Sect. 2.3.8, concentrates a weak pre-existing magnetic field in intergranular lanes. It is just a redistribution of magnetic flux from larger to smaller scales, without production of magnetic energy. On the contrary, in small-scale dynamos, a very weak seed magnetic energy is amplified exponentially by non-linear effects which occur in turbulent convection, to reach a significant fraction of the kinetic energy. However, the physical mechanisms at work here are not fully understood yet. Two different models have been developed so far, by Cattaneo and co-workers [26, 73] and by Vögler and Schüssler [74]. The Cattaneo and al's model is not realistic, because the atmosphere is incompressible and the computational box is closed; but the idealized numerical experiment helps to isolate and understand the physical processes which are involved. Vögler and Schüssler developed a more realistic model, with a compressible atmosphere and an open box, but is is difficult to interpret the results. In both simulations a weak bipolar magnetic field (10 mG) with nul net flux, is inserted when the simulated convection is well developed. Soon after, the flow acts as a turbulent dynamo, generating a small-scale disordered magnetic field, with no net flux through the box. The unsigned mean magnetic density at the level $\tau = 1$ reaches a value of about 25 G. The resulting field exhibits an intricate small-scale mixed-polarity structure. The strongest magnetic features reach occasionally vertical field strengths beyond 1 kG near $\tau = 1$ (Fig. 2.27).

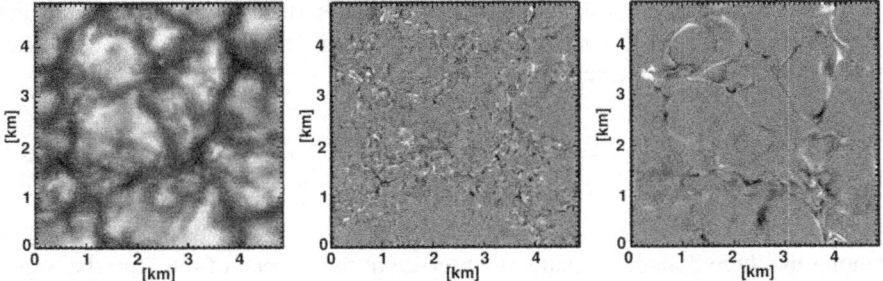

Fig. 2.27 Snapshot from a dynamo run, taken 5 h after introducing the seed field. The vertically emerging intensity (brightness, *left panel*) reveals a normal solar granulation pattern. The *other panels* show the vertical component of the magnetic field on two surfaces of constant optical depth, τ. Near the visible surface (*middle panel*, $\tau = 1$, grey scale saturating at ±250 G), the magnetic field shows an intricate small-scale pattern with rapid polarity changes and an unsigned average flux density of 25 G. About 300 km higher, at the surface $\tau = 0.01$ (*right panel*), grey scale saturating ±50 G), the unsigned average flux density has decreased to 3.2 G and the field distribution has become considerably smoother, roughly outlining the network of intergranular downflow lanes (darker area on the *left panel*) (from Vögler [74])

2.3.10 Origin of the Photospheric Network

The Quiet-Sun photospheric network decays in canceling collisions between magnetic concentrations of opposite polarities. The origin of the flux which compensates this decay is still unclear, although diffusion of magnetic flux emerged in Ephemeral Active Regions (ERs) is the most accepted one. ERs are small bipolar active regions, $10''$–$20''$ of extension and 3×10^{19}–1×10^{20} Hx of flux, which emerges frequently through the solar surface, at all latitudes. Their lifetime is of the order of one day. They are exclusively made with bright magnetic elements, without any dark sunspot nor pore. The origin of the ER population is not very clear yet. Many speculate that the ER flux originates close to the surface, forming as a consequence of the vigorous convective motions present in the near-surface layers of the convection zone [75–77]. But ERs may also be decay products of active regions, either sheared off on emergence, or in the aging process of large bipolar regions. It is also possible that both of these processes contribute to the formation of ERs, since as much flux may erupt in the form of small ERs as in large active centers. Owing to the properties of the photospheric network (fairly homogeneous distribution on the solar surface and weak dependence of the phase of the sunspot cycle), ERs appear as a better candidate to feed it. The flux emerging in this form is dispersed by supergranular advection and large scale circulation.

Alternatively, it has been suggested that the origin of the photospheric network is the IN magnetic flux generated by a local dynamo at the granulation scale of flux of the order of 10^{15} Mx per element. The IN flux is advected first to the mesogranular boundaries, where the flux is concentrated to 10^{16} Mx per element, then to the supergranular boundaries where it is concentrated to still higher flux, of the order of

10^{17} Mx. The total IN flux integrated over the solar surface represents a significant fraction of the flux in the network. In fact both sources of flux, ER and IN, can contribute to maintain the network.

2.3.11 Conclusion

Among the three dynamical patterns observed at the surface of the Sun, the origin of the granulation is well understood, on the basis of the observed properties and of the results of numerical simulations, which reproduce most of them satisfactorily, although the viscosity of the numerical plasma is much larger than the real one: it is a convective pattern driven by surface radiative cooling. Concerning the mesogranulation, the observed properties are those of convective cells. In numerical simulations, a convective flow pattern of mesogranular scale appears below the surface, which could be associated with the observed mesogranulation. Its origin could thus be convective, as a simple extension to deeper layers of the convection driven by the surface cooling. This is supported by the fact that there are no two distinct peaks corresponding to the scales of granules and mesogranules in the power spectrum of photospheric velocities. Some observational properties of the mesogranulation could suggest they are of convective origin too; this is supported by some recent numerical simulations [31, 32] The simulations, primarily devoted for investigating granulation and mesogranulation, have been extended to the size of supergranules (but at the expense of granules, which are badly resolved), and show that the scale of the developed cells increases smoothly from the granulation scale at the optical depth unity (which simulates the solar surface), to the supergranulation scale at the bottom of the simulated domain, 20 Mm below the surface. Large cells advect the smaller cells above, as observed at the real solar surface (where granules are advected by mesogranules, which are advected, in their turn, by supergranules). The mesogranulation thus seems to be of convective origin, indirectly driven by surface radiative cooling. The convective origin proposed by Leighton in the sixteen's, which was later discarded by most authors, could be the true nature of the solar supergranulation; but the driving force beeing a surface radiative cooling instead of HeII recombination as believed by Leighton. There is a problem, however, with this interpretation, because the power spectra of vertical and horizontal flows observed at the solar surface, have a distinct peak at the supergranulation scale, which is not expected in the case of a smooth size increase of the convective cells with depth. Alternatively to convection, it has been suggested that supergranules could result from non linear interactions between granules or between downflow plumes, or could be driven by the magnetic network.

Recent high spatial resolution and high polarisation sensitivity observations performed with ground based telescopes and with the Solar Optical Telescope on board the satellite HINODE, have revealed the complexity of the magnetic field in the quiet photosphere. In the same time, the development of realistic numerical simulations (except for the viscosity, which is much higher than in the solar atmosphere) allows us to better understand the physical processes which generate the magnetic field at the surface, or regenerate magnetic debris from active regions, and further concentrate the field in intergranular lanes. However, much remains to be done in order to establish quantitatively reliable results of near-surface local dynamo action. This includes studying the effects of deeper and wider computational boxes as well as variations of the boundary conditions.

Magnetic field is now found everywhere in the quiet Sun, in different forms. At the boundaries of supergranules, it is concentrated in vertical tubes of kG field strengths, and of diameters smaller, or even much smaller, than a few hundreds kilometers, located in intergranular lanes. The flux in the network is believed to be concentrated by a so-called convective collapse process. It is also ubiquitous inside supergranules, where the so-called Intranetwork field consists of small loops of lower strength (hectoGauss) and of lower flux (one or two orders of magnitude lower than in the network), closely related to granules; the vertical part of the loop is anchored in intergranular lanes and are connected through granules. The time scale of the network and internetwork fields is closely related to the time evolution of granules, which is less than 10 min. It is not clear yet whether the granular field is locally generated by a fast local dynamo action, or is emerging from deeper layers as small dipoles or as debris from decaying active regions.

The ubiquitous presence of small-scale magnetic field in surface layers may affect the propagation of surface or near-surface waves propagation and this has to be taken into account when deriving informations from the wave properties. It would also be interesting to know how much magnetic flux is contained in the deeper layers, away from active regions, and in which form.

References

1. Berger, T.E., Rouppe van der Voort, L.H.M., Löfdahl, M.G., Carlsson, M., Fossum, A, Hansteen, V.H., Marthinussen, E., Title, A.M., Scharmer, G.: A&A **428**, 613 (2004)
2. Mehltretter, J.P.: A&A **62**, 311 (1978)
3. Carlier, A., Chauveau, A., Hugon, M., Rösch, J.: Comptes rendus Acad. Sci. **266**, 199 (1969)
4. Kawaguchi, I.: Sol. Phys. **65**, 207 (1980)
5. Roudier, T., Lignières, F., Rieutord, M., Brandt, P.N., Malherbe, J.M.: A&A **409**, 299 (2003)
6. Roudier, T., Muller, R.: A&A **419**, 757 (2004)
7. Espagnet, O., Muller, R., Roudier, T., Mein, N., Mein, P.: A&A Suppl. Ser. **109**, 79 (1995)
8. November, L.J., Toomre, J., Gebbie, K.B., Simon, G.W.: ApJ **245**, L213 (1981)
9. November, L.J., Simon, G.W.: ApJ **333**, 427 (1988)
10. Roudier, T., Rieutord, M., Malherbe, J.M., Vigneau, J.: A&A **349**, 301 (1999)
11. Muller, R., Auffret, H., Roudier, T., Vigneau, J., Simon, G.W., Frank, Z., Shine, R.A., Title, A.M.: Nature **356**, 322 (1992)
12. Hart, A.B.: MNRAS **114**, 17 (1954)
13. Leighton, R.B., Noyes, R.B., Simon, G.W.: ApJ **135**, 474 (1962)
14. DeRosa, M., Duvall, T.L., Toomre, J.: Sol. Phys. **192**, 351 (2000)
15. Shine, R.A., Simon, G.W., Hurlburt, N.E.: Sol. Phys. **193**, 313 (2000)
16. Rieutord, M., Meunier, N., Roudier, T., Rondi, S., Beigbeder, F., Parès, L.: A&A **479**, L17 (2008)

17. Rast, M.P.: ApJ **597**, 1200 (2003)
18. Meunier, N., Tkaczuk, R., Roudier, T.: A&A **463**, 745 (2007)
19. Hathaway, D.H., Beck, J.G., Bogart, R.S., Bachmann, K.T., Khatri, G., Petitto, J.M., Han, S., Raymond, J.: Sol. Phys. **193**, 299 (2000)
20. Stein, R.F., Nordlund, Å.: ApJ **342**, L95 (1989)
21. Spruit, H.C., Nordlund, Å., Title, A.M.: Annual Rev. Astron. Astrophys. **28**, 263 (1990)
22. Stein, R.F., Nordlund, Å.: ApJ **499**, 914 (1998)
23. Stein, R.F., Nordlund, Å.: Sol. Phys. **192**, 91 (2000)
24. Rast, M.P.: ApJ **443**, 863 (1995)
25. Cattaneo, F., Lenz, D., Weiss, N.: ApJ **563**, L91 (2001)
26. Cattaneo, F., Emonet, T., Weiss, N.: ApJ **588**, 1183 (2003)
27. Shelyag, S., Schüssler, M., Solanki, S.K., Berdyugina, S.V., Vögler, A.: A&A **427**, 335 (2004)
28. Vögler, A., Shelyag, S., Schüssler, M., Cattaneo, F., Emonet, T., Linde, T.: A&A **429**, 335 (2005)
29. Nordlund, Å.: Sol. Phys. **100**, 209 (1985)
30. Rincon, F., Lignières, F., Rieutord, M.: A&A **430**, L57 (2005)
31. Stein, R.F., Nordlund, Å., Georgobiani, D., Benson D., Schaffenberger, W.: GONG2008 SOHO21 Conference Proceedings (2008)
32. Nordlund, Å., Stein, R.F., Asplund, M.: Living Rev. Sol. Phys. **6**, 2 (2009)
33. Rieutord, M., Roudier, T., Malherbe, J.M., Rincon, F.: A&A **357**, 1063 (2000)
34. Crouch, A.D., Charbonneau, P., Thibault, K.: ApJ **662**, 715 (2007)
35. Stenflo, J.O.: Sol. Phys. **32**, 41 (1973)
36. Dunn, R.B., Zirker, J.B.: Sol. Phys. **33**, 281 (1973)
37. Spruit, H.C.: Sol. Phys. **50**, 269 (1976)
38. Nagata, S., Tsuneta, S., Suematsu, Y., Ichimoto, K., Katsukawa, Y., Shimizu, T., Yokoyama, T., Tarbell, T.D., Lites, B.W., Shine, R.A., Berger, T.E., Title, A.M., Bellot Rubio, L.R., Orozco Suárez, D.: ApJ **677**, L145 (2008)
39. Bellot Rubio, L.R., Rodríguez Hidalgo, I., Collados, M., Khomenko, E., Ruiz Cobo, B.: ApJ **560**, 1010 (2001)
40. Socas-Navarro, H., Manso-Sainz, M.: ApJ **620**, L71 (2005)
41. Steiner, O., Grossmann-Doerth, U., Knoölker, M., Schüssler, M.: ApJ **495**, 468 (1998)
42. Muller, R., Roudier, T.: Sol. Phys. **94**, 33 (1984)
43. Muller, R., Keil, S.L.: Sol. Phys. **87**, 243 (1983)
44. Title, A.M., Berger, T.E.: ApJ **463**, 797 (1996)
45. Utz, D., Hanslmeier, A., Möstl, C., Muller, R., Veronig, A., Muthsam, H.: A&A **498**, 289 (2009)
46. Wiehr, E., Bovelet, B., Hirzberger, J.: A&A **422**, L63 (2004)
47. Muller, R.: Sol. Phys. **85**, 113 (1983)
48. Muller, R., Roudier, Th.: Sol. Phys. **141**, 27 (1992)
49. Berger, T.E., Title, A.M.: ApJ **463**, 365 (1996)
50. Berger, T.E., Löfdahl, M.G., Shine, R.S., Title, A.M.: ApJ **495**, 973 (1998)
51. Rouppe von der Voort, L.H.M., Hansteen, V.H., Carlsson, M., Fossum, A., Marthinussen, E., van Noort, M.J., Berger, T.E.: A&A **435**, 327 (2005)
52. Berger, T.E., Title, A.M.: ApJ **553**, 449 (2001)
53. Choudhuri, A.R., Auffret, H., Priest, E.R.: Sol. Phys. **143**, 49 (1993)
54. Choudhuri, A.R., Dikpati, M., Banerjee, D.: ApJ **413**, 811 (1993)
55. Muller, R., Roudier, T., Vigneau, J., Auffret, H.: A&A **283**, 232 (1994)
56. Livingston, W., Harvey J.: In: Howard, R. (ed.) IAU Symposium (Dordrecht Reidel), vol. 43, p. 51 (1971)
57. Livingston, W., Harvey, J.: BAAS **7**, 346 (1975)
58. Martin, S.F.: Solar photosphere: structure, convection and magnetic fields. In: Stenflo, J.O. (ed.) IAU Symposium, vol. 138, p. 129. Kluwer, Dordrecht (1990)
59. Lin, H., Rimmele, Th.: ApJ **514**, 448 (1999)

60. Domínguez Cerdeña, I., Sánchez Almeida, J., Kneer, F.: A&A **407**, 741 (2003)
61. Lites, B.W., Leka, D., Skumanich, A., Martínez Pillet, V., Shimizu, T.: ApJ **460**, 1019 (1996)
62. Lites, B.W., Kubo, M., Socas-Navarro, H., Berger, T.E., Frank, Z., Shine, R.A., Tarbell, T., Title, A.M., Ichimoto, K., Katsukawa, Y., Tsuneta, S., Suematsu, Y., Shimizu, T., Nagata, S.: ApJ **672**, 1237 (2008)
63. Martínez González, M.J., Collados, M., Ruiz Cobo, B., Solanki, S.K.: A&A **469**, L39 (2007)
64. Centeno, R., Socas-Navarro, H., Lites, B.W., Kubo, M., Frank, Z., Shine, R.A., Tarbell, T., Title, A.M., Ichimoto, K., Tsuneta, S., Katsukawa, Y., Suematsu, Y., Shimizu, T., Nagata, S.: ApJ **666**, L137 (2007)
65. Ishikawa, R., Tsuneta, S., Ichimoto, K., Isobe, H., Katsukawa, Y., Lites, B.W., Nagata, S., Shimizu, T., Shine, R.A., Suematsu, Y., Tarbell, T., Title, A.M.: A&A **469**, L39 (2008)
66. Orozco Suárez, D., Bellot Rubio, L.R., DelToro Iniesta, J.C., Tsuneta, S.: A&A **481**, L33 (2008)
67. Stenflo, J.O.: Sol. Phys. **80**, 209 (1982)
68. Faurobert-Scholl, M.: A&A **268**, 765 (1993)
69. Faurobert-Scholl, M., Feautrier, N., Machefert, F., Petrovay, K., Spielfiedel, A.: A&A **298**, 289 (1995)
70. Trujillo Bueno, J., Shchukina, N., Asencio Ramos, A.: Nature **430**, 326 (2004)
71. Stein, R.F., Nordlund, Å.: ApJ **642**, 1246 (2006)
72. Cheung, M.C., Schüssler, M., Moreno-Insertis, F.: A&A **467**, 703 (2007)
73. Cattaneo, F.: ApJ **515**, L39 (1999)
74. Vögler, A., Schüssler, M.: A&A **465**, L43 (2007)
75. Schüssler, M.: Second solar cycle and space weather conference. In: Sawaya-Lacoste, H. (ed.) SOLSPA, p. 3. ESA, Noodwijk (2001) (ESA SP-477)
76. Hagenaar, H.J., Schrijver, C.J., Title, A.M., Shine, R.A.: ApJ **584**, 1107 (2003)
77. Stein, R.F., Bercick, D., Nordlund, Å.: Current theoretical models and future high resolution solar observations: preparing for ATST. In: Pevtsov, A.A., Uitenbroeck, H (eds.) ASP Conf. Ser. 286, p. 121. ASP, San Francisco (2003)

Chapter 3
Modeling and Prediction of Solar Cycles Using Data Assimilation Methods

Irina N. Kitiashvili and Alexander G. Kosovichev

Abstract Variations of solar activity are a result of a complicate dynamo process in the convection zone. We consider this phenomenon in the context of sunspot number variations, which have detailed observational data during the past 23 solar cycles. However, despite the known general properties of the solar cycles a reliable forecast of the 11-year sunspot number is still a problem. The main reasons are imperfect dynamo models and deficiency of the necessary observational data. To solve this problem we propose to use data assimilation methods. These methods combine observational data and models for best possible, efficient and accurate estimates of physical properties that cannot be observed directly. The methods are capable of providing a forecast of the system future state. It is demonstrated that the Ensemble Kalman Filter (EnKF) method can be used to assimilate the sunspot number data into a non-linear $\alpha - \Omega$ mean-field dynamo model, which takes into account dynamics of turbulent magnetic helicity. We apply this method for characterization of the solar dynamo properties and for prediction of the sunspot number.

3.1 Introduction

A thoughtful investigation of a natural phenomenon consists of three basic parts: observation, construction of a model and prediction. Predictions based on a model determine correctness of our understanding of the physical processes. Because observation data contain errors, and a model constructed on their basis is characterized by some approximations, a prediction of the next set of observations will deviate from

Irina N. Kitiashvili (✉)
Center for Turbulence Research (CTR), Stanford University, Stanford CA 94305 , USA
e-mail: irinasun@stanford.edu

Alexander G. Kosovichev
Stanford University, Stanford CA 94305 USA
e-mail: sasha@sun.stanford.edu

J.-P. Rozelot and C. Neiner (eds.), *The Pulsations of the Sun and the Stars*,
Lecture Notes in Physics 832, DOI: 10.1007/978-3-642-19928-8_3,
© Springer-Verlag Berlin Heidelberg 2011

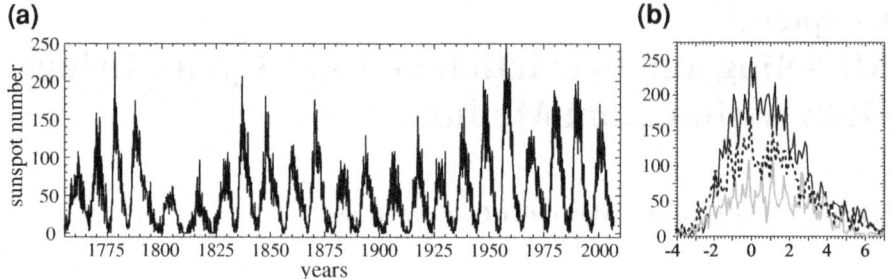

Fig. 3.1 Observed monthly sunspot number series **a** for 1755–2007 from NGDC, and **b** for three solar cycles 14 (*gray curve*), 19 (*black curve*) and 23 (*dotted curve*), which are aligned according to their maxima ($t = 0$)

the real data. Nevertheless, an estimate of the uncertainties in the model and observations allows us to correct the model solution according to the information obtained from new measurements. Thus, updated observational data and a consistent correction of the model solution allow us to more accurately describe the system's behavior and forecast its future state. This procedure also provides additional information for system parameters that are difficult to observe directly.

In this paper, we discuss an application of the Ensemble Kalman Filter method (EnKF) to a simple non-linear dynamo model for analysis of the solar activity cycles [1, 2]. One of manifestations of solar magnetic activity is the 11-year sunspot cycle (Fig. 3.1a), which is characterized by fast growth and slow decay of the sunspot number parameter (Fig. 3.1b). For modeling the solar cycle we use a non-linear $\alpha - \Omega$ dynamo model [3], which takes into account temporal variations of turbulent magnetic helicity.

3.2 Formulation of the Dynamo Models

3.2.1 Parker's Migratory Dynamo

In a kinematic approximation, the dynamo problem can be described by the induction equation [4]

$$\frac{\partial \mathbf{B}}{\partial t} = \nabla \times (\mathbf{v} \times \mathbf{B}) - \eta_m \nabla^2 \mathbf{B}, \tag{3.1}$$

where $\mathbf{B}$ is the magnetic field strength, $\mathbf{v}$ is the fluid velocity, η_m is the molecular magnetic diffusivity. Magnetic field, $\mathbf{B}$, and the fluid velocity, $\mathbf{v}$, can be separated into two components representing mean and fluctuating (turbulent) parts, or $\mathbf{B} = \langle B \rangle + \mathbf{b}$ and $\mathbf{v} = \langle \mathbf{v} \rangle + \mathbf{u}$. Here, $\langle \mathbf{B} \rangle$ represents the longitudinally averaged magnetic field, $\mathbf{b}$ is the fluctuating part of $\mathbf{B}$, $\langle \mathbf{v} \rangle$ represents mean global-scale motions in the Sun (such as the differential rotation, and $\mathbf{u}$ is the velocity of turbulent convective motions. Taking into account that the average of fluctuations is zero, $\langle \mathbf{b} \rangle = \mathbf{0}$ and $\langle \mathbf{u} \rangle = \mathbf{0}$, for the case of isotropic turbulence, we obtain the following mean-field induction equation [5]

$$\frac{\partial \langle \mathbf{B} \rangle}{\partial t} = \nabla \times (\langle \mathbf{v} \rangle \times \langle \mathbf{B} \rangle + \alpha \langle \mathbf{B} \rangle - \eta \nabla \times \langle \mathbf{B} \rangle)) \tag{3.2}$$

where η describes the total magnetic diffusion, which is the sum of the turbulent and molecular magnetic diffusivity, $\eta = \eta_t + \eta_m$ (usually $\eta_m \ll \eta_t$). Parameter α is turbulent fluid helicity. The first term of the equation describes transport of magnetic field lines with fluid, the second term describes the α-effect, and the last term determines diffusion and dissipation of the field.

For describing the average magnetic field, following [4], we choose a local coordinate system, xyz, where z represents the radial coordinate, axis y is the azimuthal coordinate and axis x coincides with co-latitude. Effects of sphericity are not included in this model. Hence, the vector of the mean field, $\langle \mathbf{B} \rangle$, can be represented as

$$\langle \mathbf{B} \rangle = B(x, y)\mathbf{e}_y + \nabla \times \left[A(x, y)\mathbf{e}_y \right], \tag{3.3}$$

where $B(x, y)$ is the toroidal component of magnetic field, $A(x, y)$ is the vector-potential of the poloidal field. Assuming that $\langle \mathbf{v} \rangle = v_y(x)\mathbf{e}_y$ (rotational component), we can write the dynamical system describing Parker's model of the $\alpha - \Omega$ dynamo [4] in the standard form:

$$\frac{\partial A}{\partial t} = \alpha B + \eta \nabla^2 A, \quad \frac{\partial B}{\partial t} = G \frac{\partial A}{\partial x} + \eta \nabla^2 B, \tag{3.4}$$

where $G = \partial \langle v_y \rangle / \partial z$ is the rotational shear.

Assuming that the coefficients are constants and seeking a solution of the model in the form $(A, B_y) \sim (A_0, B_0) \exp[i(kx - \omega t)]$, we find a well-known result that a pure periodic solution exists if $D = \alpha G/(\eta^2 k^3) = 2$, where D is the so-called "dynamo number". The solutions grow in time for $D > 2$, and decay for $D < 2$.

For the periodic dynamo solutions toroidal and poloidal field components vary in time in a sinusoidal fashion which is clearly different from the observed, asymmetric profile of the solar cycle (Fig. 3.1b). As shown in [2], in the one-mode approximation the classical Parker's dynamo model gives only periodic oscillatory solutions, and therefore cannot explain the observed variations of the sunspot number in the solar cycles. For creating chaotic variations of the magnetic field in the low-mode approximation it is necessary to add to the Parker's model a third equation describing variations of the magnetic helicity and its interaction with the large-scale magnetic field [3, 6].

3.2.2 The Kleeorin–Ruzmaikin Model

For modeling the solar cycle we choose the formulation [3], which is based on the idea of magnetic helicity conservation, and has reasonable agreement with the

observational data of solar magnetic fields [7, 8]. Due to the fact that the kinetic helicity makes the magnetic field small-scaled, the back influence on the turbulent fluid
motions can restrict the unlimited growth of the magnetic field. In the mean-field
approach the magnetic helicity is separated into large- and small-scale components.
Because of conservation of the total helicity, a growth of the large-scale magnetic
helicity due to the dynamo action is compensated by the growth of the small-scale helicity of opposite sign [8]. Thus, small- and large-scale magnetic fields grow together
and are mirror-asymmetrical. This means that the condition of magnetic helicity conservation is, perhaps, more severe for restricting the dynamo action than the condition
of the energy conservation.

The turbulent helicity can be divided into two parts: hydrodynamic and magnetic:
$\alpha = \alpha_h + \alpha_m$. The kinetic helicity, α_h, describes helical turbulent fluid motions; the
magnetic helicity, α_m, determines the order of twisted magnetic field lines:

$$\alpha_h = -\tau \langle \mathbf{u} \cdot (\nabla \times \mathbf{u}) \rangle / 3, \quad \alpha_m = \tau \langle \mathbf{b} \cdot (\nabla \times \mathbf{b}) \rangle / (12\pi\rho), \qquad (3.5)$$

where τ is the lifetime of turbulent eddies, ρ is density.

It is convenient to define the influence of the magnetic helicity on magnetic field
using spectral density χ [3]

$$\bar{\chi} \equiv \langle \mathbf{a} \cdot \mathbf{b} \rangle, \qquad (3.6)$$

where $\mathbf{a}$ is the fluctuating part of the magnetic field vector-potential, $\mathbf{A}$.

To derive an equation for the averaged helicity density we multiply the basic
induction (3.1) written without the differential rotation term by the fluctuating part
of the vector potential, $\mathbf{a}$; and also multiply the equation for the vector-potential

$$\frac{\partial \mathbf{A}}{\partial t} = \mathbf{v} \times \mathbf{B} - \eta_{\mathrm{m}} \nabla \times \nabla \times \mathbf{A}, \qquad (3.7)$$

by the fluctuating part of magnetic field, $\mathbf{b}$. Averaging the sum of (3.1) and (3.7),
and taking into account that $\mathbf{b} = \nabla \times \mathbf{a}$, after some transformations we obtain the
following expression for the helicity density [2, 3]

$$\frac{\partial \bar{\chi}}{\partial t} = \left\langle \mathbf{a} \cdot \frac{\partial \mathbf{b}}{\partial t} + \mathbf{b} \cdot \frac{\partial \mathbf{a}}{\partial t} \right\rangle = -2\langle [\mathbf{v} \times \mathbf{b}] \cdot \langle \mathbf{B} \rangle \rangle - 2\eta_{\mathrm{m}} \langle \mathbf{b} \cdot \nabla \times \mathbf{b} \rangle. \qquad (3.8)$$

Two terms, $\langle \triangle [\mathbf{a} \times [\mathbf{v} \times \langle \mathbf{B} \rangle]] \rangle$ and $\langle \triangle [\mathbf{a} \times [\mathbf{v} \times \mathbf{b}]] \rangle$, disappear as a result of volume averaging. Using the mean-field electrodynamics approximation and retaining
only the first two terms for the mean electric field [5]

$$\varepsilon \equiv \langle \mathbf{v} \times \mathbf{b} \rangle \cong \alpha \langle \mathbf{B} \rangle - \eta \left(\nabla \times \langle \mathbf{B} \rangle \right), \qquad (3.9)$$

we obtain

$$\frac{\partial \bar{\chi}}{\partial t} = 2 \left(\eta \langle \mathbf{B} \rangle \cdot (\nabla \times \langle \mathbf{B} \rangle) - \alpha \langle \mathbf{B} \rangle^2 - \eta_{\mathrm{m}} \langle \mathbf{b} \cdot \nabla \times \mathbf{b} \rangle \right). \qquad (3.10)$$

Then, the expression for variations of the magnetic helicity, α_m, in terms of the mean magnetic field is the following [3]:

$$\frac{\partial \alpha_m}{\partial t} = \frac{Q}{2\pi\rho} \left[\langle \mathbf{B} \rangle \cdot (\nabla \times \langle \mathbf{B} \rangle) - \frac{\alpha}{\eta} \langle \mathbf{B} \rangle^2 \right] - \frac{\alpha_m}{T}, \tag{3.11}$$

where coefficient $Q \sim 0.1$, T is the characteristic time for magnetic diffusion. Equation (3.11) is written for the case of uniform turbulent diffusion, and when the magnetic Reynolds number is large, $\eta \approx \eta_t$.

For further analysis of the Kleeorin–Ruzmaikin model, we transform (3.4) and (3.11) into a non-linear dynamical system in non-dimensional variables. Following the approach of [9] we average the system of (3.4) and (3.11) in a vertical layer to eliminate z-dependence of A and B and consider a single Fourier mode propagating in the x-direction assuming $A = A(t)e^{ikx}$, $B = B(t)e^{ikx}$; then we get the following system of equation

$$\frac{dA}{dt} = \alpha B - \eta k^2 A, \quad \frac{dB}{dt} = ikGA - \eta k^2 B,$$

$$\frac{d\alpha_m}{dt} = -\frac{\alpha_m}{T} - \frac{Q}{2\pi\rho} \left[-ABk^2 + \frac{\alpha}{\eta} \left(B^2 - k^2 A^2 \right) \right]. \tag{3.12}$$

This transformation allows us to investigate more easily various non-linear regimes, from periodic to chaotic, and obtain relationships of the basic properties, such as the cycle growth and decay times, duration and amplitude. Note that the formulation and the interpretation of solutions of the simplified system are not straightforward because it does not adequately describes non-linear coupling of the spatial harmonics. For simplicity we retain only the second harmonic ($k = 2$), which has the largest growth rate among the antisymmetric solutions.

To relate the dynamo model solutions to the observations we used Bracewell's definition [10, 11] of the sunspot number in the form $W \sim B(t)^{3/2}$, where $B(t)$ is the toroidal magnetic field component. We note that the solutions of the dynamical system are qualitatively similar for the different harmonics. Nevertheless, we choose the parameters, which correspond to the solar situation.

Making the following substitutions: $A = A_0 \hat{A}$, $B = B_0 \hat{B}$, $t = T_0 \hat{t}$, $k = \hat{k}/r$ (r is a layer radius), $T_0 = 1/(k^2\eta)$ and $\alpha_m = \alpha_0 \hat{\alpha}_m$, and taking into account that $A_0 = B_0 \eta k/G$, we obtain:

$$\frac{d\hat{A}}{d\hat{t}} = \hat{D}\hat{B} - \hat{A}, \quad \frac{d\hat{B}}{d\hat{t}} = i\hat{A} - \hat{B},$$

$$\frac{d\hat{\alpha}_m}{d\hat{t}} = -\nu\hat{\alpha}_m + \left[\hat{A}\hat{B} - \hat{D} \left(\hat{B}^2 - \lambda\hat{A}^2 \right) \right], \tag{3.13}$$

where $\hat{D} = D_0\hat{\alpha}$ and $\hat{\alpha} = \hat{\alpha}_h + \hat{\alpha}_m$ are the non-dimensional dynamo number and total helicity, $D_0 = \alpha_0 Gr^3/\eta^2$, $\alpha_0 = 2Qkv_A^2/G$, v_A is the Alfvén speed, ν is the ratio of the characteristic times of turbulent and magnetic diffusion [3] and $\lambda = (k^2\eta/G)^2 = Rm^{-2}$, and Rm is the magnetic Reynolds number.

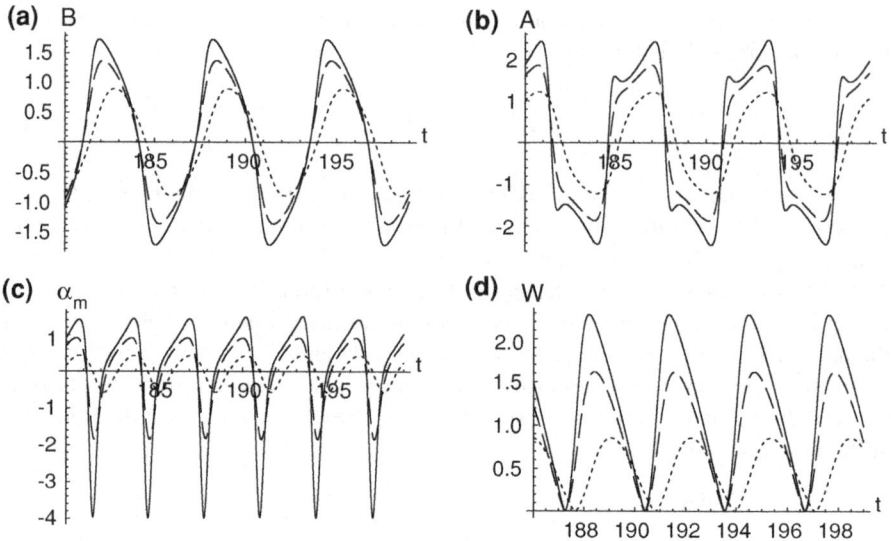

Fig. 3.2 Variations of the magnetic field for the middle convective zone $\alpha_h D_0 = -2$: $\nu = 1.28$, $\alpha_h = 2.439$, $D_0 = -0.82$ for different initial conditions: $B_0 = 4i$, $A_0 = -0.01i$ (*dotted curve*), $B_0 = 4i$, $A_0 = -i$ (*dashed curve*) and $B_0 = 1 + 4i$, $A_0 = -i$ (*solid curve*): **a** toroidal component, B; **b** vector-potential, A, of the poloidal magnetic field; **c** magnetic helicity variations, α_m; **d** evolution of the model sunspot number, W

3.2.3 Periodic and Chaotic Solutions

In order to estimate the range of parameters of the Kleeorin–Ruzmaikin model (3.12) and for modeling the solar cycle, we use the standard model of the interior structure rotation of the Sun for the top, bottom and middle areas of the convective zone [12]. The key parameter of the model is the dynamo number $D = D_0\alpha_h$, because its magnitude determines behavior of the magnetic field, which depends on the rotational velocity and magnetic field strength. According to [2] for the Kleeorin–Ruzmaikin model, given by (3.12), the linear instability condition is also $|D| \equiv |\alpha_h D_0| > 2$. However, in this case the profile of the periodic solutions is not sinusoidal, and depends on the initial conditions, A_0 and B_0. For higher initial values of these parameters the amplitude of non-linear oscillations in the stationary state is higher. However, the shapes of the oscillation profiles are similar.

Figure 3.2 illustrates solutions for the model of Kleeorin–Ruzmaikin, and the corresponding variations of the sunspot number for different initial conditions. As mentioned, different initial values for magnetic field components A_0 and B_0 lead to very similar profiles. In high amplitude cases, dual peaks may appear in variations of the vector potential, A, of the poloidal field. The evolution of the magnetic helicity shows a relatively slow growth followed by a sharp decay [2]. The helicity has maxima when the toroidal field is zero. In these calculations the value of parameter ν,

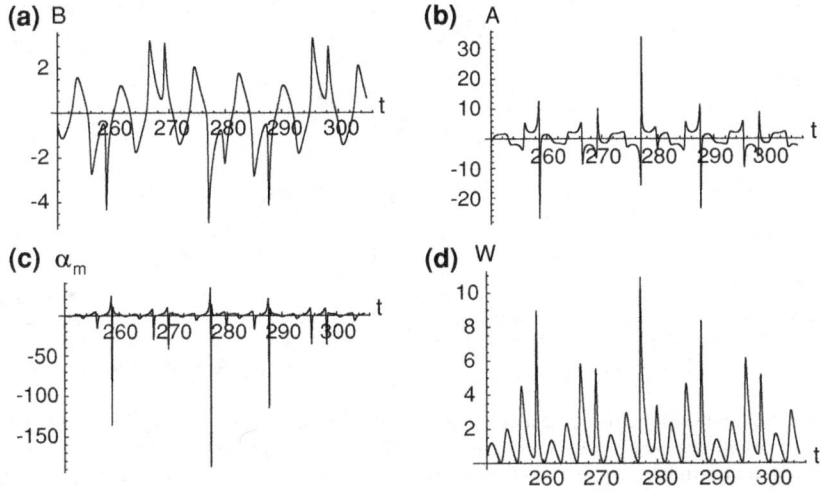

Fig. 3.3 Example of chaotic solutions for parameters of the middle convective zone: **a** toroidal component of magnetic field B, **b** vector–potential, A, **c** magnetic helicity α_m and **d** model sunspot number W

which describes the damping rate of magnetic helicity and depends on the turbulence spectrum and dissipation though helicity fluxes, is of the order of unity. Finally, the variations of the sunspot number, W, with the amplitude increase are characterized by higher peaks and shorter rising times (see Fig. 3.2d). Note that in the sunspot number profile we can recognize the well-known general properties of the sunspot number profile with a rapid growth at the beginning of the cycle and a slow decrease after the maximum.

With the increase of $|\alpha_h D_0|$ ($|\alpha_h D_0| > 2$) the profile of magnetic field variations continues to deform and can become unstable with very steep variations of the magnetic field. The solution can be stable again if we enhance the back reaction by increasing the quenching parameter. We use the following quenching formula for the kinetic part of helicity, α_h, $\alpha = \alpha_h/(1 + \xi B^2) + \alpha_m$ [6]. Thus we can always obtain periodic solutions for sufficiently strong ξ.

The transition from periodic to chaotic solutions occurs when the dynamo number, $|\alpha_h D_0|$, increases above a certain value. In the transition regime the cycle amplitude becomes modulated: it slowly increases with time, and then suddenly and very sharply declines, and then starts growing again [2].

In the case of significant deviations from the condition of linear stability, the solutions become chaotic for all variables of the dynamical system. Figure 3.3 shows an example of chaotic variations for the middle convective zone parameters: $\nu = 1.28$, $\lambda = 1.23 \times 10^{-6}$, $D_0 = -0.82$, $\alpha_h = 3.2$, $\xi = 3.9 \times 10^{-3}$, for the magnetic field components, the magnetic helicity and the sunspot number parameter. In the chaotic solutions, peaks of the toroidal magnetic field, B (Fig. 3.3a) strongly correlate with peaks of the vector-potential, A, and the magnetic helicity, α_m, (Fig. 3.3b,

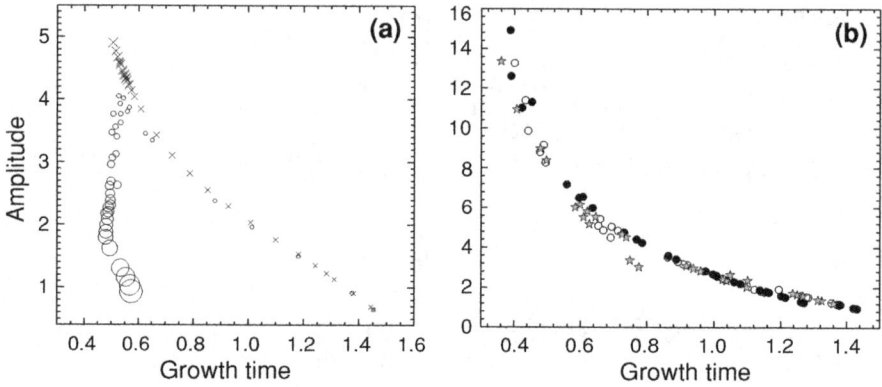

Fig. 3.4 Relationships between the amplitude of the model sunspot number and the growth time for **a** periodic solutions: the *circles* show a sequence for a fixed value of the kinetic helicity, $\alpha_h = 2.44$, and the dynamo number varying from -7 to -0.82; the crosses show the case of fixed $D_0 = -0.82$ and α_h varying from 2.44 to 3 (the size of *crosses* and *circles* is proportional to the corresponding values of $|D_0|$ and α_h); **b** chaotic solutions for $D_0 = -0.82$ and $\alpha_h = 2.8$ (*black circles*), $\alpha_h = 3$ (*empty circles*) and $\alpha_h = 3.2$ (*stars*)

Fig. 3.5 Relationships between the amplitude of the sunspot number and the growth time for the real solar cycles in 1755–2007

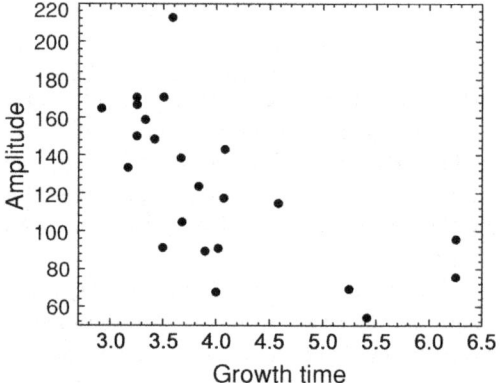

c). The growth of the toroidal field also leads to strengthening of the poloidal field and strong fluctuations of the magnetic helicity.

Now we can see from Figs. 3.2d and 3.3d that the profiles of the model sunspot number variations qualitatively describe the mean profile of the solar cycles. The next important characteristic of the solar cycles is the relationship between the amplitude and the growth time. This relationship is shown for some periodical solutions in Fig. 3.4a, four chaotic solutions in Fig. 3.4b and properties for the real 23 solar cycles in Fig. 3.5. The time scales are non-dimensional. Figure 3.5 shows the observed amplitude-growth time properties of the solar cycles of 1755–2007. Thus, all three panels demonstrate that the growth time is shorter for stronger cycles.

3.3 Data Assimilation Methods

In the previous section we obtained a solution for the dynamical system, which qualitatively reproduces the basic properties of the solar cycle. In this section we adapt the non-linear periodic solution for describing the sunspots number series using data assimilation methods.

3.3.1 Basic Formulation

The main goal of any model is an accurate description of properties of a system in the past and present times, and prediction of its future behavior. However, a model is usually constructed with some approximations and assumptions, and contains uncertainties. Therefore, it cannot describe the true condition of a system. On the other hand, observational data, d, also include errors, ϵ, which are often difficult to estimate. Data assimilation methods such as the Kalman Filter [13] allow us, with the help of an already constructed model and observational data, to determine the initial state of the model that is in agreement with a set of observations, and obtain a forecast of future observations and an error estimate [14, 15]. For instance, in our case we know from observations the sunspot number (with some errors) and want to estimate the state of solar magnetic fields, described by a dynamo model.

Generally, if the state, ψ, of a system can be described by a dynamical model $d\psi/dt = g(\psi, t) + q$, with initial conditions $\psi_0 = \Psi_0 + p$, where $g(\psi, t)$ is a non-linear vector-function, q and p are errors of the model and initial conditions. Then, the system forecast is $\psi^f = \psi^t + \phi$, where ψ^t is the true system state, and ϕ is the forecast error. The relationship between the true state and the observational data is given by a relation $d = M[\psi] + \epsilon$, where d is a vector of measurements, $M[\psi]$ is a measurement functional.

For a realization of the data assimilation procedure in the case of non-linear dynamics, it is convenient to use the EnKF method [14, 16]. The main difference of the EnKF from the standard Kalman Filter is in using an ensemble of possible states of a system, which can be generated by Monte Carlo simulations. If we have an ensemble of measurements $d_j = d + \epsilon_j$ with errors ϵ_j (where $j = 1, \ldots, N$), then we can define the covariance matrix of the measurement errors $C^e_{\epsilon\epsilon} = \overline{\epsilon\epsilon^T}$, where the overbar means the ensemble averaged value, and superscript T indicates transposition. Using a model we always can describe future states of a system, ψ^f. However, errors in the model, initial conditions and measurements do not allow the model result be consistent with observations. To take into account this deviation, we consider a covariance matrix of the first-guess estimates (our forecast related only to model calculations): $(C^e_{\psi\psi})^f = \overline{(\psi^f - \overline{\psi^f})(\psi^f - \overline{\psi^f})^T}$. Note that the covariance error matrix is calculated for every ensemble element. Then, the estimate of the system state is given by:

$$\psi^a = \psi^f + K\left(d - M\psi^f\right), \tag{3.14}$$

where $K = (C^e_{\psi\psi})^f M^T \left(M(C^e_{\psi\psi})^f M^T + C^e_{\epsilon\epsilon}\right)^{-1}$, is the so-called Kalman gain [13, 14]. The covariance error matrix of the best estimate is calculated as $(C^e_{\psi\psi})^a = \overline{(\psi^a - \overline{\psi^a})(\psi^a - \overline{\psi^a})^T} = (I - K_e M)(C^e_{\psi\psi})^f$. We can use the last best estimate obtained with the available observational data as initial conditions and make the next forecast step. At the forecast step, we calculate a reference solution of the model, according to the new initial conditions, then simulate measurements by adding errors to the model and to the initial conditions. Finally we obtain a new best estimate of the system state, which is our forecast. A new set of observations allows us to redefine the previous model state and make a correction for the predicted state.

3.3.2 Implementation of the Data Assimilation Method

For assimilation of the sunspot data into the dynamo model, we select a class of non-linear periodic solutions, which correspond to parameters of the middle convective zone and describe the typical behavior of the observed sunspot number variations (Fig. 3.2d). Implementation of the EnKF method consists of three steps [1]: preparation of the observational data for analysis, correction of the model solution according to observations, and prediction.

Step 1: Preparation of the observational data. Following [10, 11], we transform the annually averaged sunspot number for the period of 1856–2007 into the toroidal field values using the relationship $B \sim W^{2/3}$ while alternating the sign of B. We also select the initial conditions of the model such that the reference solution coincides with the beginning of the first cycle in our series, cycle 10, which started in 1856. We do not consider the previous solar cycles because of uncertainties in the early sunspot number measurements. Then we normalize the toroidal field in the model in such a way that the model amplitude of B is equal to the mean toroidal field calculated from the sunspot number. In addition, we normalize the model time scale assuming that the period of the model corresponds to the typical solar cycle duration of 11 years.

Step 2: Assimilation for the past system state. Unfortunately we do not have observations of the magnetic helicity, and the toroidal and poloidal components of the magnetic field. Therefore, in the first approximation, we generate observational data as random values around the reference solution with a standard deviation of ~12%, which was chosen to roughly reproduce the observed variations of the sunspot number. Then, we calculate the covariance error matrixes of the observations, $C^e_{\epsilon\epsilon}$, and the forecast, $(C^e_{\psi\psi})^f$. After combining the observation and model error covariances in the Kalman gain, K, we obtain the best estimate for the evolution of the system, ψ^a from (3.14) (Fig. 3.6, first half). Figure 3.7 shows the result of assimilation of the sunspot data into the dynamo model: the best EnKF estimate (black curve), the initial model (gray curve) and the actual sunspot data (circles).

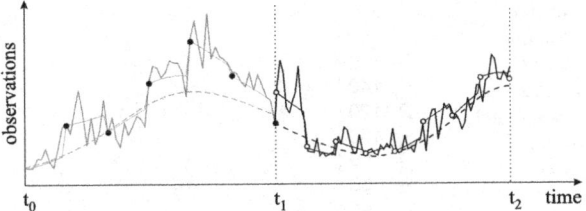

Fig. 3.6 Scheme of the data assimilation procedure. (*Dashed curves*) show exact solutions of a model, (*thin solid curves*) describe the first correction of a model according to observations, (*thick curves*) are the best estimate of an observable system state. (*Gray*) and (*black colors*) indicate estimations for past and forecast states. (*Black*) and (*empty circles*) mean real and simulated observations, respectively

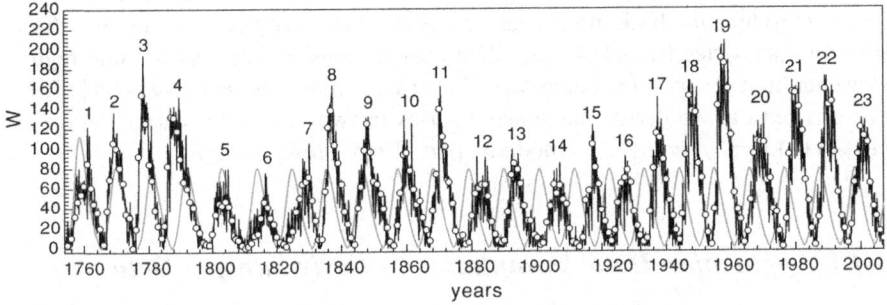

Fig. 3.7 Results of assimilation of the annual sunspot number data (*circles*) into the dynamo model. The (*gray curve*) shows the reference solution (without assimilation analysis), and the (*black curve*) shows the best EnKF estimate of the sunspot number variations, obtained from the data and the dynamo model

Step 3: Prediction. To obtain a prediction of the next solar cycle, we determine the initial conditions from the best estimated solution for the previous cycle in terms of the amplitude and phase to continue the model calculations. Then after receiving the reference solution with the new initial conditions, we simulate future observational data by adding random noise and repeat the analysis (Fig. 3.6, right). This provides the best EnKF estimate of the future state of the system (forecast).

3.4 Reproducing and Predicting Observational Data by the Ensemble Kalman Filter

As discussed early, for a successful application of the data assimilation method we need: (1) a model, which reproduces a phenomenon as accurately as possible; (2) a sufficiently long set of observational data.

Fig. 3.8 Annual sunspot data records from National Geophysical Data Center (NGDC), (*empty circles*) and corrected data (*black circles*) by Svalgaard (personal communication, [18])

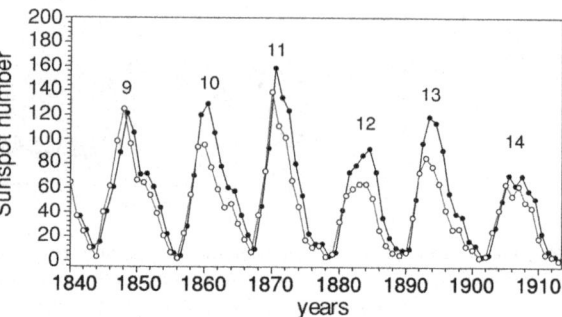

Here, we use the simplified non-linear dynamo model (3.13), which can qualitatively reproduce the basic properties of a solar cycle (see Sect. 2.3) and the Zurich sunspot data series from 1840 yr, when observations of sunspots became regular. However, first observational data strongly varied among different observers [17, 18]. For this reason, we divide the sunspot series in two parts: (1) "early" data series (1844–1915 yrs), and (2) the "modern" part (1915–present time).

3.4.1 Application Data Assimilation to Early Sunspot Data

Because of uncertainties there were many attempts to revise the early sunspot data taking into account various historical and physiological facts. For example, using information about sunspot observations, auroras, wine harvest and hailstorms, Fritz [19] estimated the solar activity for the period of 188–1638.

Here we examine the data assimilation approach for "prediction" of "early" solar cycles, when the distribution of the sunspot number is unclear. We use two annual sunspot data sets: from the National Geophysical Data Center (NGDC) and the sunspot data series corrected by Svalgaard [17, 18] for 1840–1920. In Fig. 3.8 we show differences between these data records.

Then, by applying the data assimilation method we estimate variations of the sunspot number for solar cycles 10–14. For this, we use information only for the previous cycles. Figure 3.9 (panels a–e) illustrates results of the "predictions" for the corrected data and the errors of the predicted cycle amplitudes and times of the cycle maxima (panel e). In the plot the observed sunspot maxima are aligned at $t = 0$. The corrected data (stars) give a better agreement between "predicted" and actual cycle amplitudes than the NGDC data (circles). However, the accuracy of the predictions of the solar maxima times is similar for both data sets. Note, that we get a strong over-estimation in amplitude for cycle 12 for both data set, that perhaps may indicate on missed data for this cycle.

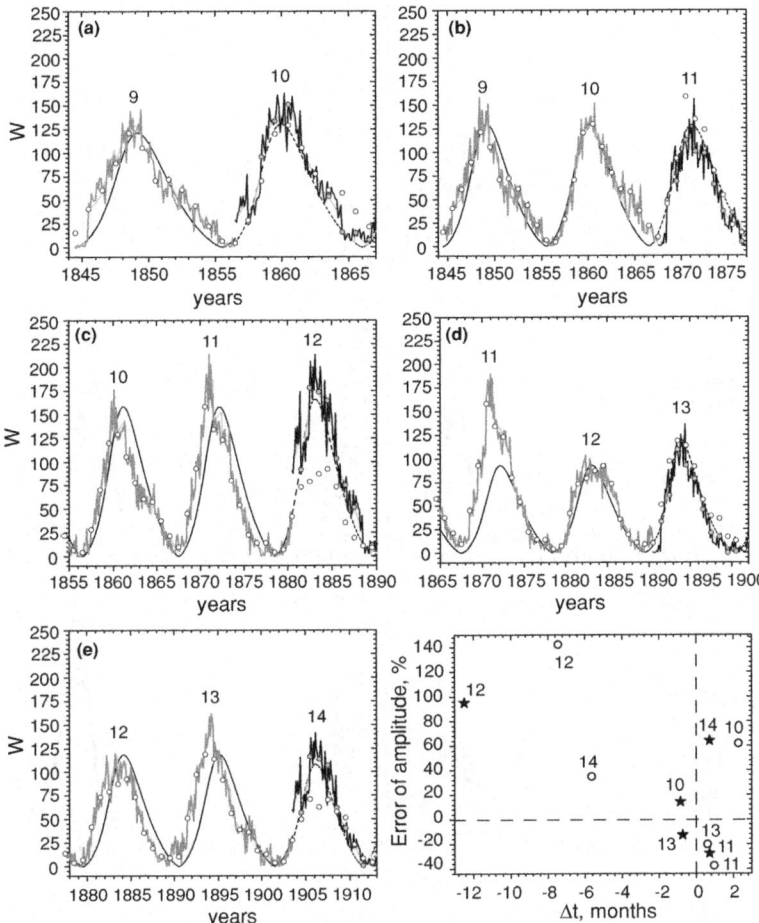

Fig. 3.9 Predictions for "early" solar cycles 10–14 (panels **a–e**). (*Grey curves*) show the assimilated model data using the previous known data sets, (*black curves*) show the forecasts. (*Filled and empty circles*) correspond to the simulated annual sunspot number and to the observed ones. Panel **f** shows the errors of the predicted amplitudes and times of the solar maxima. In this panel the *circles* show the errors for the NGDC data, and *stars* show the errors for corrected data [18]. Errors of the solar maximum times are shown as deviation from the actual maxima in months

3.4.2 Prediction of the Last Solar Cycles

The described analysis also has been tested by calculating predictions of previous "modern" cycles. Figure 3.10 shows examples of the EnKF method implementation for forecasting the sunspot number of cycles 16–23. For these forecasts, we first obtain the best estimated solutions using the observational data prior to these cycles. We then compute the model solution (black dashed curves) according to the initial

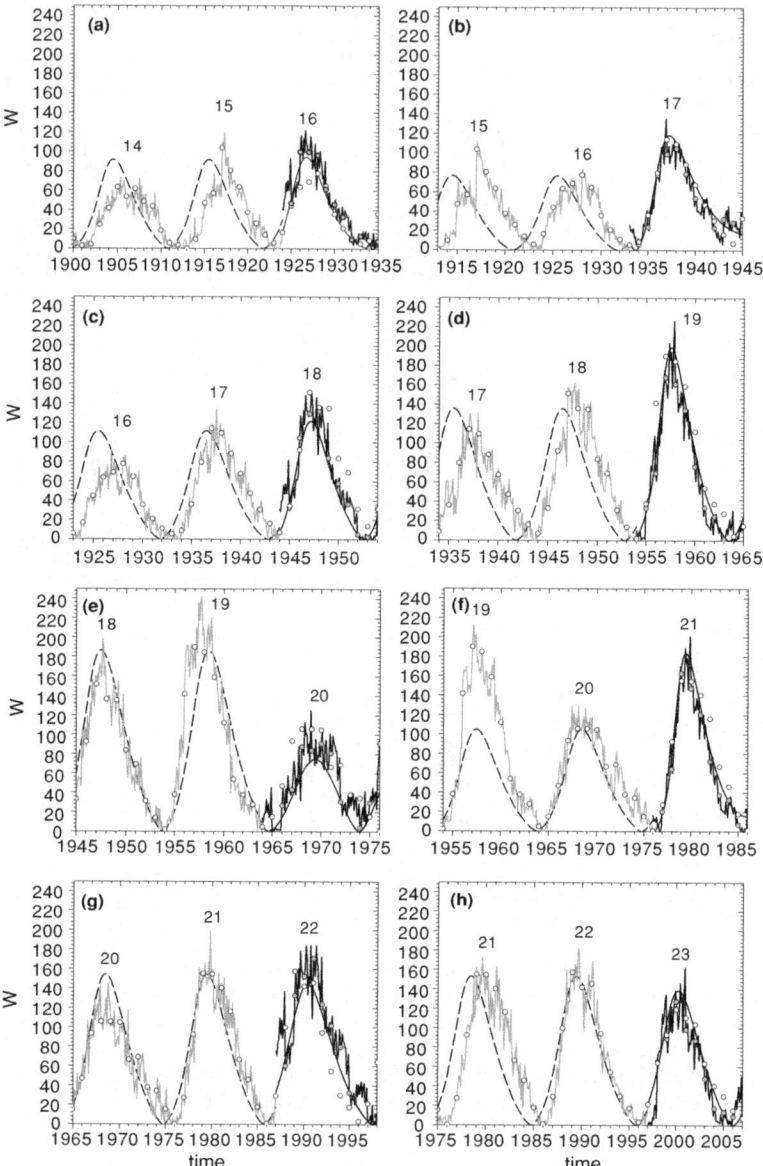

Fig. 3.10 Predictions for solar cycles 16–23. *Black dashed curves* show the model reference solution. *Gray curves* show the best estimate of the sunspot number using the observational data (*empty circles*) and the model, for the previous cycles. *Filled circles* are simulated observational data. *Black curves* show the prediction results

conditions of the time of the last measurement and simulate a new set observation by adding random noise. Then, we obtain the EnKF estimates using the simulated observations, which give us the prediction (Fig. 3.10, black curves).

Fig. 3.11 Prediction for solar cycle 24. Notations same as in Fig. 3.10

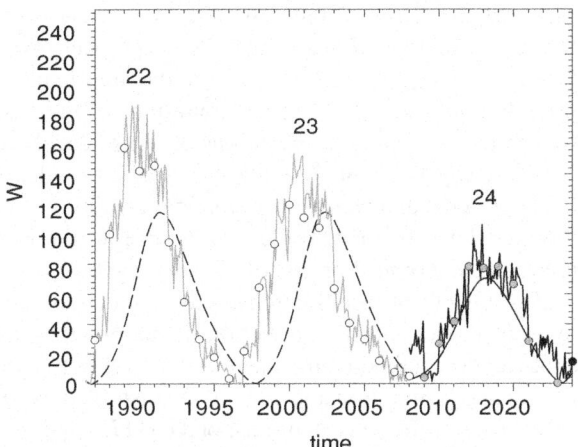

These experiments show that this approach can provide reasonable forecast of the strength of the next solar cycles. However, there are significant discrepancies. For instance, the strength of cycle 16 is over-estimated, and the strength of cycle 19 is under-estimated. The main uncertainties are caused by inaccuracies in determining the time of the end of the previous cycle from the sunspot number data, and by the incompleteness of the model and insufficiency of the sunspot number data. In particular, we found the forecast is inaccurate when the sunspot number change significantly from the value of the previous cycle [1]. Also, our forecast experiments show a strong dependence on the phase relation between the reference model solution and the observations. The phase difference appears to be due to the constant period of the model solution. Curiously, when the model phase is ahead of the solar cycle phase, adding a data point at the start of the cycle substantially improves the forecast. However, when the model phase lags, this does not happen. This effect is taken into account by correcting the phase of a reference solution that it is slightly ahead of the solar cycle phase.

The same analysis scheme is applied for predicting of the next solar cycle 24. According to this result, solar cycle 24 will be weaker than the previous cycle by approximately 30%. To test the stability of this prediction we used two other sets of initial conditions in 2008 and obtained close results (Fig. 3.11).

3.5 Discussion and Conclusions

We have presented a numerical analysis of simple dynamical models describing the non-linear behavior of two dynamo models, the classical Parker's dynamo model with the standard α-quenching and the model [3], which describes the evolution of the magnetic helicity based on the balance between the large-scale and turbulent magnetic helicities, shows the existence of non-linear periodic and chaotic solutions. Using a low-order dynamical system approach we examine the influence of the kinetic and magnetic helicities on the non-linear fluctuations of the dynamo-generated magnetic field in the conditions of the solar plasma, and compare these with the sunspot number variations observed during the solar 11-year cycles.

The analysis of the Kleeorin–Ruzmaikin model showed the existence of non-linear periodic and chaotic solutions for conditions of the solar convective zone. For this model we obtained profiles of the sunspot number variations, which qualitatively reproduce the typical profile of the solar cycles.

The results of assimilation of the annual sunspot number data into the solar dynamo model and the prediction of the previous solar cycles (Fig. 3.10) demonstrate a new method of forecasting the solar activity cycles. The application of the exam data assimilation approach to the historical data of cycles 10–14 shows a better agreement for the corrected sunspot data [17, 18] than for the standard NGDC data in terms of the cycle amplitudes. However, errors in the predictions of the cycle maxima are similar for both data sets (Fig. 3.9).

Using the EnKF method and a simple dynamo model for "modern" sunspot data record, we obtained reasonable predictions usually for the first half of sunspot cycles with an error of ~8–12%, and in some cases also for the declining phase of the cycles. This method predicts a weak solar cycle 24 with a maximum of the smoothed annual sunspot number of approximately 80 (Fig. 3.11). It is interesting to note that the simulations show that the previous cycle does not finish in 2007 as was expected, but still continues into 2008. According to the prediction, the maximum of the next cycle will be reached approximately in 2013.

The application of the data assimilation method, EnKF, for modeling and predicting solar cycles shows the power of this approach and encourages further development. It also reveals significant uncertainties in the model and the data. Among these are the uncertainties in the determination of the start of a solar cycle from the sunspot number series (in particular, when the cycles overlap), leading to the uncertainty in the phase relation between the model solution and the data. Also, there are significant uncertainties in the relationship between the sunspot number data and the physical properties of the solar magnetic field, in the absence of magnetic field and helicity data, and, of course, in the dynamo model. Our conclusion is that for more robust and accurate predictions of solar cycles, the information contained in the sunspot number data is insufficient.

Acknowledgment This work was supported by the Center for Turbulence Research (Stanford) and the International Space Science Institute (Bern).

References

1. Kitiashvili, I.N., Kosovichev, A.G.: ApJL. **688**, L49 (2008)
2. Kitiashvili, I.N., Kosovichev, A.G.: Geophys. Astrophys. Fluid Dyn. **103**, 53 (2009)
3. Kleeorin, N.I, Ruzmaikin, A.A.: Magnetohydrodynamics **18**, 116 (1982)

4. Parker, E.N.: ApJ **122**, 293 (1955)
5. Moffatt, H.K.: Magnetic Field Generation in Electrically Conducting Fluids. Cambridge University Press, New York (1978)
6. Kleeorin, N., Rogachevskii, I., Ruzmaikin, A.: Astron. Astrophys. **297**, 159 (1995)
7. Kleeorin, N., Kuzanyan, K., Moss, D., Rogachevskii, I., Sokoloff, D., Zhang, H.: Astron. Astrophys. **409**, 1097 (2003)
8. Sokoloff, D.: Plasma Phys. Control Fusion **49**, B447 (2007)
9. Weiss, N.O., Cattaneo, F., Jones, C.A.: Geophys. Astrophys. Fluid Dyn. **30**, 305 (1984)
10. Bracewell, R.N.: Nature **171**, 649 (1953)
11. Bracewell, R.N.: Mon. Not. R. Astr. Soc. **230**, 535 (1988)
12. Schou, J. et al.: ApJ. **505**, 390 (1998)
13. Kalman, R.E.: J Basic Eng. **82**(series D), 35 (1960)
14. Evensen, G.: Data Assimilation. The Ensemble Kalman Filter, Springer (2007)
15. Kitiashvili, I.: ASP. Conf. Ser. **383**, 55 (2008)
16. Evensen, G.: J. Geophys. Res. **99**(C5), 10143 (1994)
17. Svalgaard, L., Cliver, E.W.: J. Geophys. Res. **110**(A12), A12103 (2005)
18. Svalgaard, L.: Recalibration of the sunspot number and consequences for predictions of future activity and reconstructions of past solar behavior. Report at conference "Solar activity during the Onset of Solar Cycle 24", 2008, Napa, California, USA. Available at http://www.leif.org/research/Napa%20Solar%20Cycle%2024.pdf
19. Fritz, H.; Verzeichnis beobachteter Polarlichter. Gerold & Sohn, Wien (1873)

Chapter 4
Amplitudes of Solar Gravity Modes

K. Belkacem

Abstract Solar gravity modes are mainly trapped inside the radiative region and are then able to provide information on the properties of the central part of the Sun granulation. However, there is no consensus on the detection of solar gravity modes which remains a major challenge. In this paper, we discuss the underlying driving and damping processes of solar g modes, and review the quantitative estimates of their amplitudes. This issue is important since a theoretical determination of mode amplitudes may help to design the track for gravity modes.

4.1 Motivations

Identification of the solar 5-min oscillations as global acoustic standing waves (p modes) by Ulrich [1] and Leibacher and Stein [2] led to major improvements of the knowledge of the Sun internal structure. Contrary to classical pulsators for which only a few modes had been detected from the ground, the Sun pulsates with million of modes. The resulting rich frequency spectrum allows to probe the internal structure by means of inversion of the seismic data [3]. For instance, the sound speed profile or the rotation profile were inferred, from the knowledge of frequencies, in the whole Sun except near the core. Solar acoustic modes do not permit to get access to the inner most region since their amplitudes in those layers are very small.

The potential of gravity modes in doing so had been recognized for many years. Such modes are mainly trapped in the radiative region and are thus able to provide information on the properties of the central part of the Sun ($r < 0.3R_\odot$) [4]. Hence, the track for the detection of g modes began more than 30 years ago. The first claims of detection of solar gravity modes started with the works of Severnyi et al. [5]

K. Belkacem (✉)
Institut d'Astrophysique et Géophysique, Université de Liège,
Allée du 6 Août 17-B 4000 Liège, Belgium
e-mail: Kevin.Belkacem@ulg.ac.be

J.-P. Rozelot and C. Neiner (eds.), *The Pulsations of the Sun and the Stars*,
Lecture Notes in Physics 832, DOI: 10.1007/978-3-642-19928-8_4,
© Springer-Verlag Berlin Heidelberg 2011

and Brookes et al. [6]. None of them were confirmed even after more than 10 years
of observations from SOHO [7]. Hence, the detection and identification of gravity
modes mode [7–11] to determine the rotation profile in the whole radiative core from
the tachocline to the nuclear region [12–14], remains one of the key issue in solar
physics.

As g modes are evanescent in the convective region, their amplitudes are expected
to be very low at the photosphere and above, where observations are made, making
their detection a challenge. In this framework, the theoretical determination of g
mode amplitudes is an important task since it can gives some indications about
the needed observational threshold to achieve. In addition, the investigation of g
mode amplitudes and the underlying involved physical mechanisms, i.e. driving
and damping processes, gives informations on the dynamical properties of the Solar
convective region. Amplitudes of g modes, as of p modes, are believed to result from
a balance between driving in the solar convection zone and damping processes. Two
major processes have been identified as stochastically driving the resonant modes
in the stellar cavity. The first is related to the Reynolds stress tensor, the second is
caused by the advection of turbulent fluctuations of entropy by turbulent motions.
Theoretical estimations based on stochastic excitation have been obtained by Gough
[15], Kumar et al. [16], and Belkacem et al. [17]. We also note that penetrative
convection is thought to be an another possible excitation mechanism [18, 19], as
well as other mechanisms such as mode coupling [20–24], or excitation by magnetic
torques [25].

This lecture aims at explaining the main features of mode driving and damping
of solar gravity modes by turbulent convection. In Sect. 4.2, the basic properties
of solar g modes, which are of interest in regard to their amplitudes, are outlined.
Section 4.3 is dedicated to the presentation of the driving by turbulent convection
and damping processes while Sect. 4.4 emphasizes quantitative results obtained for
g mode amplitudes. In Sect. 4.5 we discuss another likely excitation mechanism, i.e.
penetrative convection. Section 4.6 is dedicated to conclusions.

4.2 Basic Properties of Solar Gravity Modes

In this first section, we present some basic properties of solar g modes. In doing so,
we emphasize their peculiar behavior, which is an advantage since they probe the
internal layers of the Sun but also a difficulty for their detection.

4.2.1 Dispersion Relation of Gravito-Acoustic Modes

Let us write the governing equations of the fluid motion, neglecting the effect of
molecular viscosity, rotation as well as magnetic field

$$\frac{\partial \rho}{\partial t} + \nabla \cdot (\rho \boldsymbol{u}) = 0, \tag{4.1}$$

$$\frac{\partial \rho \boldsymbol{u}}{\partial t} + \nabla : (\rho \boldsymbol{u} \boldsymbol{u}) = \rho \boldsymbol{g} - \nabla p, \tag{4.2}$$

$$\rho T \frac{\mathrm{d}s}{\mathrm{d}t} = -\nabla \cdot \boldsymbol{F} + \epsilon \tag{4.3}$$

where ρ is the density, $\boldsymbol{u}$ the velocity field, p the pressure, $\boldsymbol{g}$ the gravitational acceleration, T temperature, s the specific entropy, ϵ an external source of energy, and F the radiative flux.

To establish the equations governing the oscillatory motion, the scalar and vectorial fields are split into an equilibrium and a fluctuating part associated with the oscillations, such as

$$y = y_0 + y' \tag{4.4}$$

$$|y'| \ll |y_0| \tag{4.5}$$

with $y = \{p, \rho, T, s, \boldsymbol{u}, \boldsymbol{F}, \boldsymbol{g}\}$, and the equilibrium state defined by

$$\nabla p_0 = \rho_0 g_0 \tag{4.6}$$

$$\epsilon + \nabla \cdot \boldsymbol{F}_0 = 0 \tag{4.7}$$

Equation (4.1) are linearized and the variables (y') expanded onto the spherical harmonics (Y_ℓ^m), for instance the Lagrangian displacement associated with the oscillations is expanded such as

$$\boldsymbol{\xi}(r) = \sum_{\ell,m} (\xi_r \boldsymbol{e}_r + \xi_H \nabla_H) Y_{\ell,m} e^{i\sigma t} \tag{4.8}$$

with $\sigma = \omega + i\eta$, ω the pulsation frequency and η the damping rate, and $\boldsymbol{u}' = \sigma \boldsymbol{\xi}$.

Further assuming that the specific entropy is conserved during the oscillations and using (4.4) and (4.6) into (4.1), one gets [26]

$$\frac{1}{r^2}\frac{\mathrm{d}}{\mathrm{d}r}\left(r^2 \xi_r\right) - \frac{g}{c_s^2}\xi_r + \left(1 - \frac{S_\ell^2}{\omega^2}\right)\frac{p'}{\rho c_s^2} = 0, \tag{4.9}$$

$$\frac{1}{r}\frac{\mathrm{d}p'}{\mathrm{d}r} + \frac{g}{\rho c_s^2}p' + (N^2 - \omega^2)\xi_r = 0, \tag{4.10}$$

where the fluctuations of the gravitational acceleration associated to the oscillations have been neglected for sake of simplicity, c_s is the sound speed, S_ℓ is the lamb frequency, and N is the buoyancy frequency defined such as

$$N^2 = g \left(\frac{1}{\Gamma_1} \frac{d \ln p_0}{dr} - \frac{d \ln \rho_0}{dr} \right) \qquad (4.11)$$

where $\Gamma_1 = (\partial \ln p_0 / \partial \ln \rho_0)_s$, p_0 the mean pressure, and ρ_0 the mean density

Combining (4.9) and (4.10) into a second order differential equation and further assuming a local solution of the form $e^{ik_r r}$, one can derive the dispersion relation o gravito-acoustic waves

$$k_r^2 = \frac{\omega^2}{c_s^2} \left(1 - \frac{N^2}{\omega^2} \right) \left(1 - \frac{S_\ell^2}{\omega^2} \right) \qquad (4.12)$$

From (4.12), one can distinguish two limits, namely

- $\omega^2 \gg N^2$, S_ℓ^2 corresponds to acoustic modes (p modes), and the dispersion rela tion reduces to

$$k_r^2 = \frac{\omega^2}{c_s^2} \qquad (4.13)$$

- $\omega^2 \ll N^2$, S_ℓ^2 corresponds to the limit of gravity modes (g modes)

$$k_r^2 = \left(\frac{N^2}{\omega^2} - 1 \right) k_h^2 \qquad (4.14)$$

with $k_h^2 = \ell(\ell + 1)/r^2$ is the local horizontal wave number.

Note that modes are gravito-acoustic waves; depending on their frequency and the location in the Sun their properties are a mixture between pure gravity and pure acoustic modes. Equations (4.13) and (4.14) are only asymptotic limits but are useful to understand the main properties of the solar cavity. In the following, we restrict our discussion to gravity modes, for which (4.14) will be used.

4.2.2 Some g Mode Properties Near the Sun Surface

A key feature of solar g modes is the behavior of their displacement across the convective region and near the surface. The upper part of the Sun is convective such that the buoyancy frequency vanishes,[1] i.e. the restoring force no longer exist. From (4.14) the radial wave number is imaginary since $k_r^2 < 0$. This permits to conclude that gravity modes are evanescent in the convective region.

In order to understand the consequences of the evanescent nature of g modes on their displacement, let us use an asymptotic solution of the wave equation for ξ_r. Following Unno et al. [26]

---

[1] More precisely, $N^2 < 0$ in a convective region that is not strictly adiabatic. Nevertheless, the departure from adiabaticity is small for the problem we are considering.

Fig. 4.1 Mode velocity as a function of the radius, normalized so that the kinetic energy equals to unity (see (4.17)) for a solar g mode of radial order $n = 10$ ($\nu \approx 60\,\mu\text{Hz}$) in *solid line* and a solar p mode of radial order $n = 3$ ($\nu \approx 600\,\mu\text{Hz}$) in *dashed line*

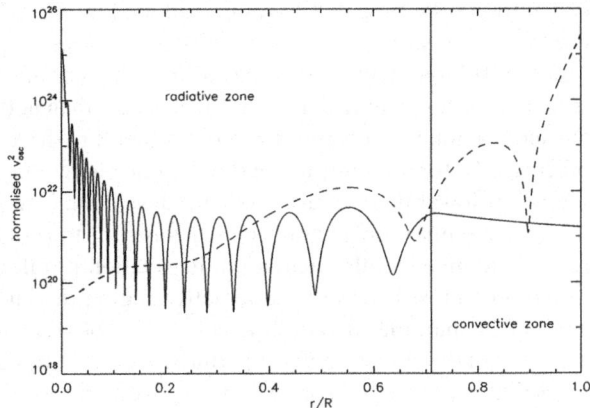

$$\xi_r \propto \frac{1}{\sqrt{\kappa}} \exp\left(-\int_{r_b}^{r} \kappa(r')dr'\right) \tag{4.15}$$

for $r \gg r_b$, where r_b is the bottom of the convective region, and $\kappa^2 = -k_r^2 > 0$. From (4.15) and (4.14), one clearly see that ξ_r decreases exponentially through the convective region. The mode is evanescent and consequently will exhibit a small amplitude at the Sun surface (compared to the amplitude at $r = r_b$). This is one of the main difficulty in the observational track for solar g modes. This property is emphasized in Fig. 4.1 for a g mode of radial order $n = 10$. One can see the dichotomy between the radiative region, where the mode oscillates, and the convective region, where the mode displacement exponentially decreases toward the surface. Further assuming that $N^2 = 0$, one can also infer the dependence of the mode amplitude with the angular degree since in that case (4.15), with the help of (4.14), becomes

$$\xi_r \propto \frac{1}{\sqrt{\kappa}} \left(\frac{r}{r_b}\right)^{-\sqrt{\ell(\ell+1)}} \tag{4.16}$$

which permits to point out that the higher the angular degree the higher the decrease of mode amplitude through the convection region. Hence, low-angular degree modes are expected to present higher amplitudes at the Sun surface. This is one of the reason why most of observational efforts focused on low-degree g modes.

In addition to the evanescent nature of g modes in the convection region, there is an other effect that tends to make the mode amplitude very small at the solar surface. Equations (4.9) and (4.10) are solved with an arbitrary amplitude so we choose to normalize the modes such that its total energy is

$$E = \mathcal{M} v_{\text{osc}}^2 \equiv 1 \tag{4.17}$$

where $\mathcal{M}$ is the mode mass that is the normalized inertia such that $\mathcal{M} = I/|\boldsymbol{\xi}(r)|^2$, and v_{osc} the mode velocity. The squared mode velocity can then be computed using

(4.17) for a solar model. The result is plotted in Fig. 4.1 for a p and g mode. It turns out that at the surface, a g mode presents a lower velocity than the p mode as the result of its larger inertia. It comes from the nature of a g mode that probes the inner layers of the Sun where density is much larger than at the surface. Consequently, for the same amount of energy it is more difficult to drive a g mode than a p mode. In other words, for the same amount of kinetic energy the displacement associated to a g mode is lower than for a p mode because there is more "mass" to move.

As a summary, solar gravity modes permit to probe the solar core so that they are able to give us information on the rotation profile, the sound speed profile etc. In the innermost layers of the radiative region the buoyancy frequency is positive and higher than the mode frequency $N \approx 500 \, \mu\text{Hz}$ such that the buoyancy force acts efficiently to restore wave perturbations toward equilibrium, thus allowing for an oscillatory motion. Hence, g modes mainly probe the inner layers of the Sun, as displayed in Fig. 4.1. However, this advantage is also a source of difficulties since their high-inertia and evanescent nature tend to make their surface amplitudes very small. Nevertheless, a definite answer on g-mode surface amplitudes needs to take the energetic aspects into account and it is the object of the following sections.

4.3 Energetic of Gravity Modes: Driving and Damping Processes

Mode amplitude is a balance between driving and damping processes, thus in this section we aim at explaining the main features of mode driving and damping mechanisms under the assumption that driving is dominated by turbulent convection and the mode is stable. We first establish the equation governing is the mode kinetic energy so as to emphasize the way both the driving and damping contribute to the mode energy.

4.3.1 Principle: Forced and Damped Oscillator

The mode total energy is by definition the quantity

$$E_{\text{osc}}(t) = \int dm \, |v_{\text{osc}}|^2 \, (\mathbf{r}, t) \tag{4.18}$$

where v_{osc} is the mode velocity at that radius.

We assume that the mode damping to occur over at a time-scale much longer than that associated with the driving. Indeed, for solar p modes the damping is proportional to the mode line-width that is of the order of several μHz while the mode driving is dominated by eddies at the top of the convection zone with a frequency of several mHz. Accordingly, damping and driving can be completely decoupled in time. Let us define $\mathcal{P}$ to be the amount of energy injected per unit time into a mode by an arbitrary source of driving (which nevertheless acts over time scale much shorter than $1/\eta$). Then, as described by Samadi (2009, this volume)

$$\frac{dE_{osc}}{dt} + 2\eta E_{osc} = \mathcal{P}. \tag{4.19}$$

The mode kinetic energy is then a balance between the driving and damping[2] processes. For solar g modes one has to provide a physical modeling of both processes to determine their amplitude since contrary to solar p modes they are not observed and the damping can not be inferred from the observations. An extended discussion on p modes and comparison with observations can be found in Samadi (2009).

4.3.2 Driving Mechanisms

A complete description of mode excitation by turbulent convection as well as a derivation of the equations governing the driving is addressed in Samadi (2009, this volume). Here, we recall the main features and discussed the particular case of solar g modes.

The inhomogeneous wave equation governing mode excitation is

$$\left(\frac{\partial^2}{\partial t^2} - \mathbf{L}\right) \mathbf{v}_{osc} + \mathcal{C}_{osc} = \mathbf{\mathcal{S}}_t, \tag{4.20}$$

where $\mathbf{L}$ is the linear operator [26], the operator $\mathcal{C}_{osc}$ involves both turbulent and pulsational velocities and contributes to the linear dynamical damping [27].

Finally the $\mathcal{S}_t$ operator contains the source terms, given by

$$\mathcal{S}_t = \mathcal{S}_R + \mathcal{S}_E + \mathcal{S}_\Omega + \mathcal{S}_M + \mathcal{L}_t \tag{4.21}$$

with

• the Reynolds stress contribution

$$\mathcal{S}_R = -\frac{\partial}{\partial t}\nabla : (\rho_0 \mathbf{u}_t \mathbf{u}_t), \tag{4.22}$$

where $\mathbf{u}_t$ is the turbulent velocity field and ρ_0 the mean density. This term is the dominant term (Samadi 2009, same volume), it corresponds to the generation of acoustic noise by turbulence. It scales as the square of the Mach number [27].
• the entropy contribution

$$\mathcal{S}_E = \nabla(\alpha_s \mathbf{u}_t \cdot \nabla s_t), \tag{4.23}$$

---

[2] Note that $\eta > 0$ corresponds to the damping in the following.

with $\alpha_s = (\partial \ln p/\partial \ln s)_\rho$, and s_t the turbulent fluctuations of entropy. The entropy contribution is related to the advection of turbulent fluctuations of entropy by the turbulent field. It scales as the Mach number to the third and its contribution is small compared to the Reynolds one. More precisely, it is dominant in the uppermost part of the Sun, the super-adiabatic region, but globally remains small compared to the Reynolds contribution. Consequently, the entropy term is negligible for solar gravity modes.

- the rotational contributions

$$S_\Omega = -\frac{\partial}{\partial t}\rho_t \left[\Omega\frac{\partial}{\partial \phi}u_t - 2\Omega \times u_t - r \sin \theta u_t \cdot \nabla\Omega e_\phi \right], \qquad (4.24)$$

where $\Omega = \Omega e_\phi$ is the rotational frequency. Those contributions are related to rotation, they scale as the Mach number to the third. In addition, they are proportional to the ratio Ω/ω, which is very small for slow rotators such as the Sun. Hence, those terms are negligible for solar g modes.

- terms involving the second order mass flux

$$S_M = \frac{\partial}{\partial t}(\rho_t g) + \nabla\left[c_s^2 \nabla \cdot (\rho_t u_t) \right] - g\nabla \cdot (\rho_t u_t) - \frac{\partial^2}{\partial t^2}(\rho_t u_t) \qquad (4.25)$$

All those contributions are negligible because they scale as $\mathcal{M}^3$.

- $\mathcal{L}_t$ contains linear terms in term of turbulent fluctuations. [27] have shown that those terms do not contribute to the excitation.

As the sources are random, the mode amplitude (A) can only be calculated in square average, $\langle |A|^2 \rangle$. From Samadi and Goupil [27], one finds

$$\langle |A|^2 \rangle = \frac{C^2}{8\eta(\omega_0 I)^2}, \qquad (4.26)$$

with

$$C^2 \equiv \int d^3 x_0 \int_{-\infty}^{+\infty} d^3 r d\tau e^{-i\omega_0\tau} \langle (\xi \cdot S_R)_1 (\xi \cdot S_R)_2 \rangle \qquad (4.27)$$

where η is the mode damping rate, and the subscripts 1,2 denote two different spatial and temporal locations.

Eventually, it is possible to express the excitation rates from (4.26) such as [28]

$$\mathcal{P} = \eta\langle |A|^2 \rangle \omega_0^2 I = \frac{C^2}{8I} \qquad (4.28)$$

As expected in Sect. 4.2, the excitation rate is inversely proportional to the mode inertia.

Note also that $\mathcal{P}$ is independent of the mode damping and will be determined by the correlation product between the source term, here the divergence of the Reynolds

Fig. 4.2 Integrant of the mode excitation rates a function of the radius. The *dashed line* corresponds to the mode $\ell = 1$, $n = 30$ and the *solid line* to $\ell = 1$, $n = 2$. Note that $r/R \approx 0.7$ corresponds to the base of the solar convection zone

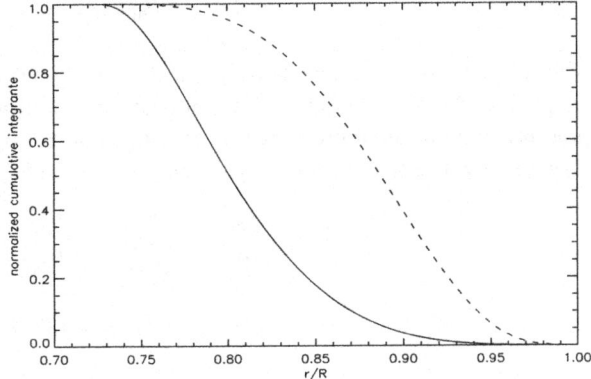

stress, and the eigenfunction as shown by (4.27). Figure 4.2 displays the integrant of g mode excitation rates for two modes, namely a low- radial-order and high-radial-order modes. It turns out, as a result of the balance between mode shape and turbulent kinetic energy flux, that g modes are excited in the deeper layers of the solar convective region in sharp contrast with p modes.

4.3.3 Sources of Damping (η)

The physics of mode damping is unclear and many uncertainties remain for solar-like oscillations. In this section we attempt to provide a short description of the major physical mechanisms involved and we discuss the particular case of g modes.

The oscillation exchanges energy with its surrounding medium through many processes, which can be split into two groups

$$\eta = \eta_{\mathrm{p}} + \eta_{\mathrm{other}} \qquad (4.29)$$

where η_{p} represents the contributions associated to pressure fluctuations and η_{other} are extra-contributions, usually neglected [29].

To investigate the mode damping processes, it is mandatory to take the energy equation into account. Some insight into the different physical mechanisms involved in mode damping can be obtained when writing the integral equation for η. This equation is established by considering the equation of kinetic energy of the oscillation. This yields [26]

$$\eta_{\mathrm{p}} = \frac{1}{2\omega I} \int\limits_0^M \mathcal{I}m \left(\frac{\delta\rho}{\rho_0}^* \frac{\delta P}{\rho_0} \right) dm \qquad (4.30)$$

with

$$\delta P = \delta P_{\text{turb}} + \delta P_g \tag{4.31}$$

where ρ_0 denotes the mean density, I is the mode inertia, ω the real part of the mode frequency, $\delta\rho, \delta P_g, \delta P_{\text{turb}}$ the Lagrangian perturbations of the mode density, gas pressure and turbulent pressure (i.e. the diagonal part of the Reynolds stress tensor), respectively. Using the following thermodynamic relations

$$\frac{\delta P_g}{P_0} = \chi_\rho \frac{\delta S}{c_v} + \Gamma_1 \frac{\delta\rho}{\rho_0} \tag{4.32}$$

$$\chi_\rho = (\Gamma_3 - 1)c_v\rho_0 T_0/P_0 \tag{4.33}$$

$$(\Gamma_3 - 1) = \left(\frac{\partial \ln T_0}{\partial \ln \rho_0}\right)_s \tag{4.34}$$

one can express the damping rate more explicitly as

$$\eta = \frac{1}{2\omega I} \int_0^M \mathcal{I}m \left[\left(\frac{\delta\rho}{\rho_0}^* T_0 \delta S\right)(\Gamma_3 - 1) + \left(\frac{\delta\rho}{\rho_0}^* \frac{\delta P_{\text{turb}}}{\rho_0}\right) \right] dm \tag{4.35}$$

where δS is the perturbation of entropy, T_0 the mean temperature, and the star denotes the complex conjugate.

For sake of simplicity, we further restrict our discussion to the radial case[3] so that the perturbed wave energy equation reads

$$i\sigma T_0 \delta S = -\frac{d\delta L_r}{dm} - \frac{d\delta L_c}{dm} + \delta\epsilon_t \tag{4.36}$$

with $\delta L_r, \delta L_c$ the perturbations of the radiative and convective fluxes, respectively, and $\delta\epsilon_t$ the perturbation of the dissipation rate of turbulent kinetic energy into heat. The set of (4.30–4.36) eventually permit to identify the different sources of damping related to pressure fluctuations. One can write

$$\eta_p = \eta_{\text{turb}} + \eta_{\text{conv}} + \eta_{\text{rad}} + \eta_{\text{dissipation}} \tag{4.37}$$

where

---

[3] Even if non-consistent with the non-radial nature of g modes, this assumption is sufficient for the discussion. Indeed, this assumption is justified when the perturbation of the horizontal component of the radiative flux can be neglected in front of the radial one, i.e. when

$$\frac{1}{T_0}\frac{dT_0}{dr} \gg \frac{\ell(\ell+1)}{r}$$

This is verified for low-degree asymptotic solar g modes.

- η_{turb} is the contribution due to the perturbation of the turbulent pressure, its expression is

$$\eta_{\text{turb}} = \frac{1}{2\omega I} \int_0^M \mathcal{I}m \left(\frac{\delta\rho^*}{\rho_0} \frac{\delta P_{\text{turb}}}{\rho_0} \right) dm \qquad (4.38)$$

The oscillation loses (or gains) part of its energy by producing a work $\delta P_{\text{turb}} dV$, where the variation of volume dV induced by the oscillation is related to the mode compressibility $\nabla \cdot \boldsymbol{\xi} = -\delta\rho/\rho$. These losses of energy are mainly controlled by the phase differences between $\delta\rho$ and δP_{turb}. As shown by (4.38), if those two quantities are in phase, there is no damping.

- η_{rad} is the contribution to the damping rate due to the perturbation of the radiative flux, it reads

$$\eta_{\text{rad}} = \frac{1}{2\omega^2 I} \int_0^M \mathcal{R}e \left(\frac{\delta\rho^*}{\rho_0} \frac{d\delta L_r}{dm} \right) (\Gamma_3 - 1) dm \qquad (4.39)$$

This contribution contains two dominant terms, namely the opacity effect that is responsible for the instability of modes in classical pulsators but negligible in solar-type stars [30, 31]. The other contribution is related to the temperature fluctuations δT. In the diffusion approximation, the radiative flux is approximated by a Fourier law, hence in the perturbed energy equation the divergence of $\mathbf{F}_r$ introduces the second derivatives of δT. The former then introduces a factor k_r^2 (where $k_r \approx \sqrt{\ell(\ell+1)} N/(\omega_0 r)$ is the vertical local wavenumber in the g mode cavity). Accordingly, the higher the mode radial order the higher η_{rad}. Indeed, we also note that the higher the mode angular degree the higher η_{rad}, since ℓ is nothing but the horizontal equivalent of n.

- η_{conv} is the damping associated to the perturbation of the convective heat flux

$$\eta_{\text{conv}} = \frac{1}{2\omega^2 I} \int_0^M \mathcal{R}e \left(\frac{\delta\rho^*}{\rho_0} \frac{d\delta L_c}{dm} \right) (\Gamma_3 - 1) dm \qquad (4.40)$$

This contribution is certainly the more complex to evaluate since it strongly depends on how convection and oscillations are coupled and consequently it depends on the modeling of convection. It is necessary to take η_{conv} into account when the convective turn-over time-scale is of the same order as the modal period. This is typically the case for the solar 5 min oscillations for which the eddies near the photosphere have a time-scale of several minutes. In contrast, i.e. when the time-scales are uncorrelated ($\tau_{\text{conv}} \gg P_{\text{osc}}$, with τ_{conv} the convective turn-over time-scale and P_{osc} the modal period), the frozen convection assumption can be adopted and η_{conv} no longer plays significant role [32].

- $\eta_{\text{dissipation}}$ is the contribution to the damping associated with the perturbation of the dissipation rate of turbulent kinetic energy into heat. This contribution was introduced by Ledoux and Walraven [33] and more recently by Grigahcène et al. [34], it partly compensates the effect of turbulent pressure (η_{turb}) and in the limit of a fully ionized gas in which radiative pressure can be ignored the sum $\eta_{turb} + \eta_{dissipation}$ vanishes.

The contributions to the damping rate, i.e. η_{turb}, η_{rad}, η_{conv}, and $\eta_{dissipation}$, are often thought to be dominant in the Solar case. Nevertheless, other contributions had been proposed [η_{other} in (4.29)], namely

- $\eta_{viscturb}$ is the contribution to the damping rate related to turbulent viscosity

$$\eta_{viscturb} = \frac{1}{2I} \int_0^M v_t \left| r \frac{\partial}{\partial r} \left(\frac{\xi_r}{r} \right) \right|^2 dm \qquad (4.41)$$

where v_t is the turbulent viscosity. Turbulent viscosity is an effective viscosity by analogy to the molecular case in which one considers eddies as "particles". In kinetic gas theory, molecular viscosity corresponds to a perpendicular (in respect to the direction of the fluid motion) transfer of impulsion between particles of the fluid. In fact, it is the off-diagonal terms of the Reynolds stress tensor that act to damp the modes. While the diagonal part is related to the mode compression (see (4.38)), these off-diagonal contributions are related to the shear of the mode.

- $\eta_{surface}$ is a contribution corresponding to the losses of energy at the star upper boundary

$$\eta_{surface} = -\frac{2}{\omega_0^2 I} \int_V \mathcal{R}e \left(\nabla \cdot (\delta P \boldsymbol{v}_{osc}) \right) dV = -\frac{2}{\omega_0^2 I} \int_S \mathcal{R}e \left(\delta P \boldsymbol{v}_{osc} \cdot dS \right)$$

$$(4.42)$$

One immediately sees that the term $\delta P \boldsymbol{v}_{osc}$ corresponds to the wave flux. In the ideal case, the wave flux of a mode is zero since a standing wave is composed of two progressive waves traveling in the opposite direction. Both waves induce a wave flux but the sum vanishes. Nevertheless, in the non-ideal case, at the upper boundary waves are reflected due to a density drop but there is still part of the wave energy that tunnels over this barrier. A full discussion of this mechanism can be found in [35].

- $\eta_{scattering}$ is a contribution to the damping rate that was originally considered by Goldreich and Murray [36] in the context of solar p modes. The principle can be sketched as follows: a progressive wave that travels a turbulent media will be affected by random phase shifts. Now considering a standing wave, i.e. a mode, it is a sum of two progressive waves and consequently it will also be affected. However, modes are affected in a different way since scattering couple modes. Eddies can modify the phase of a mode α so that the final state corresponds to a mode β, thus energy is transmitted from α to β. This is possible under the condition [36]

$$\left| \omega_\beta - \omega_\alpha \right| < \frac{1}{\tau} \qquad (4.43)$$

where ω_β, ω_α are the frequency of the mode β and α, respectively. τ is the turn-over time-scale of the energy bearing eddies.

- Some other contributions due to non-linear effects have also been investigated, for instance non-resonant three mode coupling (see [37] and references therein) or saturation of over-stable modes [20].

Many works attempted to model the damping rates of solar modes. For instance, Balmforth [38], Dupret et al. [39] based on a perturbed mixing-length approach of Gough [40] and Grigahcène et al. [34], respectively. One can also mention the work of Xiong et al. [41] based on a Reynolds stress approach that consists in averaging the Navier–Stockes equation and in resolving the high-order moment of this equation.

For solar p modes, the dominant contribution to the damping rates is still a mystery. For instance, Goldreich and Kumar [42] have shown that both the radiative contribution (η_{rad}) and the one associated with turbulent viscosity (η_{viscturb}) are of the same order of magnitude. In contrast, Gough [43] and Balmforth [38] found that the damping is dominated by the modulation of turbulent pressure (η_{turb}) while the result of Dupret et al. [39] suggests the perturbation of the convective heat flux (η_{conv}) is dominant. Such disagreements are mainly related to the strong coupling between convection and oscillation which makes the problem difficult when the characteristic times associated with the convective motions are of the same order as the oscillation periods.

For low-order solar g modes, the situation is similar. Recent computation [17] have shown that those modes remain sensitive to the upper layers of the Sun and to the treatment of non-local convection. In contrast, Belkacem et al. [17] have shown that high-order gravity modes, i.e. low-frequency modes, are insensitive to this treatment and that their damping is dominated by radiative damping (see Fig. 4.3). This is the result of the large-amplitude of these modes in the inner part of the radiative zone and of their high-radial wave number (k_r). Note that such a result is in agreement with the computation of Kumar et al. [16] in this particular frequency range.

4.4 Theoretical Estimates on g Mode Amplitudes

Keeley [44] was the first to investigate the hypothesis that g modes are excited by turbulent convection. This author tried to explain the observational results of Severnyi et al. [5], Brookes et al. [6], and Delache and Scherrer [45], which claim to detect a solar g modes with a period near $2^\mathrm{h}40^\mathrm{m}$. He concluded that excitation by turbulent convection is unlikely to excite only one mode but rather the whole spectrum of gravity modes. However, those detections have not been confirmed [7]. Hence, while the stochastic excitation of solar p modes by turbulent convection was privileged, it becomes the more likely driving mechanism also for g modes.

Fig. 4.3 Contributions to the work from the radial radiative flux variation (*solid line*), the transverse radiative flux variation (*dotted line*), and the time-dependent convection terms (*dashed line*), for the mode $\ell = 1$, g_{10} (*top panel*) and $\ell = 1$, g_{32} (*bottom panel*). dW_{FRr} corresponds to the contribution of the radial component of the radiative flux, dW_{FRh} to the horizontal component and dW_c to the contribution associated to convection and turbulent pressure

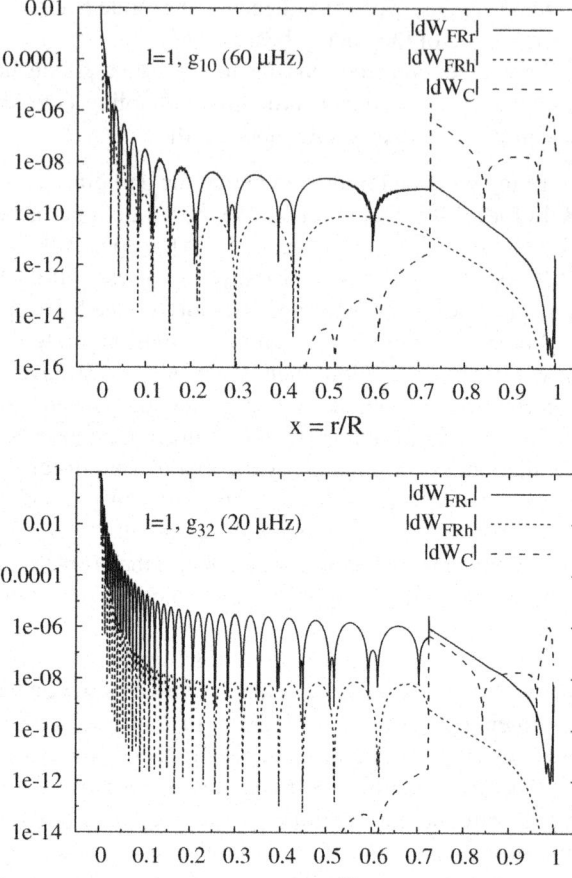

4.4.1 Computation Based on the Equipartition Assumption

A simple way to evaluate mode amplitudes, without computation of both excitation and damping processes, is to use the *equipartition assumption*. It consists in equating the mode energy with the kinetic energy of resonant eddies whose lifetimes are close to the modal period.

This assumption has been theoretically justified for p modes, by Goldreich and Keeley [46] assuming that the modes are damped by eddy viscosity. They found that the modal energy is inversely proportional to the damping rate, η, and proportional to an integral involving the term $E_\lambda v_\lambda \lambda$ where $E_\lambda \equiv (1/2) m_\lambda v_\lambda^2$ is the kinetic energy of an eddy with size λ, velocity v_λ and mass $m_\lambda = \rho \lambda^3$ (see Eq. (46) of Goldreich and Keeley [46]). Using a solar model, they assume that the damping rates of solar p modes are dominated by turbulent viscosity and that accordingly the damping rates

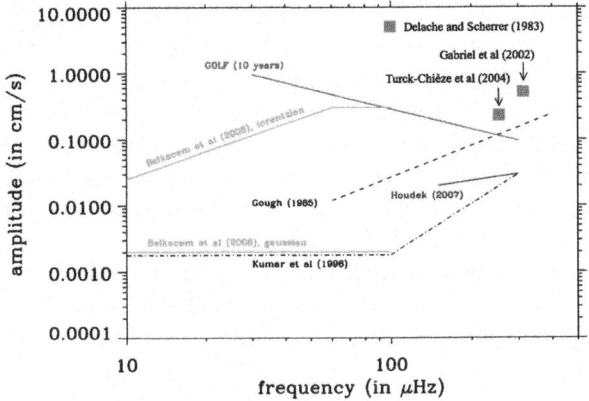

Fig. 4.4 Schematic mode surface velocity as function of frequency. Theoretical computations by Gough [15], Kumar et al. [16], and Belkacem et al. [17] are provided in *dashed line*, *dot-dashed line* and *solid line*, respectively. The GOLF observational threshold associated with ten years of observation is in *solid line* and observational results obtained by Gabriel et al. [8], Turck-Chièze et al. [9], Delache and Scherrer [45] are represented by squares. Courtesy: T. Appourchaux

are proportional to the eddy-viscosity, that is $\eta \propto v_\lambda \lambda$ (see (6) of Goldreich and Keeley [46]). Hence, after some simplifying manipulations, Goldreich and Keeley [46] found the modal energy to be (see their Eq. (52))

$$E_{\mathrm{osc}} \approx 0.26 E_\lambda = 0.13 m_\lambda v_\lambda^2. \tag{4.44}$$

This assumption was used by Christensen-Dalsgaard and Frandsen [47] for p modes and Gough [15] and Berthomieu and Provost [48] for solar g modes. The results obtained by Gough [15] is presented in Fig. 4.4. The author found a maximum of velocity of about $0.5\,\mathrm{mm\,s^{-1}}$ for the $\ell = 1$ mode at $\nu \approx 100\,\mu\mathrm{Hz}$. For the modes of angular degree $\ell = 2$, the amplitudes are found to linearly increase with frequency to reach values around $5\,\mathrm{mm\,s^{-1}}$ for low-order g modes ($\nu \approx 400$–$500\,\mu\mathrm{Hz}$). However, the result strongly depends on the way the modes are damped, and for g modes there is no evidence that they are mainly damped by turbulent viscosity.

4.4.2 Full Computation of Mode Driving and Damping

Kumar et al. [16], motivated by a claim of g-mode detection in the solar wind [49], performed the first theoretical estimate of g-mode amplitudes that is based on a computation of both excitation and damping rates.

Computations were performed using the Goldreich et al. [50] formalism for the excitation rates. In this formulation, Kumar et al. [16] assume a simplified description of turbulence in which the kinetic energy of the driving eddies scales according to the Kolmogorov spectrum. Also of particular interest is the way the eddies and the standing waves are temporally-correlated. The Goldreich and Keeley [46] approach, upon which the Goldreich et al. [50] and Kumar et al. [16] formulations are based, assume that the time-correlation between eddies is Gaussian. As we will see in the following sections, the way this function is chosen can lead to very different estimation of g-mode amplitudes . Concerning the damping rates, both turbulent and radiative contributions to the damping rates were included as derived by Goldreich and Kumar [42]. The full computation, as performed by Kumar et al. [16], shows that the mode life-time is around 10^6 years.

Eventually, the computation of both excitation and damping rates led to a maximum surface velocity $v_s \approx 1\,\mathrm{mm\,s}^{-1}$ near $v = 200\,\mu\mathrm{Hz}$ for $\ell = 1$ modes. The authors also found very low velocities ($10^{-2}\,\mathrm{mm\,s}^{-1}$) for $v < 100\,\mu\mathrm{Hz}$ (see Fig. 4.2).

4.4.3 The Special Case of Asymptotic g Modes

Recently, Belkacem et al. [17] investigated the particular case of the amplitudes of asymptotic g modes. The formalism used by Belkacem et al. [17] to compute excitation rates of non-radial modes were developed by Belkacem et al. [51] who extend to non-radial modes the work of Samadi and Goupil [27], Samadi et al. [52], and Samadi et al. [53].

In order to compute the excitation rates, one needs to determine the kinetic energy spectrum (E_k) as well as the eddy-time correlation function (χ_k). Both are derived from 3D numerical simulations see Belkacem et al. [17], for details. The Lorentzian function (χ_k) is found to better reproduce the eddy-time correlation function from the 3D numerical simulation than a Gaussian function in the frequency range $v \in [20\,\mu\mathrm{Hz}; 110\,\mu\mathrm{Hz}]$. In addition, the eddy-time correlation function is poorly represented by a Gaussian function, which underestimates χ_k by many order of magnitudes (see Fig. 4.5).

Computation of the excitation rates, with the input of 3D numerical simulations from the ASH code [54], permits to define two regimes. As shown by Fig. 4.5, at lower frequencies ($v < 100\,\mu\mathrm{Hz}$), the excitation rates (P) reach values about $10^{20-21}\,\mathrm{erg\,s}^{-1}$. At higher frequencies ($v > 100\,\mu\mathrm{Hz}$), the excitation rates (P) are found smaller, with values around $10^{19}\,\mathrm{erg\,s}^{-1}$. This can be explained by considering the balance between two contributions to the excitation rate (P), which are the mode inertia I and mode compressibility. Mode inertia decreases with frequency since the higher the frequency, the higher up the mode is confined in the upper layers. This then tends to decrease the efficiency of the excitation of low-frequency modes and seems to favor high-frequency g modes. On the other hand, mode compressibility is to be considered. It is minimum for frequencies near the fundamental mode (f mode) and increases for asymptotic p and g modes [17, 51]. It is found that it competes and dominates over the effect of mode inertia. In the asymptotic regime ($v < 100\,\mu\mathrm{Hz}$), the modes are compressible explaining the efficiency of the excitation.

Damping rates are computed with a fully non-radial non-adiabatic pulsation code MAD [39, 55, 56]. It takes into account the role played by the variations of the

Fig. 4.5 *Top*: Eddy-time correlation function versus frequency (χ_k). Crosses represent $\chi_k(\omega)$ obtained from the 3D simulation at the wave number k_0 that corresponds to the maximum of $E(k)$, and at the radius $r/R_\odot = 0.89$. Data (from the numerical simulation) are obtained with a time series of duration ~4.68 days with a sampling time of 800 s. The theoretical curves are normalized so that their integrals over-frequency equal that of the simulated data. *Solid line* corresponds to a Lorentzian function, *dashed line* to a Gaussian one and *dashed-dot line* to a combination of Lorentzian and Gaussian functions [17]. *Bottom*: Rate (P) at which energy is supplied to the modes versus the frequency for modes with angular degree $\ell = 1, 2,$ and 3, using a Lorentzian eddy-time correlation function

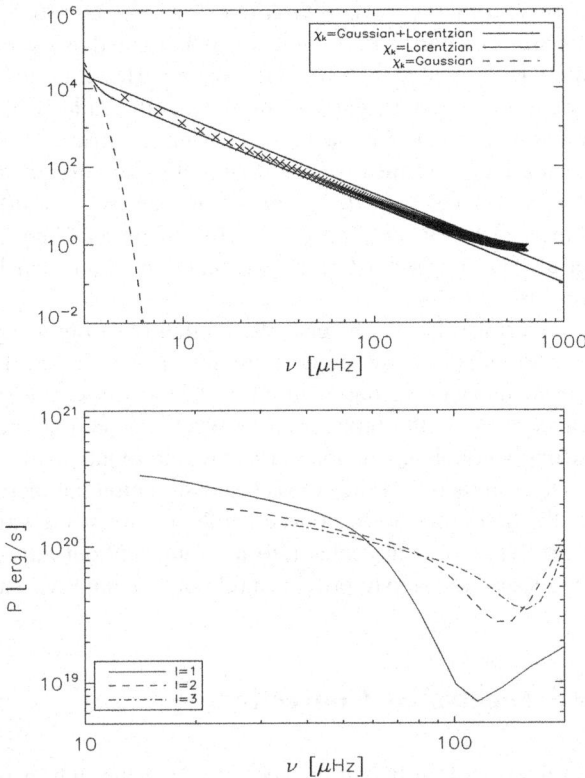

convective flux, the turbulent pressure and the dissipation rate of turbulent kinetic energy (see Sect. 4.3.3 for details). Belkacem et al. [17] have found that for high-frequency g modes ($\nu > 110 \, \mu$Hz), the work integrals and thus the damping rates are sensitive to the convection/pulsation interactions because the role of the surface layers in the work integrals becomes important. In contrast, for low-frequency g modes ($\nu < 110 \, \mu$Hz), the work integral and then the damping rates are found *insensitive* to the convection/pulsation interactions as well as the non-local parameters. Eventually, the damping rates dominated by radiative losses behave as ν^{-3}.

Figure 4.4 presents *intrinsic* values of the velocities. Taking visibility factors as well as the limb-darkening into account, Belkacem et al. [17] finally found that the maximum of apparent surface velocities of asymptotic g-modes is $\approx 3 \, \text{mm s}^{-1}$ for $\ell = 1$ at $\nu \approx 60 \, \mu$Hz and $\ell = 2$ at $\nu \approx 100 \, \mu$Hz. Those results then put the theoretical g-mode amplitudes near the GOLF observational threshold.

4.4.4 Discussion

Amplitudes found by Gough [43], Kumar et al. [16] and Belkacem et al. [17] differ by orders of magnitude. Gough [43] estimates were based on an equipartition principle derived from the work of Goldreich and Keeley [46, 57] and designed for p modes. Its use for asymptotic g modes is not adapted as the damping rates of these modes are not dominated by turbulent viscosity. However, for low-order g modes, the main contribution is not clearly identified and the validity of the equipartition assumption remains unclear. Kumar et al. [16] have carried another investigation of g mode amplitudes, and most of the quantitative disagreement with Belkacem et al. [17], for asymptotic g modes, comes from the use of a different eddy-time correlation function. Indeed, Kumar et al. [16] assumed a Gaussian function while Belkacem et al. [17] choose a Lorentzian function, motivated by 3D numerical simulation results.

A critical issue concerns the modeling of damping rates, which is challenging in particular for low-order g modes and p modes. The strong coupling between convection and oscillation in solar-like stars makes the problem difficult. A predictive description of the interaction between convection and oscillations when both are strongly coupled, is mandatory. It would require a sophisticated analytical or semi-analytical theory of the convection-oscillation interaction, which will not be limited to the first order in the convective fluctuations and which will take the contribution of different spatial scales into account, without adjusting free parameters. It is a necessary first step to provide a reliable quantitative estimate of low-order g modes.

4.5 Penetrative Convection

Excitation by turbulent convection is not the only way to generate gravity modes. Penetrative convection is also thought to be an efficient mechanism to excite internal waves as it has been known for many years for geophysical flows [58]. In the Sun, turbulent plumes are created at the upper boundary of the convection zone, where radiative cooling becomes dominant and where the flow reaches the stable atmosphere. In this region, the updrafts become cooler than their environment and stop their ascent. This cool flow is then denser than its environment and it triggers the formation of turbulent descending plumes [59]. When plumes fall down through the convection zone, they entrain the surrounding flow at their edge. It is the *entrainment hypothesis*, first introduced by G.I. Taylor and supported by observations in geophysical flows (for a review see [60]). This leads to the formation of large-scale downwelling turbulent structures that reach the stably stratified radiative zone below and that penetrate over some distance releasing its kinetic energy into internal waves and presumably into modes.

Andersen [18] proposed a theoretical estimation of g mode amplitudes based an 2D cartesian numerical simulations. The authors computed the attenuation factors between the bottom and top of the simulated convective region, that correspond to the exponential decreases of g modes in a convective region. Using order of magnitude energetic considerations and the attenuation factors they extrapolated an estimation of mode amplitude at the Sun surface, from 0.01 to 5 mm s^{-1}. Nevertheless, those assumptions were based on rather crude estimates such as the number of modes involved.

More recently, Dintrans et al. [19] investigated the internal wave generation by penetrative convection by projecting the wave flux on g-modes of the simulated cavity. The authors then studied the relation between penetration and g mode amplitudes and have shown that up to 40% of the convective kinetic energy can reside into g modes. In addition, their results show that the g-mode mean life-time is about twice the modal period. Those promising results are based on 2D polytropic cartesian simulations, which can not permit to infer g-mode amplitudes in the Sun. However, this work presents the advantage of giving a quantitative study of g mode amplitudes by penetrative convection.

Pushing realism one step further consists in performing 2D numerical simulations in spherical geometry with a realistic stratification. Such a work has been performed by Rogers and Glatzmaier [61]. A striking result is that using a quasi-linear simulation Rogers and Glatzmaier [61] have shown that standing waves, i.e. g modes, are excited by penetrative convection. However, full non-linear simulations do not exhibit modes but only low-frequency progressive waves. Non-linear wave-wave interaction is thought to be responsible for this result. A similar mechanism was proposed by Kumar and Goldreich [62] to explain p mode damping. The authors demonstrated that the dominant non-linear effect couples three waves: two trapped p modes from which a propagative wave drain its energy. Nevertheless, the high-thermal diffusivities used in the simulation imply an over-estimation of the solar convective flux by a factor 10^5. Hence, one can expect that the wave flux is over-estimated by a similar factor, which calls for caution since non-linear effects depend on mode amplitudes .

4.6 Concluding Remarks

In this lecture, we discussed the excitation and for damping mechanisms of solar g modes and reviewed the results obtained their amplitudes.

Concerning the damping rates, the non-detection of these modes favors that they are linearly stable. In the asymptotic regime, radiative damping is thought to be the dominant contribution [16, 17]. For low-order g modes the situation is less clear since those modes are sensitive to the interaction with convection. For p modes there is no consensus about the dominant contribution of the mode line-width [29], so does for low-order g modes. This issue is critical in the sense that it prevents an unambiguous theoretical determination of low-order g mode amplitudes .

For the driving mechanism, turbulent convection is thought to be responsible for g mode excitation [16, 17, 43]. Quantitative estimates differ from each other by orders of magnitude and those discrepancies have been partly explained by Belkacem et al. [17]. Their results tend to show that high-order g mode amplitudes are higher than previous findings and are close to the actual observational limit from the GOLF instrument. For low-order g modes, further progress on the determination of g mode amplitudes is strongly related to the understanding of damping processes.

Other processes have also been proposed to drive g modes, for instance penetrative convection which can be studied with the development of numerical simulations. Nevertheless, a quantitative estimate is not yet available since it would require realistic 3D numerical simulations that computer resources do not yet permit.

Acknowledgment I am grateful to organizers of the CNRS school of St-Flour for their invitation and to Thierry Appourchaux for kindly providing Fig. 4.2. Financial support from Liège University through the Subside Fédéral pour la Recherche is acknowledged.

References

1. Ulrich, R.K.: Ap. J. **162**, 993 (1970)
2. Leibacher, J.W., Stein, R.F.: Astrophys. Lett. **7**, 191 (1971)
3. Thompson, M.J.: Challenges to theories of the structure of moderate-mass stars. In: Gough, D., Toomre, J. (eds.) Lecture Notes in Physics, vol. 388, p. 61. Springer, Berlin (1991)
4. Turck-Chièze, S., Couvidat, S., Kosovichev, A.G., et al.: Ap. J. **555**, L69 (2001)
5. Severnyi, A.B., Kotov, V.A., Tsap, T.T.: Nature **259**, 87 (1976)
6. Brookes, J.R., Isaak, G.R., van der Raay, H.B.: Nature **259**, 92 (1976)
7. Appourchaux, T., Fröhlich, C., Andersen, B., et al.: Ap. J. **538**, 401 (2000)
8. Gabriel, A.H., Baudin, F., Boumier, P., et al.: A&A **390**, 1119 (2002)
9. Turck-Chièze, S., García, R.A., Couvidat, S., et al.: Ap. J. **604**, 455 (2004)
10. García, R.A., Turck-Chièze, S., Jiménez-Reyes, S.J., et al.: Science **316**, 1591 (2007)
11. García, R.A., Jiménez, A., Mathur, S., et al.: Astronomische Nachrichten **329**, 476 (2008)
12. Mathur, S., Turck-Chièze, S., Couvidat, S., García, R.A.: Ap. J. **668**, 594 (2007)
13. Mathur, S., Eff-Darwich, A., García, R.A., Turck-Chièze, S.: A&A **484**, 517 (2008)
14. García, R.A., Mathur, S., Ballot, J.: Sol. Phys. **251**, 135 (2008)
15. Gough, D.O.: Theory of Solar Oscillations, Tech. rep. (1985)
16. Kumar, P., Quataert, E.J., Bahcall, J.N.: Ap. J. **458**, L83 (1996)
17. Belkacem, K., Samadi, R., Goupil, M.J., et al.: A&A **494**, 191 (2009)
18. Andersen, B.N.: A&A **312**, 610 (1996)
19. Dintrans, B., Brandenburg, A., Nordlund, Å., Stein, R.F.: A&A **438**, 365 (2005)
20. Dziembowski, W.: Sol. Phys. **82**, 259 (1983)
21. Guenther, D.B., Demarque, P.: Ap. J. **277**, L17 (1984)
22. Ando, H.: Astrophys. Space Sci. **118**, 177 (1986)
23. Wentzel, D.G.: Ap. J. **319**, 966 (1987)
24. Wolff, C.L., O'Donovan, A.E.: Ap. J. **661**, 568 (2007)
25. Dziembowski, W.A., Paterno, L., Ventura, R.: A&A **151**, 47 (1985)
26. Unno, W., Osaki, Y., Ando, H., Saio, H., Shibahashi, H.: Nonradial oscillations of stars, 2nd edn. University of Tokyo Press, Tokyo (1989)
27. Samadi, R., Goupil, M.: A&A **370**, 136 (2001)
28. Samadi, R., Goupil, M., Lebreton, Y.: A&A **370**, 147 (2001)
29. Houdek, G.: Stochastic excitation and damping of solar-like oscillations. In: Proceedings of SOHO 18/GONG 2006/HELAS I, Beyond the spherical Sun. ESA Special Publication, vol. 624. http://adsabs.harvard.edu/abs/2006ESASP.624E..28H (2006) (Provided by the SAO/NASA Astrophysics Data System)
30. Pamyatnykh, A.A.: Acta Astronomica **49**, 119 (1999)
31. Houdek, G., Balmforth, N.J., Christensen-Dalsgaard, J., Gough, D.O.: A&A **351**, 582 (1999)
32. Grigahcène, A., Dupret, M.-A., Garrido, R.: Time Dependent Convection vs. Frozen Convection Approximations. In: Straka, C.W., Lebreton, Y., Monteiro, M.J.P.F.G. (eds.) in EAS Publications Series, vol. 26, pp. 137–144. EAS Publications Series (2007)

http://adsabs.harvard.edu/abs/2007EAS....26..137G, (Provided by the SAO/NASA Astrophysics Data System). doi 10.1051/eas:2007132
33. Ledoux, P., Walraven, T.: Handbuch der Physik **51**, 353 (1958)
34. Grigahcène, A., Dupret, M.-A., Gabriel, M., Garrido, R., Scuflaire, R.: A&A **434**, 1055 (2005)
35. Balmforth, N.J., Gough, D.O.: Ap. J. **362**, 256 (1990)
36. Goldreich, P., Murray, N.: Ap. J. **424**, 480 (1994)
37. Kumar, P., Goldreich, P., Kerswell, R.: Ap. J. **427**, 483 (1994)
38. Balmforth, N.J.: MNRAS **255**, 603 (1992)
39. Dupret, M.A., Barban, C., Goupil, M.-J., et al.: Beyond the spherical Sun. In: Proceedings of SOHO 18/GONG 2006/HELAS I, vol. 624, ESA Special Publication (2006)
40. Gough, D.O.: Ap. J. **214**, 196 (1977)
41. Xiong, D.R., Cheng, Q.L., Deng, L.: MNRAS **319**, 1079 (2000)
42. Goldreich, P., Kumar, P.: Ap. J. **374**, 366 (1991)
43. Gough, D.: Nonradial and nonlinear stellar pulsation. In: Hill, H.A., Dziembowski, W.A. (eds.) Lecture Notes in Physics, vol. 125, pp. 273–299. Springer, Berlin (1980)
44. Keeley, D.: Nonradial and nonlinear stellar pulsation. In: Hill, H.A., Dziembowski, W.A. (eds.) Lecture Notes in Physics, vol. 125, pp. 245–25. Springer, Berlin (1980)
45. Delache, P., Scherrer, P.H.: Nature **306**, 651 (1983)
46. Goldreich, P., Keeley, D.A.: Ap. J. **212**, 243 (1977)
47. Christensen-Dalsgaard, J., Frandsen, S.: Sol. Phys. **82**, 469 (1983)
48. Berthomieu, G., Provost, J.: A&A **227**, 563 (1990)
49. Thomson, D.J., Maclennan, C.G., Lanzerotti, L.J.: Nature **376**, 139 (1995)
50. Goldreich, P., Murray, N., Kumar, P.: Ap. J. **424**, 466 (1994)
51. Belkacem, K., Samadi, R., Goupil, M.-J., Dupret, M.-A.: A&A **478**, 163 (2008)
52. Samadi, R., Nordlund, Å.., Stein, R.F., Goupil, M.J., Roxburgh, I.: A&A **403**, 303 (2003)
53. Samadi, R., Nordlund, Å., Stein, R.F., Goupil, M.J., Roxburgh, I.: A&A **404**, 1129 (2003)
54. Miesch, M.S., Brun, A.S., DeRosa, M.L., Toomre, J.: Ap. J. **673**, 557 (2008)
55. Dupret, M.-A., Samadi, R., Grigahcene, A., Goupil, M.-J., Gabriel, M.: Commun. Asteroseismol. **147**, 85 (2006)
56. Dupret, M.-A., Goupil, M.-J., Samadi, R., Grigahcène, A., Gabriel M.: In: Proceedings of SOHO 18/GONG 2006/HELAS I, vol. 624, Beyond the spherical Sun, ESA Special Publication (2006)
57. Goldreich, P., Keeley, D.A.: Ap. J. **211**, 934 (1977)
58. Stull, R.B.: J. Atmospheric Sci. **33**, 1279 (1976)
59. Stein, R.F., Nordlund, A.: Ap. J. **499**, 914 (1998)
60. Turner, J.S.: J. Fluid Mech. **173**, 431 (1986)
61. Rogers, T.M., latzmaier, G.A.: MNRAS **364**, 1135 (2005)
62. Kumar, P., Goldreich, P.: Ap. J. **342**, L558 (1989)

Chapter 5
Unveiling Stellar Cores and Multipole Moments via their Flattening

Jean-Pierre Rozelot, Cilia Damiani, Ali Kilcik, Berrak Tayoglu and Sandrine Lefebvre

Abstract Rotation, and more precisely differential rotation, has a major impact on the internal dynamics of stars (and the Sun) and induces many instabilities driving the transport of angular momentum. In this chapter we shall consider these effects on the shape of shelllular layers, and to first order, those concerning the apparent oblateness. Thanks to the advent of interferometry techniques, stellar shapes can now be measured with a great accuracy. We will review here some main results obtained so far on different stars and we will give their main physical parameters taking into account differential rotation. We will discuss how the core density can be reached. Gravitational moments are presented for these observed flattened stars, and for the Sun, for which some conflicting results are presented.

Keywords Sun · Stars · Oblateness

Jean-Pierre Rozelot (✉)
Université de Nice-Sophia-Antipolis, OCA-FIZEAU (UMR CNRS 6525), Av. Copernic 06130, Grasse (France)
e-mail: e@obs-azur.fr

Cilia Damiani
INAF- Osservatorio Astrofisico di Catania, Via S. Sofia 78 95123, Catania (Italy)
e-mail: damiani@oact.inaf.it

Ali Kilcik
Big Bear Solar Observatory (BBSO), 40386 North Shore Lane Big Bear City, CA 92314-9672 USA
e-mail: kilcik@bbso.njit.edu

Berrak Tayoglu
Kandilli Observatory and E.R.I., Bogazici University, Cengelkoy 34684, Istanbul Turkey

Sandrine Lefebvre
Université Paris 6 (UPMC–LATMOS), Tour 45–46 - 4ème étage - Case 102, 4 place Jussieu 75 252, PARIS Cedex 05 France
e-mail: Sandrine.Lefebvre@aero.jussieu.fr

J.-P. Rozelot and C. Neiner (eds.), *The Pulsations of the Sun and the Stars*,
Lecture Notes in Physics 832, DOI: 10.1007/978-3-642-19928-8_5,
© Springer-Verlag Berlin Heidelberg 2011

5.1 Relating Rotation and Measured Flattening

It has been demonstrated that a sphere is the unique solution to the problem of hydrostatic equilibrium for a fluid mass at rest in tridimensional space. The problem complicates when the mass is rotating. In stars, axial rotation modifies the shape of equilibrium by adding a centrifugal acceleration term to the total potential, breaking the spherical symmetry. The sphere becomes an oblate figure, and we have no a priori knowledge of its stratification, boundary shape, planes of symmetry, transfer of angular momentum in differentially rotating body, etc. When the velocity rotation rate is non constant, in depth and in latitude, the surface changes from a spherical to a spheroidal shape (and even more complex figures if asymmetry exists, due to the presence of magnetic fields for instance). We will argue here that, if *the stellar geometrical deformation can be accurately observed, and measured, one should be able to deduce informations on the stellar rotation, and density distribution, down to the core*.

For example, let us consider the case of a mass of polytropic gas of index n, rotating at a constant angular velocity Ω. The equilibrium configuration and shape of such a body is known since the works of [1, 2]. By writing the mechanical equilibrium equations and seeking a solution in the form of a perturbed case of the non-rotating configuration, neglecting furthermore the effects arising from Ω^4, the boundary of the star, defined for instance by a constant null density can be obtained. The surface oblateness f is given by an equation of the type:

$$f = \upsilon \frac{\Omega^2}{G\rho_c},$$

where G is the constant of gravitation, ρ_c the density of the core, and υ a term depending on the polytropic index chosen. Extensive computations can be found in [2]; and for the case $n = 3$:

$$f = \left(0.5 + 0.856\frac{\rho_m}{\rho_c}\right)\frac{\Omega^2 R_{eq}}{g} \tag{5.1}$$

where ρ_m/ρ_c, is the ratio of mean to central density, R_{eq} the equatorial radius of the star, and g the gravity at surface. Even if such a formalism can be now considered as outdated, it could be noticed that the approximation is still rather good and good enough for non polytropic structures with discontinuous variation of density, such as the Earth.

In the solar case, taking $\rho_c/\rho_m = 107.168$, $\Omega = 2.85 \times 10^{-6}$ rad/s, $R_{eq} = R_{\odot} = 6.955080 \times 10^{10}$ cm and $g = 2.74 \times 10^4$ cm/s^2 (values taken in [3]), it follows that $f = 1.04 \times 10^{-5}$, in satisfying agreement with the best up-to-date determination of 8.55×10^{-6}.

Conversely, if the flattening is accurately measured, the mass determined by different means, and the mean density computed (the volume can be known through the measured two radii—equatorial and polar one), thus the core density can be reached.

The model described by (5.1) is a crude simplification of the actual configuration of the Sun and stars, but it serves as an illustration to link the flattening with the rotation velocity and density distribution.

Since there is no simple analytical description of the mechanical equilibrium of a non-uniformly dense star, with varying rotation velocity with depth and latitude (except for the Sun), we shall limit the present study to the description of the relationship between the flattening and the main parameters of the star according to different assumptions.

After reviewing the current observational knowledge on solar and stellar flattenings, we will detail different formalisms to link the main parameters of the star to the observed flattening.

5.2 Measurements of the Flattening of the Sun and Stars

5.2.1 Flattening of the Sun

Solar flattening has been identified as a key parameter for astrophysics as early as the late nineteenth century when Newcomb [4] demonstrated that if the difference between equatorial and polar radii, $\Delta r = R_{eq} - R_{pol} = 500$ mas, it would explain the discrepancy between the prediction of Newtonian gravitational theory and the perihelion advance of Mercury observed by Le Verrier in 1859. However measurements soon ruled this hypothesis out, for example Auwers [5] found $\Delta r = 38 \pm 23$ mas. In modern times, even though general relativity had given a satisfactory prediction of Mercury's perihelion, the argument was once again debated after Dicke's historical measurement of $\Delta r = 41.9 \pm 3.3$ mas [6]. We know today that such measurements were inaccurate; nevertheless they have been a source of progress. To summarize the present knowledge of the flattening of the Sun obtained by different experiments, we show in Fig. 5.1 the values obtained since 1993, comparing both ground-based and space measurements. The actual value of the flattening of the Sun is still debated but at the moment a commonly accepted value is $\Delta r = 8.6$ mas [3]. Regarding the high correlation found between the measured oblateness values and the facular index data, we recommend to use:

$$\Delta r = 8.21 \text{ mas}.$$

Further details are given in [7–9].

Figure 5.1 shows that the solar flattening might be time dependent, the physical mechanisms involved being likely an exchange of the first two multipole moments with time. The first one (J_2), which carries the oblateness, is predominant in period of higher activity to the detriment of the second one (J_4), the mechanisms inverting in periods of lower activity [10, 11]. Those considerations are beyond the scope of this chapter and require a fine understanding of the Sun's interior to be tackled. Nonetheless, even to first order, the flattening is a consequence of rotation and its

Fig. 5.1 Recent flattening measurements: ground-based measurements obtained by means of the Pic du Midi Heliometer (*black diamonds*), space-born RHESSI's result (*violet holed square*), balloon-borne Solar Disk Sextant (SDS) experiment (*blue diamonds*) and MDI instrument on board SOHO (*red squares*). The *green curve* (*right scale*) is the faculae index which is a good indicator of the solar activity cycle. All errors are $\pm 1\sigma$

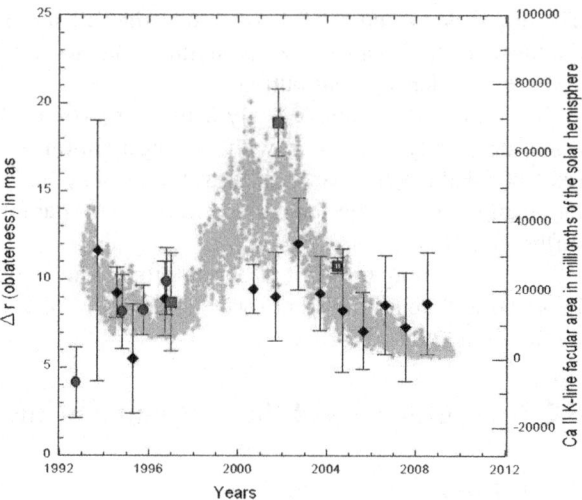

value depends on the inner parameters of the star. In this regard, what has been learned in the case of the Sun can be transposed to other stars, and mainly what is known from the multipole moments.

5.2.2 Flattening of Stars

The recent advent in interferometric techniques has provided us with a possibility to measure the flattening of distant stars. Surprisingly enough, the absolute accuracy of such measurements is better than the one presently reached for the Sun. For instance, the difference between the equatorial and polar radius of Altair (measured on the projected sky-plane) is $2 \times \Delta r_{\text{Altair}} = 0.424 \pm 0.079$ mas ($2R_{\text{eq}} = 3.461 \pm 0.038$ and $2R_{\text{pol}} = 3.037 \pm 0.069$ mas) [12], an accuracy better than for the Sun. And that of Achernar is no more than 3 % only as $R_{\text{eq}} = 12.0 \pm 0.4\ R_\odot$ and $R_{\text{pol}} = 7.7 \pm 0.2\ R_\odot$ [13]. Some years ago the measurements were mainly available for fast rotators; today at the current level of the technique, a more number of stars will be accessible to oblateness measurements. We give in Table 5.1 a list of stars for which both the polar and equatorial radii has been already measured, together with some of their main physical parameters.

5.3 Different Models of Uniformly Rotating Body

In this section, we will derive the main parameters of a star via their flattening always considering that the star rotates rigidly, i.e. with constant angular velocity in latitude and depth.

Table 5.1 List of stars for which the oblateness have been measured through different interferometric techniques. The last one is derived from spectroscopic and photometric observations. The table is not exhaustive. Notations: Sp for Spectral; T_p: polar temperature; T_{eq}: equatorial temperature; M: mass of the star; R_{eq} and R_{pol}: equatorial and polar radii; v_{eq}: velocity at the equator; i: axis star inclination; v_c: critical velocity

Star	Sp Type	T_p (K)	T_{eq} (K)	M ($M_\odot$)	R_{eq}/R_{pol}	R_{eq} ($R_\odot$)	v_{eq} (km/s)	i (°)	v_{eq}/v_c	Ref.
Achernar	B3Vpe	20000	9500	6.07	1.450	12.0	292	50.0	0.96	[14]
Regulus	B7V	15400	10314	3.04	1.325	4.16	317	90.0	0.86	[15]
Vega	A0V	10150	7900	2.303	1.230	2.78	270	4.7	0.75	[16]
		9988	7557		1.246	2.87	270	4.5	0.77	[17]
Alderamin	A7IV-V	8440	7486	2.0	1.298	2.82	283	88.2	0.83	[18]
Altair	A7IV-V	8500	6509	1.8	1.237	2.12	277	55.0	0.76	[19]
		8740	6890		1.215	1.99	273	63.9	0.73	[20]
		8710	6850		1.217	2.02	271	62.7	0.73	[21]
		8450	6860		1.221	2.03	286	57.2	0.75	[21]
Rasalhague	A5 III	9300	7460	2.10	1.201	2.871	237	87.7	0.89	[22]
α Cep		8863	6707	1.92	1.246	2.739	262	64.9	0.93	[22]
v Cygni	B2Ve	22200		6.97	1.352	4.80	453	23	0.95	[23]

5.3.1 Model for Centrally Condensed Body

The total external potential of a star of mass M rotating at constant angular velocity Ω can be expressed, assuming symmetry about the rotation axis, as the sum of a gravitational and a centrifugal term, Φ_g and Φ_c, both depending on the radius r and the colatitude θ

$$\Phi_{tot}(r, \theta) = \Phi_g(r, \theta) + \Phi_c(r, \theta) \tag{5.2}$$

In hydrostatic equilibrium, the equipotential surfaces are simultaneously surfaces of constant pressure p and constant density ρ [1] In particular, $p = \text{const} = 0$ represents the surface of the star and its figure is wholly determined by the form of the equipotential surfaces.

The radius of the star's surface can be then expressed in the general form

$$r(\theta) = s \left[1 + \sum_{n=0}^{\infty} s_{2n}(s) P_{2n}(t) \right] \tag{5.3}$$

where the coefficients s_{2n} characterize the shape of the surface, and are therefore called sometimes "shape coefficients", $P(t)$ are the usual Legendre's polynomials

---

[1] The Von Zeipel's [24] theorem stipulates that contours of temperature, density, or pressure should be nearly coincident near the surface. Differential rotation, magnetic fields and turbulent pressure are the largest local acceleration sources that may violate this theorem. It has been generalized to account for differential rotation in the case of a "shellular" rotation law by Maeder [36].

with $t = \cos\theta$ and s is the mean radius of the equipotential surface under consideration, defined as the radius of a sphere of equivalent volume. It can be shown [25] that the first coefficients of (5.3) can be written as[2]

$$
s_0 = -\frac{4}{45}\varepsilon^2 - \frac{52}{567}\varepsilon^3 - \frac{32}{315}\varepsilon k,
$$

$$
s_2 = -\frac{2}{3}\varepsilon - \frac{23}{63}\varepsilon^2 - \frac{8}{21}k - \frac{4}{27}\varepsilon^3 + \frac{2}{21}h - \frac{152}{315}\varepsilon k,
$$

$$
s_4 = \frac{12}{35}\varepsilon^2 + \frac{32}{35}k + \frac{4}{11}\varepsilon^3 + \frac{192}{385}h + \frac{32}{105}\varepsilon k,
$$

$$
s_6 = -\frac{40}{231}\varepsilon^3 - \frac{80}{231}h. \tag{5.4}
$$

where $\varepsilon = (R_{eq} - R_{pol})/R_{eq}$ is the oblateness (R_{eq} and R_{pol} are the equatorial and polar radii of the surface), k and h characterize the difference between the equilibrium spheroid and the ellipsoid of rotation in second-order and third-order oblateness, respectively; k and h are functions of s (see Annex 1). One can then readily obtain expressions relating the equatorial radius to the mean radius:

$$
R_{eq} = s\left(1 + \frac{1}{3}\varepsilon + \frac{2}{9}\varepsilon^2 + \frac{8}{15}k + \frac{14}{81}\varepsilon^3 + \frac{26}{105}h + \frac{16}{63}\varepsilon k\right), \tag{5.5}
$$

or conversely

$$
s = R_{eq}\left(1 - \frac{1}{3}\varepsilon - \frac{1}{9}\varepsilon^2 - \frac{8}{15}k - \frac{5}{81}\varepsilon^3 - \frac{26}{105}h + \frac{32}{315}\varepsilon k\right). \tag{5.6}
$$

The external gravitational potential of the star is in the form (see [26])

$$
\Phi(r, t) = -\frac{GM}{r}\left[1 - \sum_{n=1}^{\infty}(R_{eq}/r)^{2n} J_{2n} P_{2n}(t)\right] \tag{5.7}
$$

where J_{2n} are the gravitational moments of the star. The coefficients J_{2n} are linked with ε, k and h in (5.4) when taking (5.7) at the surface. It can be easily found through some algebra that:

$$
J_2 = +\left[\frac{2}{3}\varepsilon - \frac{1}{3}q - \frac{1}{3}\varepsilon^2 + \frac{8}{21}k(1+q) + \frac{3}{7}\varepsilon q + \frac{40}{147}\varepsilon k \right.
$$
$$
\left. -\frac{50}{294}\varepsilon^2 q - \frac{2}{21}h\right],
$$

$$
J_4 = -\left[-\frac{4}{5}\varepsilon^2 - \frac{32}{35}k + \frac{4}{5}\varepsilon^3 - \frac{50}{49}\varepsilon^2 q + \frac{3616}{2695}\varepsilon k + \frac{4}{7}\varepsilon q \right.
$$
$$
\left. +\frac{208}{385}qk - \frac{192}{385}h\right],
$$

$$
J_6 = +\left[\frac{8}{7}\varepsilon^3 - \frac{20}{21}\varepsilon^2 q - \frac{160}{231}qk + \frac{128}{77}\varepsilon k + \frac{80}{231}h\right]. \tag{5.8}
$$

---

[2] Note the relation $-s_0 = \frac{1}{5}s_2 + \frac{2}{105}s_2^3$.

where q is the dimensionless square of the angular velocity of rotation of the star defined as

$$q = \Omega_{eq}^2 R_{eq}^3 / GM \tag{5.9}$$

In the same way, it will be useful to use χ defined as

$$\chi = \Omega_{eq}^2 s^3 / GM \tag{5.10}$$

(Note that q and χ are related in terms of each other and are expressed in terms of $s_0, s_2, s_4 \ldots$, as: $q/\chi = (R_{eq}/s)^3 = 1 - \frac{3}{2}s_2 + (\frac{3}{4}s_2^2 + 3s_0 + \frac{9}{8}s_4) + \cdots ,$).

To first approximation, one can take $s = R_{sp}$, the radius of the best sphere passing through R_{eq} and R_{pol}:

$$s = (R_{eq}^2 R_{pol})^{1/3} \tag{5.11}$$

It results from this formalism that

$$-s_2 = J_2 + \frac{1}{3}\chi + \frac{11}{7}J_2^2 + \frac{19}{21}\chi J_2 + \frac{8}{63}\chi 2$$
$$+ \frac{25}{4}J_2^3 + \frac{162}{35}\chi J_2^2 + \frac{73}{60}\chi^2 J_2 + \frac{116}{945}\chi^3$$
$$+ \frac{27}{28}J_2 J_4 + \frac{2}{21}\chi J_4,$$

$$-s_4 = J_4 + \frac{6}{35}(6J_2^2 - \chi - J_2\chi) + \frac{1548}{385}J_2^3$$
$$+ \frac{564}{385}\chi J_2^2 - \frac{34}{77}\chi^2 J_2 - \frac{62}{385}\chi^3$$
$$+ \frac{274}{77}J_2 J_4 + \frac{328}{231}\chi J_4,$$

$$-s_6 = J_6 + \frac{30}{11}J_2 J_4 + \frac{18}{11}J_2^3 + \frac{12}{77}\chi^2 J_2$$
$$+ \frac{6}{77}\chi J_2^2 + \frac{5}{33}\chi J_4 + \frac{8}{77}\chi^3. \tag{5.12}$$

This set of equations can be inverted. If one is able *to measure accurately the shape coefficients, thus the successive gravitational moments can be estimated*:

$$-J_2 = s_2 + \frac{1}{3}\chi + \frac{1}{7}(11s_2 + \chi)s_2 + \frac{27}{28}s_2 s_4 + \frac{19}{84}\chi s_4$$
$$+ \frac{1311}{980}s_2^3 + \frac{109}{980}\chi s_2^2,$$

$$+J_4 = s_4 + \frac{6}{37}\left(\chi + \frac{6}{5}s_2\right)s_2 + \frac{274}{77}s_2 s_4 + \frac{1548}{539}s_2^3$$

$$+ \frac{5106}{2695} \chi s_2^2 - \frac{18}{77} \chi s_4,$$

$$-J_6 = s_6 + \frac{30}{11} \left(s_2 s_4 + \frac{3}{7} s_2^3 + \frac{5}{18} \chi s_4 + \frac{4}{7} \chi s_2^2 \right). \tag{5.13}$$

In the same way, k and h can be determined. One gets:

$$k = -\frac{7}{8}\varepsilon^2 + \frac{5}{8}\varepsilon q - \frac{35}{32} J_4 - \frac{53}{20}\varepsilon^3 + \frac{11}{4}\varepsilon^2 q - \frac{15}{16}\varepsilon q^2$$
$$- \frac{89}{16}\varepsilon J_4 + \frac{15}{16} q J_4 - \frac{63}{40} J_6,$$

$$h = \frac{9}{10}\varepsilon^3 - 2\varepsilon^2 q + \frac{5}{4}\varepsilon q^2 + \frac{21}{4}\varepsilon J_4 - \frac{35}{16} q J_4 + \frac{231}{80} J_6. \tag{5.14}$$

5.3.2 Model for a Body of Constant Density

Such a model is a limiting case, and has the greatest gravitational moments (in absolute magnitude). In real stars, the density is not constant with depth, generally increasing, so that the formalism is equivalent to transferring mass to the interior, where the level surfaces are more spherical.

After some algebra, one get

$$J_2 = -\frac{1}{2}\chi(1 - \frac{5}{14}\chi + \frac{25}{98}\chi^2),$$

$$J_4 = \frac{15}{28}\chi^2(1 - \frac{5}{17}\chi),$$

$$J_6 = -\frac{125}{168}\chi^3. \tag{5.15}$$

5.3.3 The Generalized Roche Model: Stars with a Convective Zone and Massless Envelope

Let us consider a star with a convective zone (CZ) of radius β in such a manner that the dimensionless density $\delta = \rho/\overline{\rho}$ is

$$\delta(\beta) = \begin{cases} \delta_c = \beta_c^{-3}, \ 0 < \beta < \beta_c \\ 0 \qquad\quad \beta_c < \beta < 1 \end{cases} \tag{5.16}$$

where β is the normalized mean radius of the equisurface, or (s/R_{eq}) [see (5.6)].

Within the CZ we get

$$\varepsilon(\beta) = \varepsilon_c = \frac{5}{4}\chi\beta_c^3 \left(1 + \frac{15}{56}\chi\beta_c^3 + \frac{925}{1568}\chi\beta_c^6 \right) \tag{5.17}$$

Table 5.2 Moments of inertia J_2 for the eight stars given in Table 5.1, computed through the generalized Roche model

Star	J_2
(1) Achernar	-1.24×10^{-4}
(2) Regulus	-1.42×10^{-5}
(3) Vega	-5.54×10^{-6}
id	-5.62×10^{-6}
(4) Alderamin	-4.85×10^{-6}
(5) Altair	-2.25×10^{-6}
id	-2.24×10^{-6}
id	-2.45×10^{-6}
id	-2.28×10^{-6}
(6) Rasalhague	-5.67×10^{-6}
(7) α Cep	-4.07×10^{-6}
(8) α Cygni	-4.76×10^{-6}

Note that (5.19) gives a two orders of magnitude less.

and the corresponding gravitational moments are given by:

$$J_2 = -\frac{1}{2}\chi\beta_c^5\left[1 - \frac{1}{3}\chi\left(1 - \frac{10}{7}\beta_c^3 + \frac{3}{2}\beta_c^5\right)\right.$$
$$-\chi^2\left(\frac{23}{180} + \frac{10}{63}\beta_c^3 + \frac{1}{30}\beta_c^5 + \frac{925}{882}\beta_c^6\right.$$
$$\left.\left.-\frac{10}{21}\beta_c^8 - \frac{5}{140}\beta_c^{10}\right)\right],$$
$$J_4 = +\frac{15}{28}\chi^2\beta_c^{10}\left[1 - \frac{2}{3}\chi\left(1 - \frac{10}{7}\beta_c^3 + \frac{3}{2}\beta_c^5\right)\right],$$

$$J_6 = -\frac{125}{168}\chi^3\beta_c^{15}. \tag{5.18}$$

We give in Table 5.2 the computed values of J_2 obtained in the Roche model for stars given in Table 5.1.

5.3.4 Polytropic Model of Unit Index n

The gravitational moments have been calculated in [27] and are of the form:

$$J_2 = -0.173273\,q + 0.197027\,q^2 - 0.15q^3,$$
$$J_4 = +0.081092\,q^2 - 0.15\,q^3, \tag{5.19}$$
$$J_6 = -0.056329\,q^3.$$

Table 5.3 Central density for the stars given in Table 5.1 taking into account their oblateness

	(1)	(2)	(3)	(4)	(5)	(6)
ρ_c (kg/m^3)	132985	128305	123374	122521	127811	123521
	(7)	(8)	(9)	(10)	(11)	(12)
ρ_c (kg/m^3)	119885	120783	122317	122546	121889	125217

Formulas are accurate up to $n = 3$. They can be easily transformed in terms of χ. As was shown in the introduction, the polytropic model allows to reach the central density of the star using the flattening. We give in Table 5.3 the corresponding values calculated for the stars in Table 5.1.

5.4 Introducing Differential Rotation

5.4.1 Solar Case

The usual centrifugal potential Φ_c must be rewritten to take into account the differential rotation, in this case, the problem is no longer conservative, and the star is baroclinic. However we can postulate that the centrifugal force must derive from a potential, in such a way that it must be possible to find a function U which satisfies [28]:

$$\vec{F}_{\text{centrifugal}} = -\vec{\nabla} U$$

At a depth r_p, one can use an equation of the form

$$\Omega = \Omega_{\text{pol}} \left[1 + \sum_{i=1}^{\infty} a_{2i} r_p^{2i} \cos^{2i}(\theta) \right]^{1/2} \tag{5.20}$$

which derives from a potential.

Using the solar Greenwich database that records the sunspots position as a function of time, [29] have computed a_i for $i = 1$ and $i = 2$, which are:

$a_2 = +0.442$, $a_4 = +0.056$ at the surface ($r_p = 1$) and $\omega_{\text{pol}} = 2.399 \, \mu$rad/s.

The reader will be able to verify the perfect adjustment of the two curves given by (5.21) and (5.20), using the numerical values given here. Figure 5.2 shows the rotation rate with depth (from $r_p = 1$ down to 0.75 $R_\odot$). The inversion of the radial gradient rotation rate can be seen at $\theta = 37°$ of latitude, within the leptocline. This mechanism signs the main difference with a stellar structural approach and put in evidence the key role of the solar gravitational moments J_2 and J_4.

Those values are in agreement with the usual rotational law taken for the Sun under the form

Fig. 5.2 Differential rotation (velocity rate in nHz versus the heliographic latitude), for the solar case and according to a law deriving from a potential. The different depth are listed in the right box. One can perfectly see the inversion of the radial gradient of rotation at $\theta =$ 37° of latitude (Influence of the leptocline at 0.99 $R_\odot$. See [30]

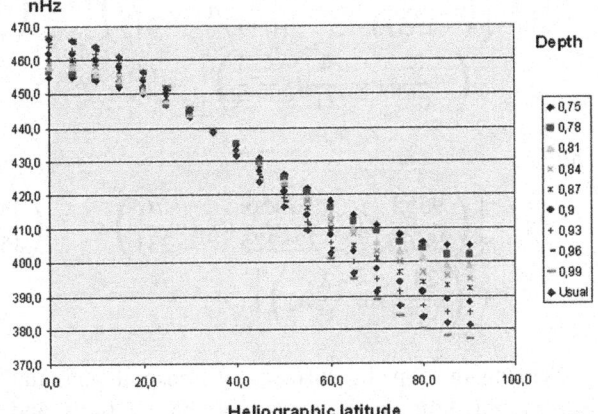

$$\Omega(\theta) = A + B\sin^2(\theta) + C\sin^4(\theta). \qquad (5.21)$$

The coefficient A represents the equatorial rotation rate and the coefficients B and C measure the latitudinal gradient in the rotation rate, with B representing mainly low latitudes whereas C represents largely higher latitudes [31]. Best known estimates of A, B and C, according to several authors, are given in Table 5.2 published in [26]. The following two examples differ from the way sunspots and faculae rotation were analyzed (coefficients in μrad/s):

$A = 2.913$, $B = -0.283$ and $C = -0.269$ [32],

$A = 2.82$, $B = -0.33$ and $C = -0.53$ [33].

Numerical values are of importance, as a small change could have a relatively large effect on the implied multipolar moments.[3]

The rotational potential Φ_c can then be expressed in terms of Legendre's polynomials

$$\Phi_c = -\chi \frac{GM}{R_{\rm sp}} [A_0 + A_2 P_2 + A_4 P_4]$$

with

---

[3] Helioseismic inversions provide a more realistic rotation profile, albeit less practical, that we shall recall here. Assuming a solid rotation below 0.66, and a differential rotation above the interface, one can write:

$$\Omega(r, \theta) = \Omega_c + \frac{1}{2}[1 + \text{erf}(2 * \frac{r - r_c}{d_1}] \times (\Omega_{Eq} + a_2\cos^2\theta + a_4\cos^4\theta - \Omega_c), \qquad (5.22)$$

with $\Omega_{Eq} = 1$, $\Omega_c = 0.93944$, $r_c = 0.7$, $d_1 = 0.05$, $a_2 = -0.136076$ and $a_4 = -0.145713$; "erf" is the error function. With this profile, the radial shear is maximal at the tachocline. Using this profile and numerically integrating the equipotentials from the core to the surface leads to a J_2 lying between $1.60 \ 10^{-7}$ and $2.20 \ 10^{-7}$ [34].

$$A_2 = \left[\left(-\frac{46832}{135135}a_4 - \frac{3064}{10395}a_2 - \frac{82}{315}\right)\varepsilon^2 + \left(-\frac{160}{693}a_4 - \frac{16}{63}a_2 - \frac{20}{63}\right)\varepsilon \right.$$
$$\left. + \left(-\frac{8}{63}a_4 - \frac{4}{21}a_2 - \frac{1}{3}\right)\right]$$

(5.23)

and

$$A_4 = \left[\left(\frac{9552}{25025}a_4 + \frac{77456}{225225}a_2 + \frac{76}{231}\right)\varepsilon^2 + \left(\frac{96}{455}a_4 + \frac{256}{1155}a_2 + \frac{8}{35}\right)\varepsilon \right.$$
$$\left. + \left(\frac{24}{385}a_4 + \frac{2}{35}a_2\right)\right]$$

(5.24)

Writing that on the surfaces of constant potential Φ must not depend on the heliographic latitude, it comes after computations, and accurate up to $\mathcal{O}(\varepsilon^3)$

$$J_2 = \frac{2}{3}\varepsilon - \frac{1}{3}\varepsilon^2 + \chi A_2 - \frac{26}{21}\chi A_2\varepsilon$$

(5.25)

and

$$J_4 = -\frac{4}{5}\varepsilon^2 + \chi A_4 - \frac{36}{35}\chi A_2\varepsilon - \frac{502}{231}\chi A_4\varepsilon$$

(5.26)

where A_2 and A_4 are determined by the a_i.

With the numerical values already given, one obtains $A_2 = -0.42(4638)$ and $A_4 = +0.028(751)$ at $r_p = 1\ R_\odot$.[4] It can be seen from (5.26) that the differential rotation increases J_4.

5.4.2 Differential Rotation on Stars

The latitudinal dependence of the angular velocity on stars can be approximated by:

$$\Omega^2 = \Omega_{eq}^2\left[1 + \overline{\alpha}_k\, l^k\right], \quad \text{where } \overline{\alpha}_k = \alpha_k s^{-k}, \quad l = (r/s)\cos(\frac{\pi}{2} - \theta), \quad (5.27)$$

l being in units of the mean star radius s. New observations by means of interferometry will permit to determine the α_k. Further details can be found in [35].

Another approach is to develop the solar rotation described by (5.21), by means of a set of disc-orthogonal functions

$T_1^0(\sin\theta) = 1$,
$T_2^1(\sin\theta) = 5\sin^2\theta - 1$, and
$T_4^1(\sin\theta) = 21\sin^4\theta - 14\sin^2\theta + 1$,

---

[4]   There is a sign mistake in [7] p. 23; the slight difference in the estimates is due to the different values used for the solar radius.

which leads to the following expansion:

$$\Omega(\theta) = \bar{A} + \bar{B}(5\sin^2\theta - 1) + \bar{C}(21\sin^4\theta - 14\sin^2\theta + 1). \tag{5.28}$$

The coefficients $\bar{A}$, $\bar{B}$, and $\bar{C}$ are free of crosstalk, $\bar{A}$ represents the 'rigid body' (or 'mean') component in the rotation, $\bar{B}$ and $\bar{C}$ are the components of the differential rotation. If the polynomial expansion is terminated at $\bar{C}$, the coefficients $\bar{A}$, $\bar{B}$, and $\bar{C}$, are related to the standard A, B, and C coefficients as follows:

$\bar{A} = A + (1/5)B + (3/35)C,$
$\bar{B} = (1/5)B + 2/15)C,$
$\bar{C} = (1/21)C.$

Using Legendre polynomials P_0, P_2 and P_4 as a set of orthogonal functions, the differential rotation can be described by:

$$\Omega(\theta) = DP_0 + EP_2(\cos\theta) + FP_4(\cos\theta). \tag{5.29}$$

where θ is the co-latitude.

If the expansion is truncated at the third term, the coefficients D, E, and F are related to the coefficients A, B, C in (5.29) as follows:

$D = A + (1/3)\ B + (1/5)\ C,$
$E = (2/3)\ B + (4/7)\ C,$
$F = (8/35)\ C.$

5.5 Moments of Inertia

The principal moments of inertia of a rotation star along the x and y axis with respect to the z axis are:

$$C(s) = \int_\tau (x^2 + y^2)\rho d\tau \quad \text{and} \tag{5.30}$$

$$B(s) = \int_\tau (y^2 + z^2)\rho d\tau \tag{5.31}$$

where $\rho(r)$ represents the density distribution inside the star (at the surface $\rho(s) \approx 0$). Equation (5.30) can be rewritten as:

$$C(s) = \frac{4\pi}{15} \int_0^s \rho dr \int_{-1}^1 [1 - P_2(t)]\, r^5(t)dt \tag{5.32}$$

which is also after some algebra:

$$C(s) = \frac{8\pi}{15} \int_0^s \rho d \left[r^5 \left(1 + \frac{2}{3}\varepsilon + \frac{5}{9}\varepsilon^2 + \frac{8}{21}k - \frac{2}{21}h + \frac{40}{81}\varepsilon^3 + \frac{88}{63}\varepsilon k \right) \right].$$

(5.33)

Similarly, one get

$$B(s) = \frac{8\pi}{15} \int_0^s \rho d \left[r^5 \left(1 - \frac{1}{3}\varepsilon + \frac{7}{18}\varepsilon^2 - \frac{4}{21}k + \frac{1}{21}h + \frac{1078}{3969}\varepsilon^3 + \frac{4}{63}\varepsilon k \right) \right].$$

(5.34)

The dynamical flattening (which intervenes in the precession constant) is given by

$$H = (C - B) / C \tag{5.35}$$

Introducing the mass of the star given by

$$M(a) = \frac{4\pi}{3} \int_0^a \rho d[a^3(1 - \varepsilon)] \quad (a = R_{eq}). \tag{5.36}$$

it results that to first approximation

$$J_2 = \frac{B - C}{M R_{eq}^2},$$

a relation which holds true whatever model chosen for the star's interior. Putting

$$\eta_1 = \frac{5}{2}\frac{\chi}{\varepsilon} - 2, \tag{5.37}$$

one gets

$$H = -\frac{\varepsilon - \chi/2}{1 - \frac{2}{5}\sqrt{1 + \eta_1}}. \tag{5.38}$$

Mathematically speaking this equation is an approximate first integral of the well known Clairaut's equation.[5]

---

[5]

given by $\dfrac{d^2 f}{dq^2} + \dfrac{6}{q}\dfrac{\rho}{D}\dfrac{f}{q} - \dfrac{6}{q^2}\left(1 - \dfrac{\rho}{D}\right)f = 0,$ where D is $\dfrac{3}{q^3}\displaystyle\int_0^q \rho q^2 dq.$

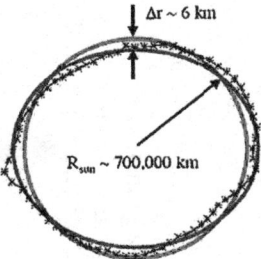

Fig. 5.3 Cross-section of the Sun through its rotation axis, showing its basically circular shape (upper curve passing below the vertical descending arrow), with a small distortion induced by the rotation (lower curve passing up the ascending arrow). The radial scale has been magnified enormously as indicated. The points show actual RHESSI data, which indicates a complex outer shape. (After [9]. See also [36])

5.6 Conclusion

All the formalism described above allows to compute the gravitational moments for stars when the shape coefficients are measured, or at least their flattening ε. The mean density of each star listed in Table 5.1 can be computed, and it is thus possible to go back to the central density ρ_c. Finally, the angular momentum A can be derived as it is directly linked to the gravitational moment of order 2:

$$A_{star} = C * \overline{\Omega}, \text{ which is also:}$$

$$A_{star} = \frac{8\pi}{3} \times \Omega_{eq} \int_0^{R_\odot} \rho r^4 dr \qquad (5.39)$$

For the solar case, results are given in Paragraph 7 and are confronted to estimates deduced by other authors. One of the major result is that $A_\odot$ constraints C and J_2.

Stellar rotation is one of the major topic of astrophysics today, and much of the research is an attempt to adjust theory and observations in order to bring the two in closer agreement. The preliminarily results presented here, show that the oblateness cannot be ignored and is crucial to constraint coherent stellar models. As discrepancies between models and observations have already been noticed, the study of stellar shapes cannot be bypassed anymore as first pointed out by [35, 37].

The differential rotation, as well as the non homogeneous mass and angular velocity distributions of stars, including the Sun, modify their outer shape. Up to a recent date, this departure to sphericity has been considered only as a second order effect in theories of stellar structure. However, the oblateness gets more important as the differential rotation gets stronger, and this encourage accurate observations of the stellar shapes.

Concerning the Sun, the equatorial radius is greater than the polar one of some 8.21 mas (i.e. 5950 m or ≈ 6 km). The differential rotation is not only surfacic, but is

Fig. 5.4 Rotation rate in the solar case from the tachocline to the leptocline. The gravitational moments J_2 and J_4 are critically dependent on this differential rotation

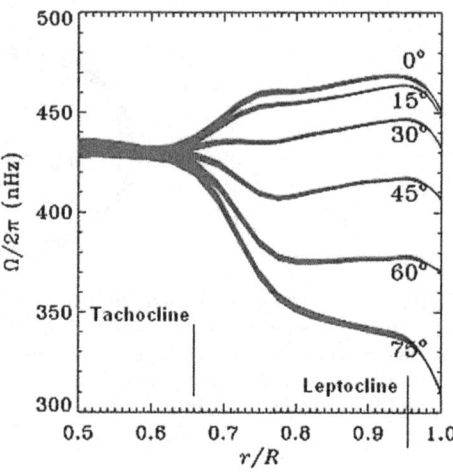

anchored deeper in the Sun, so that the successive shells which compose the Sun are not spherical, the outer shape being rather complex as seen in Fig. 5.3. It results that the interior of the Sun can be described as going from the surface to the core (i) a sub-surfacic thin layer (the leptocline), which is the seat of solar asphericities, radius variations with the 11-year cycle and the cradle of complex physical processes: partial ionisation of the light elements, opacities changes, superadiabaticity, strong gradient of rotation and pressure; (ii) a convective zone that incurs latitudinal shear, with little radial shear; (iii) a thin transition layer separating the convective zone from the deeper radiative zone, the solar tachocline, which would experience a rapidly quenched radial shear that would lead it to be prolate; (iv) a radiative zone rotating nearly uniformly, with an angular velocity of about 93.5 % of the equatorial angular velocity; (v) a core rotating at a nearly uniform velocity rate of about twice the equatorial rotation of the surface. One understand that the observed shape of our star is far from being spherical, hence the need to accurately measure the solar shape coefficients (Fig. 5.4).

For the solar case, the results show that the exact shape critically depends on the rotation of the external layers. Moreover, the oblateness owing to differential rotation is larger when the radial gradient of rotation $(d\Omega/dr)$ is <0 and decreases when this gradient is >0. The inversion is produced at around 40° of latitude [30]. Analytical differential rotation models lead to a significantly lower value of the flattening in comparison to a uniform rotation model; this can only be interpreted in terms of a positive outward rotation gradient in the subsurface (otherwise the differential rotation increases the flattening [38].) We want to emphasize here that J_2 and J_4 are very sensitive to differential rotation, both at the surface and in depth. From our analysis, we believed that the order of magnitude of J_4 is less than those deduced from equations of stellar structure and more in accordance with the values deduced from observations by [39], i.e.: $J_2 = 1.84 \times 10^{-7}$ and $J_4 = 9.83 \times 10^{-7}$ (correct order of magnitude). We must add that ε is time dependent, so that J_n must be also time dependent.

Furthermore, results show that it is impossible to conciliate other estimates than those founded there, for Δr, J_2, $A_\odot$ and J_4, taking into account accepted values of s and Ω_{eq}. Conflicting results are obtained when changing these last values, especially when computing moments of inertia, H, B and C. Is it for this reason that no estimates of these parameters are given in the literature?

We encourage observations of oblate stars—to transpose what is observed within the Sun—in order to determine their equatorial and polar radius, through existing facilities such as the CHARA array, the Keck interferometer, the Navy prototype Interferometer (NPOI) or the Palomar Testbed Interferometer PTI). A catalogue of 67 prospective rotationally distorted stars has been given by Van Belle et al. (2006) who gave a rough estimate of the ratio R_{pol}/R_{eq} based upon a simplification of an expression describing self-gravitating rotationally distorted gaseous masses:

$$v\sin i \approx (2GM/R_{pol} \times (1 - R_{pol}/R_{eq}))^{0.5}.$$

In the case of Altair, the approximation gives 1.14 instead of 1.16 observed. Further prospects concern the way to take into account magnetic fields.

5.7 Computations for the Solar Case

Masse $M = 1.989 \times 10^{33}$ g (Allen, 2000) [6]
   $G = 6.67259850 \times 10^{-11} \mathrm{m^3 kg^{-1} s^{-2}}$

$\Omega_{eq} = 2.850\ \mu\mathrm{rads}^{-1}$
$\Omega_{pol} = 2.399\ \mu\mathrm{rads}^{-1}$
$R_{eq} = 695509.9835$ km
$R_{pol} = 695504.0331$ km
$\Delta r = 5950.(336)$ m
$R_{sp} = 695508.0003$ km
$\varepsilon = 8.56 \times 10^{-6}$
β (surface) = 0,999997148206
$\chi = 2.05904734 \times 10^{-5}$
$q = 2.05906496 \times 10^{-5}$

5.7.1 Results

Constant density model; this model gives an upper limit for the gravitational moments

---

[6]   Or 1.9891×10^{33} g ± 0.0004 (Cohen and Taylor, 1984).

J_2: -1.03×10^{-5}
J_4: $+2.27 \times 10^{-10}$
J_6: -6.50×10^{-15}

Polytrope of index n = 3

J_2: -3.57×10^{-6}
J_4: $+7.68 \times 10^{-12}$
J_6: -4.92×10^{-16}

Convective zone and massless envelope

	Surface	Core ($r = 0.485$) ($M = 87.7\%$)	Core ($r = 0.22$) ($M = 39.9\%$)
β:	0.999997148	0.999999545	0,999999868
J_2:	-1.030×10^{-5}	-1.215×10^{-6}	-2.47×10^{-7}
J_4:	$+2.271 \times 10^{-10}$	$+2.936 \times 10^{-12}$	$+1.21 \times 10^{-13}$
J_6:	-6.495×10^{-15}	-1.067×10^{-17}	-8.96×10^{-20}

Differential rotation

χ: 1.631×10^{-5} (with $\Omega = 403.701 nHz$)
J_2: $-1.219(5) \times 10^{-6}$(†)
J_4: $+4.68(9) \times 10^{-7}$(†)

(†) $J_2 = -2.613 \times 10^{-7}$ and $J_4 = +6.29 \times 10^{-7}$: [40].
$J_2 \approx +2.22 \times 10^{-7}$, $J_4 = \approx -4.44 \times 10^{-9}$ and $J_6 = \approx -2.79 \times 10^{-10}$ [41], using a model of the interior structure and of solar rotation obtained from helioseismic inversions. The value of J_2 for non-uniform rotation ($\approx +2.21 \times 10^{-7}$) found by [41] for $\Omega = 435$ nHz is close to the value obtained by Pijpers (1998, MNRAS, 297, 76) ($\approx +2.18 \times 10^{-7}$) using a seismically determined rotation profile $\Omega(r, \theta)$, and also in close agreement with the value obtained by [42] ($\approx +2.15 \times 10^{-7}$).

However, [43] using a vector harmonic solution for the total potential found $J_2 = -2.22 \times 10^{-7}$ and $J_4 = +3.84 \times 10^{-9}$.

The difference in the sign is not *only a question of convention*. Results strongly depends on the rotation. Other perturbations in the rotation profile are needed to reduce the theoretical multipolar moments deduced from the equation of stellar structures to match the observations. Without additional constraints there appears to be no unique solution for the interior rotation that recovers both the oblateness and hexadecapole shape. We must emphasize that s_2 is always <0, s_4 always >0, so that J_2 must be <0 and $J_4 > 0$.

Thus, with our values, it comes:

$H = -1.6718 \times 10^{-5}$
$C = 7.01(9) \times 10^{+46}$ kg m^2

$A = 1.92(3) \times 10^{+48}$ g cm^{-2} s^{-1}. It must be noted[7]

with $\overline{\Omega} = 436.07$ nHz (2.74 rd/s)$^{(\ddagger)}$. One can verify that $\varepsilon = 8.47 \times 10^{-6}$ (= $\frac{3}{2} J_2 +$ $\frac{1}{2}\chi$, with $\chi = 2.05904734 \times 10^{-5}$, to first order).

($\ddagger$)Other values are as followed:

$1.94 \times 10^{+48}$: [44];
$1.91 \times 10^{+48}$: [45].

It is not possible to match both C, A and J_2 with the Roxburg's estimates. In addition, $\varepsilon_{Sun} = 1.02953 \times 10^{-5}$ leads to J_2 strictly $= 0$. It results that, ε_{Sun} being less than the former estimate just mentioned, J_2 must be < 0.

Annex 1: Equation of the Spheroid

The equation of an ellipsoid of revolution is given by:

$$\frac{r^2(\theta)\cos^2\theta}{R_{eq}^2(1-\varepsilon)^2} + \frac{r^2(\theta)\sin^2\theta}{R_{eq}^2} = 1 \tag{5.40}$$

Expanding $r(\theta)$ in powers of ε, we obtain

$$r(\theta) = R_{eq}\left[1 - \varepsilon\cos^2\theta - \frac{3}{2}\varepsilon^2(\sin^2\theta\cos^2\theta) + \frac{1}{8}\varepsilon^3(1-5\sin^2\theta)\sin^22\theta + \cdots\right] \tag{5.41}$$

Designing by $k(R_{eq})$ and $h(R_{eq})$ the second-order and third-order corrections, (5.41) becomes[8]

$$r(\theta) = R_{eq}\left[1 - \varepsilon\cos^2\theta - (\frac{3}{8}\varepsilon^2 + k)\sin^22\theta \right.$$
$$\left. + \frac{1}{4}(\frac{1}{2}\varepsilon^3 + h)(1-5\sin^2\theta)\sin^22\theta + \cdots\right]. \tag{5.42}$$

Inserting this development of the radius vector $r(\theta)$ in the expression of the total potential one obtains

$$\Phi_{tot} = -\frac{GM}{R_{eq}}[1 + a(\varepsilon,\theta) + b(\varepsilon,\theta)J_2 + c(\varepsilon,\theta)J_4 + \cdots] \tag{5.43}$$

J_n represents the gravitational moments(see 5.7). On the equilibrium surface, Φ_{tot} must be constant (i.e. independent from θ), so that the J_n coefficients must vanish (see 5.14).

---

[7] The estimate based on surface rotation alone [46] is $1.63 \times 10^{+48}$, which is about 15% smaller. The average angular momentum as a function of radius (and latitude) basically determined by the solar model that provides the density as a function of radius.

[8] Truncated at the order 3, and putting $h = 0$, (5.42) represents the so-called Darwin-de Sitter spheroid equation.

Annex 2: Roche's Model

The stellar equipotential surfaces of a body of mass M rotating at constant angular velocity Ω are described by

$$\Phi(\theta) = \frac{\Omega^2 R^2(\theta)\sin^2(\theta)}{2} + \frac{GM}{R(\theta)} = \frac{GM}{R_{pol}} \tag{5.44}$$

Introducing the degree of sphericity $D = R_{pol}/R_{eq}$, which relates the inverse of the oblateness to the polar radius, (5.44) can be rewritten as

$$r^3(\theta) - r(\theta)\left(\frac{1}{1-D}\right)\frac{1}{\sin^2(\theta)} + \left(\frac{D}{1-D}\right)\frac{1}{\sin^2(\theta)} = 0 \tag{5.45}$$

where $r(\theta)$ designs the normalized radius $\equiv R(\theta)/R_{eq}$. Solution of (5.45) is obtained through hypergeometric series of argument ς given by $\varsigma^2 \equiv 2\frac{(1-D)}{D}(\frac{3D}{2})^3\sin^2\theta$ [47], which determine the stellar shape. Lastly, it can be useful to remember the quantities relating critical and non-critical parameters. The critical or break-up velocity for the Roche's model is attained when the centrifugal and gravitational forces are equal. this leads to the following set of equations:

$$\frac{R_c}{R_p} = \frac{3}{2} \equiv D_c$$

$$v_c = \Omega_c R_c = \sqrt{\frac{GM}{R_c}} = \sqrt{GM\left(\frac{2}{3R_p}\right)}$$

$$\frac{v_{eq}}{v_c} = \sqrt{3(1-D)} = \frac{\Omega}{\Omega_c}\frac{R_{eq}}{R_c} \equiv \omega\frac{R_{eq}}{R_c}$$

$$\Omega_c = \frac{v_c}{R_c} = \sqrt{\frac{8GM}{27R_p^3}}$$

$$\frac{\Omega}{\Omega_c} = \omega = \sqrt{3(1-D)}\left(\frac{3}{2}D\right)$$

One will note that the angular (Ω) and linear (v) velocities are linked by $\Omega = v/R(\sin i)$.

References

1. Milne, E.A.: MNRAS **83**, 118 (1923)
2. Chandrasekhar, S.: MNRAS **93**, 390 (1933)

3. Allen, C.W., Cox, A.N.: Allen's Astrophysical Quantities, 4th edn, p. 719. Springer, New York (2000)
4. Newcomb, S.: Fundamental Constants of Astronomy, p. 111. US GPO, Washington (1865)
5. Auwers, A.: Astron. Nach. **128**, 367 (1891)
6. Dicke, R.H., Goldenberg, H.M.: Phys. Rev. Lett. **18**, 313 (1967)
7. Rozelot, J.P., Damiani, C., Lefebvre, S.: JASTP **71**, 1683 (2009)
8. Damiani, C., Rozelot, J.P., Lefebvre, S., Kilcik, A., Kosovichev, A.K. JASTP **73**(2–3), 241–250 (2010). doi:10.1016/j.jastp.2010.02.021
9. Hudson, H., Rozelot, J.P. (2010) RHESSI science nugget. http://sprg.ssl.berkeley.edu/tohban/wiki/index.php/HistoryofSolarOblateness
10. Emilio, M., Bush, R.I., Kuhn, J., Scherrer, P.: ApJ **660**, L161 (2007)
11. Damiani, C., Rozelot, J.P., Pireaux, S.: ApJ **703**(2), 1791 (2009)
12. van Belle, G.T., Ciardi, D.R., Thompson, R.R., Akeson, R.L., Lada, E.A.: ApJ **559**, 1155 (2001)
13. Jackson, S., MacGregor, K.B., Skumanich, A.: ApJ **606**, 1196 (2004)
14. Domiciano de Souza, A., Kervella, P., Jankov, S., et al.: A&A **407**, L47 (2003)
15. Mc Alister, H.A., ten Brummelaar, T.A., Gies, D.R. et al.: ApJ **628**, 439 (2005)
16. Aufdenberg, J.P., Mérand, A., Coudédu Foresto, V. et al.: ApJ **645**, 664 (2006)
17. Peterson, D.M., Hummel, C.A., Pauls, T.A. et al.: . Nature **440**, 896 (2006)
18. van Belle, G.T., Ciardi, D.R., ten Brummelaar, T. et al.: ApJ **637**, 494 (2006)
19. Domiciano de Souza, A., Kervella, P., Jankov, S. et al.: A&A **442**, 567 (2005)
20. Peterson, D.M., Hummel, C.A., Pauls, T.A. et al.: ApJ **636**, 108 (2006)
21. Monnier, J.D., Zhao, M., Pedretti, E. et al.: Science **317**, 342 (2007)
22. Zhao, M., Monnier, J.D., Pedretti, E. et al.: ApJ **701**, 209 (2009)
23. Neiner, C., Floquet, M., Hubert, A.M. et al.: A&A **437**, 257 (2005)
24. von Zeipel, H.: MNRAS **84**, 665 (1924)
25. Zharkov, V.N., Turbitsyn, V.P.: Sov. Phys. **13**, 981 (1970)
26. Rozelot, J.P., Lefebvre, S. The Sun's Surface and Subsurface, Lecture Notes in Physics, **599**, 4 (2003)
27. Hubbard, W.B.: Astron. J. **51**, 1052 (1974)
28. Lefebvre, S. (2003) : Déformées solaires, diamètre et structure interne. Ph D. Thesis, Nice University
29. Javaraiha J., Rozelot J.P.: Technical note. Observatoire de la Côte d'Azur, Nice University (2002)
30. Lefebvre, S., Kosovichev, A.K., Rozelot, J.P.: ApJ **658**, L135 (2007)
31. Javaraiah, J.: Solar Phys. **257**, 61 (2009)
32. Kuiper, J. The Sun, Chicago University Press, Chicago (1972)
33. Howard, R., Boyden, J.E., LaBonte, B.: Sol. Phys. **66**, 167 (1980)
34. Godier, S., Rozelot, J.P.: A& A **350**, 310–317 (1999)
35. Meynet, A., Milne, E.A.: Physics of rotation in stellar models. In: Rozelot, J.P., Neiner, C. (eds.) The Rotation of Sun and Stars, vol. 765, pp. 139. Springer, Heidelberg (2009)
36. Fivian, M.D. et al.: Science **322**, 560 (2008)
37. Deupree, R.G. CNRS St Flour School, this proceedings (2010)
38. Maeder, A.: A & A **347**, 185 (1999)
39. Lydon, T.J., Sofia, S.: Phys. Rev. Lett. **76**, 177 (1996)
40. Ajabshirizadeh, A., Rozelot, J.P., Fazel, Z.: Contribution of the solar magnetic field on gravitational moments. Scientia Iranica, **15** (1), 144–149 (2008)
41. Roxburgh, I.W.: A& A **377**, 688–690 (2001)
42. Paternó, L., Sofia, S., DiMauro, M. P.: The rotation of the Sun's core. Astron. & Astrophys., **314**, 940–946 (1996)
43. Armstrong, J., Kuhn, J. R.: (1999). Interpreting the Solar Limb Shape Distortions. Astrophys. Jour., **525**, 533 (1999)

44. Komm, R., Howe, R., Durney, B. R., Hill, F .: Temporal Variation of Angular Momemtum in the Solar Convection Zone. Astrophysical Journal. **586**, 650–662 (2003)
45. Antia, Chitre, & Thompson 2000, A & A, 306,335
46. Livingston, W. C. 2000, in Allen's Astrophyscial Quantities, ed. A. N. Cox (New York: Springer), p. 340.
47. Kopal, Z.: Astrophys. Space Sci. **133**, 157 (1987)

Part III
From Heliosismology to Asterosismology

S. Vauclair and M. Soriano (LATT-Toulouse University)

Chapter 6
From Helioseismology to Asteroseismology: Some Recent Developments

Sylvie Vauclair and Mélanie Soriano

Abstract While variable stars have been observed and analyzed for more than two centuries, the study of solar-type oscillations is new and presently blooming. In this introductory presentation, we first recall the general basis for asteroseismology, the so-called "asymptotic theory" of stellar oscillations and discuss important deviations from this theory. Then we focus on solar-type stars and give two examples for which it was possible to derive precise stellar parameters from seismology, specially the helium abundance, which is not obtained in these stars from spectroscopy alone. The potentiality of asteroseismology for a better knowledge of stellar structure and evolution is huge, and many new results are expected in the near future.

6.1 Introduction

If we define asteroseismology as the general study of stellar oscillations, we must recall that this thematic began long before the discovery of the solar oscillations. At that time, astronomers spoke of "variable stars" or "pulsating stars", and they only detected large amplitude oscillations.

The first Cepheid was discovered as soon as 1786 by the English astronomer John Goodricke. One century later, Henrietta Swan Leavitt, a lady who worked at the Cambridge Harvard College Observatory under the supervision of the astronomer Edward Pickering, discovered many of these variable stars and found a clear relation between their intrinsic luminosities and their periods. Henrietta Leavitt understood the potentiality of this relation to determine stellar distances, but she was not allowed by her boss to go on in her way. Some decades later, the period-luminosity relation of Cepheids and the luminosities of RR Lyrae stars were used by Harlow Shapley, as

Sylvie Vauclair (✉) and Mélanie Soriano
Laboratoire d'Astrophysique de Toulouse-Tarbes,
Université de Toulouse - CNRS 14, Avenue Edouard Belin 31400 Toulouse, France
e-mail: sylvie.vauclair@ast.obs-mip.fr

J.-P. Rozelot and C. Neiner (eds.), *The Pulsations of the Sun and the Stars*,
Lecture Notes in Physics 832, DOI: 10.1007/978-3-642-19928-8_6,
© Springer-Verlag Berlin Heidelberg 2011

well as many other astronomers, to determine stellar distances. It is still one of the most important distance estimators in the Universe.

Nowadays, variable stars are known all other the HR diagram. They can be classified according to:

- the type of waves which leads to their oscillations, either pressure of gravity waves or both
- their amplitudes
- their excitation mechanisms

Before the first discovery of solar oscillations, solar type stars were not supposed to be variable, as the acoustic waves are damped out. We now know that stochastic excitation induced by convective motions leads to permanent destabilisation so that these stars behave like resonant cavities in spite of the waves damping.

The first report of a periodic solar velocity field was given by Leighton et al. [1]. They wrote that: "The vertical velocities exhibit a striking repetitive time correlation, with a period $T = 296 \pm 3$ s. This quasi-sinusoidal motion has been followed for three full periods in the line Ca λ 6103, and is also clearly present in Fe λ 6102, Na λ 5896, and other lines. The energy contained in this oscillatory motion is about $160 \, \mathrm{J \, cm^{-2}}$; the "losses" can apparently be compensated for by the energy transport ... A similar repetitive correlation, with nearly the same period, seems to be present in the brightness fluctuations observed on ordinary spectroheliograms ... "

Evidences of the five minute oscillations were later confirmed by Ulrich [2] and Leibacher and Stein [3]. Some ten millions p-modes are observed, with frequencies around 2–4 mHz, velocity amplitudes about $1 \, \mathrm{cm \, s^{-1}}$ (max $20 \, \mathrm{cm \, s^{-1}}$), relative variations of brilliance 10^{-7}, mode lifetimes of a few hours up to a few months.

We will not present here detailed studies of helioseismology, which are out of the scope of this introductory paper and will be discussed elsewhere, neither will we give a complete review of the asteroseismology of solar type stars. In the following, we first recall the basics of stellar oscillations (a more complete discussion may be found in "Lectures notes on stellar oscillations", by Christensen–Dalsgaard). Then we will discuss the so-called "asymptotic theory" and important deviations from it. Finally we will give some examples of the asteroseismology of recently observed main sequence stars.

6.2 Basics of Stellar Oscillations

The oscillations are considered as the propagation of a perturbation in a fluid. The gas is treated as a continuum, its properties are function of the displacement $\vec{r}$ and time t. The basic equations of hydrodynamics are used and linear perturbations are performed. Some manipulations of equations and some approximations finally lead to the wave equation.

6.2.1 Some Hydrodynamics

- Equation of continuity

$$\frac{\partial \rho}{\partial t} + \mathrm{div}(\rho \vec{v}) = 0 \tag{6.1}$$

- Equation of motions

$$\rho \left(\frac{\partial \vec{v}}{\partial t} \right) + (\vec{v} \, \vec{\nabla}.\vec{v}) = -\vec{\nabla} P + \rho \vec{g} \tag{6.2}$$

where the viscosity is neglected.

The vector $\vec{g}$ is the gravitational acceleration that can be written as:

$$\vec{g} = -\vec{\nabla} \Phi, \tag{6.3}$$

where Φ is the gravitational potential that satisfies the Poisson's equation:

$$\nabla^2 \Phi = 4\pi G \rho \tag{6.4}$$

- Energy equation

$$\frac{dq}{dt} = \frac{dE}{dt} + P \frac{dV}{dt}, \tag{6.5}$$

where E is the internal energy per unit of volume. This equation can be rewritten with thermodynamic variables:

$$\frac{dq}{dt} = C_V \left[\frac{dT}{dt} - (\Gamma_3 - 1) \frac{T}{\rho} \frac{d\rho}{dt} \right] \tag{6.6}$$

$$= C_P \left[\frac{dT}{dt} - \frac{\Gamma_2 - 1}{\Gamma_2} \frac{T}{P} \frac{dP}{dt} \right] \tag{6.7}$$

$$= \frac{1}{\rho(\Gamma_3 - 1)} \left[\frac{dP}{dt} - \frac{\Gamma_1 P}{\rho} \frac{d\rho}{dt} \right] \tag{6.8}$$

where C_V and C_P are the specific heat capacities, respectively at constant volume and under constant pressure. The adiabatic indices are defined by:

$$\Gamma_1 = \left(\frac{\partial \ln P}{\partial \ln \rho} \right)_{ad}, \quad \frac{\Gamma_2 - 1}{\Gamma_2} = \left(\frac{\partial \ln T}{\partial \ln P} \right)_{ad}, \quad \Gamma_3 - 1 = \left(\frac{\partial \ln T}{\partial \ln \rho} \right)_{ad} \tag{6.9}$$

In the adiabatic case, (6.5) becomes:

$$\frac{dP}{dt} = \frac{\Gamma_1 P}{\rho} \frac{d\rho}{dt} \tag{6.10}$$

6.2.2 Linear Perturbations

Stellar oscillations have very low amplitudes compared to the characteristic scales of a star. They can be treated as small perturbations around an equilibrium state. For a variable x we have, with an eulerian description:

$$x(\vec{r}, t) = x_0(\vec{r}) + x'(\vec{r}, t) \tag{6.11}$$

and with a lagrangian description:

$$\delta x(\vec{r}) = x(\vec{r_0} + \delta\vec{r}) - x_0(\vec{r_0}) \tag{6.12}$$

$$= x'(\vec{r_0}) + \delta\vec{r} \cdot \vec{\nabla} x_0 \tag{6.13}$$

These perturbations are introduced in (6.1), (6.2), (6.4), and (6.10). The equation of continuity (6.1) becomes:

$$\delta\rho + \rho_0 \vec{\nabla} \cdot \delta\vec{r} = 0 \tag{6.14}$$

The equation of motions becomes:

$$\rho_0 \frac{\partial^2 \delta\vec{r}}{\partial t^2} = -\vec{\nabla} P' + \rho_0 \vec{g'} + \rho' \vec{g_0} \tag{6.15}$$

where $\vec{g'} = -\vec{\nabla}\Phi'$ and:

$$\nabla^2 \Phi' = 4\pi G\rho'. \tag{6.16}$$

Finally, the equation of energy becomes:

$$\frac{\partial \delta q}{\partial t} = \frac{1}{\rho_0(\Gamma_3 - 1)} \left(\frac{\partial \delta P}{\partial t} - \frac{\Gamma_{10} P_0}{\rho_0} \frac{\partial \delta\rho}{\partial t} \right) \tag{6.17}$$

If we consider adiabatic motions, the heat term is neglected, and this equation turns to:

$$\frac{\partial \delta P}{\partial t} - \frac{\Gamma_{1,0} P_0}{\rho_0} \frac{\partial \delta\rho}{\partial t} = 0 \tag{6.18}$$

6.2.3 Separation of Variables

The displacement $\vec{\delta r}$ is separated in a radial and an horizontal component:

$$\vec{\delta r} = \xi_r \vec{a_r} + \vec{\xi_h}. \tag{6.19}$$

We first consider the perturbed equation of motions (6.17). If we compute the horizontal divergence of its horizontal component, we find a first equation:

$$\rho_0 \frac{\partial^2}{\partial t^2} \left(\vec{\nabla}_h \cdot \vec{\xi}_h \right) = -\nabla_h^2 P' - \rho_0 \nabla_h^2 \Phi' \tag{6.20}$$

The equation of continuity can be rewritten as:

$$\rho' = -\text{div}(\rho_0 \vec{\delta r}) \tag{6.21}$$

$$= -\frac{1}{r^2} \frac{\partial}{\partial r} (\rho_0 r^2 \xi_r) - \rho_0 \vec{\nabla}_h \cdot \vec{\xi}_h, \tag{6.22}$$

We introduce this new expression in (6.20) and we obtain:

$$-\frac{\partial^2}{\partial t^2} \left[\rho' + \frac{1}{r^2} \frac{\partial}{\partial r} (r^2 \rho_0 \xi_r) \right] = -\nabla_h^2 P' - \rho_0 \nabla_h^2 \Phi' \tag{6.23}$$

The radial component of the equation of motions is:

$$\rho_0 \frac{\partial^2 \xi_r}{\partial t^2} = -\frac{\partial P'}{\partial r} - \rho' g_0 - \rho_0 \frac{\partial \Phi'}{\partial r} \tag{6.24}$$

The Poisson's equation becomes:

$$\frac{1}{r^2} \frac{\partial}{\partial r} \left(r^2 \frac{\partial \Phi'}{\partial r} \right) + \nabla_h^2 \Phi' = 4\pi G \rho' \tag{6.25}$$

The solutions are expressed with:

$$\xi_r(r, \theta, \phi, t) = \sqrt{(4\pi)} \tilde{\xi}_r(r) Y_l^m(\theta, \phi) \exp(-i\omega t) \tag{6.26}$$

$$P'(r, \theta, \phi, t) = \sqrt{(4\pi)} \tilde{P}'(r) Y_l^m(\theta, \phi) \exp(-i\omega t) \tag{6.27}$$

$$\cdots$$

$\tilde{\xi}_r, \tilde{P}', \ldots$, are the amplitudes of the solutions. The function Y_l^m is a spherical harmonics defined by:

$$Y_l^m(\theta, \phi) = (-1)^m \sqrt{\frac{(2l+1)(l-m)!}{4\pi(l+m)!}} P_l^m(\cos\theta) \exp(im\phi), \tag{6.28}$$

where P_l^m are Legendre polynomials.

Equations (6.23), (6.24), and (6.25) become:

$$\omega^2 \left[\tilde{\rho}' + \left(\frac{1}{r^2} \frac{\partial}{\partial r} (r^2 \rho \tilde{\xi}_r) \right) \right] = \frac{\ell(\ell+1)}{r^2} (\tilde{P}' + \rho_0 \tilde{\Phi}') \tag{6.29}$$

$$-\omega^2 \rho_0 \tilde{\xi}_r = -\frac{d\tilde{P}'}{dr} - \tilde{\rho}' g_0 - \rho_0 \frac{d\tilde{\Phi}'}{dr} \tag{6.30}$$

$$\frac{1}{r^2} \frac{\partial}{\partial r} \left(r^2 \frac{d\tilde{\Phi}'}{dr} \right) - \frac{\ell(\ell+1)}{r^2} \tilde{\Phi}' = 4\pi G \tilde{\rho}' \tag{6.31}$$

6.2.4 Non-radial Adiabatic Oscillations

To solve the fourth order system more easily, the perturbations of the gravitational potential are neglected (Cowling's approximation). The system is then reduced to:

$$\frac{d\xi_r}{dr} = - \left(\frac{2}{r} + \frac{1}{\Gamma_1} H_P^{-1} \right) \xi_r + \frac{1}{\rho c^2} \left(\frac{S_\ell^2}{\omega^2} - 1 \right) P' \tag{6.32}$$

$$\frac{dP'}{dr} = \rho \left(\omega^2 - N^2 \right) \xi_r - \frac{1}{\Gamma_1} H_P^{-1} P', \tag{6.33}$$

where $H_P^{-1} = -d \ln P / d \ln r$ is the pressure height scale.

In this expression, two characteristic frequencies have been introduced:

• The lamb frequency S_ℓ^2 defined as:

$$S_\ell^2 = \frac{\ell(\ell+1)c^2}{r^2} = k_h^2 c^2, \tag{6.34}$$

where $c^2 = \Gamma_1 P / \rho$ is the squared value of the acoustic sound speed.

• The Brunt–Väisälä frequency N^2 defined as:

$$N^2 = g \left(\frac{1}{\Gamma_1 P} \frac{dP}{dr} - \frac{1}{\rho} \frac{d\rho}{dr} \right) \tag{6.35}$$

For high radial order oscillations, the derivatives of the equilibrium quantities can be neglected. The system of (6.34) and (6.35) becomes:

$$\frac{d\xi_r}{dr} = \left(\frac{S_\ell^2}{\omega^2} - 1 \right) \frac{P'}{\rho c^2} \tag{6.36}$$

$$\frac{dP'}{dr} = \rho \left(\omega^2 - N^2 \right) \xi_r \tag{6.37}$$

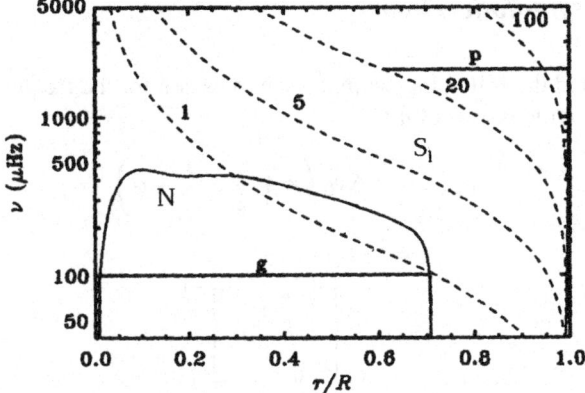

Fig. 6.1 Brunt–Väisälä (*solid line*) and lamb frequencies for various values of ℓ (*dashed lines*). The *horizontal lines* represent the propagation zones for a $100\,\mu$Hz g-mode, and for a $1000\,\mu$Hz p-mode with $\ell = 20$

which can be combined in:

$$\frac{d^2\xi_r}{dr^2} = \frac{\omega^2}{c^2}\left(1 - \frac{N^2}{\omega^2}\right)\left(\frac{S_\ell^2}{\omega^2} - 1\right)\xi_r \qquad (6.38)$$

This equation describes the global properties of the oscillation modes and can provide a good determination of their frequencies. The two characteristic frequencies play an important role in the behaviour of the oscillations. The evolution of these frequencies with the radius of the star is shown in Fig. 6.1.

The solution is oscillatory when

$$|\omega| > |N| \text{ and } |\omega| > |S_\ell|$$

or

$$|\omega| < |N| \text{ and } |\omega| < |S_\ell|$$

This induces two classes of modes:

- The high-frequency modes, or p-modes, characterized by: $|\omega| > |N|$ and $|\omega| > S_\ell$
- The low-frequency modes, or g-modes, characterized by: $|\omega| < |N|$ and $|\omega| < S_\ell$

6.3 The Asymptotic Theory and Deviations from it

It is possible to compute a more precise second order differential equation in ξ_r, which leads to asymptotic expressions of the frequencies and the eigenfunctions. This widely used asymptotic theory is discussed below.

6.3.1 The Asymptotic Theory

Tassoul [4] found the following asymptotic expression for the frequency, at the first order, valid for large values of n :

$$\nu_{n,\ell} \simeq 2\pi \Delta \nu_0 \left(n + \frac{\ell}{2} + \frac{1}{4} + \alpha \right) \tag{6.39}$$

with

$$\Delta \nu_0 = \left[2 \int_0^R \frac{1}{r} dr \right]^{-1}, \tag{6.40}$$

which is the inverse of twice the time needed by the sound waves to travel from the stellar surface down to the center. This quantity is named the mean large separation, α is a parameter which depends on the atmospheric structure.

This asymptotic theory gives us a first approximation of the oscillation frequencies. In a real star, there are deviations from this theory and these deviations can provide information about specific regions of the star.

Large separations and echelle diagram. From the (6.39), we find that two p-modes of successive radial order n, with the same degree ℓ are approximately separated by $\Delta \nu_0$. We define the large separation as:

$$\Delta \nu_{n,\ell} = \nu_{n+1,\ell} - \nu_{n,\ell} \tag{6.41}$$

In the framework of the asymptotic theory, we should have $\Delta \nu_{n,\ell}$ equal to $\Delta \nu_0$. But deviations from the asymptotic theory induce deviations of $\Delta \nu_{n,\ell}$ from $\Delta \nu_0$.

A useful representation of the oscillation frequencies is the echelle diagram. In ordinates are plotted the frequencies and in abscissae, the same frequencies modulo the mean large separation. According to the expression found for the frequency in (6.39), we should obtain vertical lines corresponding to each value of the degree ℓ. In a real star, there are deviations from the asymptotic theory which are characteristic of the inner stellar structure.

Small separations. They are defined as:

$$\delta \nu_{n,\ell} = \nu_{n,\ell} - \nu_{n-1,\ell+2} \tag{6.42}$$

In the framework of the first order of the asymptotic theory, we have $\delta \nu_{n,\ell} \simeq 0$. But in a real star, there are deviations from this theory and the small separations are not equal to zero. At second order, Tassoul [4] gives:

$$\delta \nu_{n,\ell} \simeq - (4\ell + 6) \frac{\Delta \nu_0}{4\pi^2 \nu_{n,\ell}} \int_0^R \frac{1}{r} \frac{dc}{dr} dr \tag{6.43}$$

These quantities are very sensitive to the deep stellar interior (Gough, Roxburgh and Vorontsov [5, 6]) and can give us interesting information about the stellar core.

6.3.2 Deviations from the Asymptotic Theory

Soriano et al. [7] derived a different asymptotic expression by doing similar computations as Tassoul [4], but taking into account the fact that different modes do not always travel in the same stellar regions: the modes $\ell = 0$ travel from the center of the star ($r = 0$) to the surface, but the modes $\ell \neq 0$ are trapped between the surface and their internal turning point r_t, whose position depends on the degree of the mode.

$$\nu_{n,\ell} \simeq \left(n + \frac{\ell}{2} + \frac{1}{4} + \alpha\right) \Delta\nu - \frac{\ell(\ell+1)\Delta\nu}{4\pi^2\nu_{n,\ell}} \left[\frac{c(R)}{R} - \int_{r_t}^{R} \frac{1}{r}\frac{dc}{dr}dr\right] - \delta\frac{\Delta\nu^2}{\nu_{n,\ell}},$$

(6.44)

where α is a surface phase shift, δ a function of the parameters of the equilibrium model and $\Delta\nu$ is defined as follows:

$$\Delta\nu = \frac{1}{2\int_{r_t}^{R}\frac{dr}{c}}$$

(6.45)

In this framework, several frequency combinations may be used to probe the stellar structure. Soriano and Vauclair [8] gave a second order asymptotic expression for the small separations, which becomes, for the degrees $\ell = 0$–$\ell = 2$ and $\ell = 1$–$\ell = 3$:

$$\delta\nu_{02} \simeq \left(n + \frac{1}{4} + \alpha\right)(\Delta\nu_0 - \Delta\nu_2) + I(r_t)\left[\frac{6\Delta\nu_2}{4\pi^2\nu_{n-1,2}}\right]$$

(6.46)

$$\delta\nu_{13} \simeq \left(n + \frac{3}{4} + \alpha\right)(\Delta\nu_1 - \Delta\nu_3) + I(r_t)\left[\frac{\Delta\nu_1}{2\pi^2\nu_{n,1}} - \frac{6\Delta\nu_3}{2\pi^2\nu_{n-1,3}}\right]$$

(6.47)

where

$$I(r_t) = \int_{r_t}^{R} \frac{1}{r}\frac{dc}{dr}dr,$$

(6.48)

Contrary to what was assumed in Tassoul [4] work, these quantities can become negative [7, 8]. This specific behaviour is related to the presence of a convective core or to a helium core with sharp edges. We can use this phenomenon to characterize helium-rich cores and to give strong constraints on the possible overshooting Fig. 6.2.

6.3.3 Second Differences

In regions with an important gradient of the sound velocity, like the boundary of the convective zone or the HeII ionization zone, there are partial reflexions of the waves

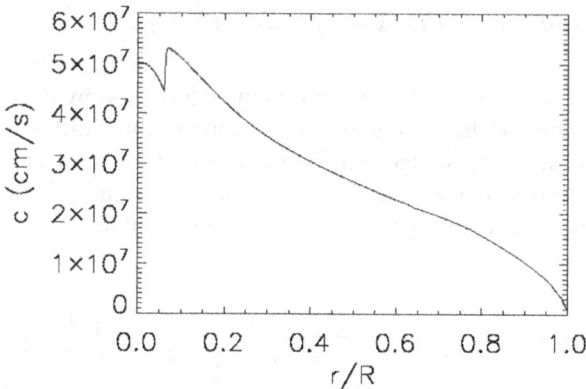

Fig. 6.2 Sound velocity profile in a 1.1 $M_\odot$ star at the end of the main sequence: the signature of the helium core clearly appears as a depression in the sound velocity . The $\ell = 0$ waves travel down to the center while the $\ell = 2$ waves have an internal turning point which depends on the mode frequency. The modes for which the turning point is at the edge of the core present a special behavior and may be the limit above which the small separation become negative

that create modulations on the frequencies. These modulations clearly appear in the so-called second differences (e.g. [9, 10]) which are defined by:

$$\delta_2 v = v_{n+1,\ell} + v_{n-1,\ell} - 2v_{n,\ell} \tag{6.49}$$

The modulation period of the oscillations is equal to twice the acoustic depth, which is the time needed for the sound waves to travel from the considered region to the stellar surface:

$$t_s = \int_{r_s}^{R} \frac{dr}{c(r)} \tag{6.50}$$

$c(R)$ is the sound velocity at the radius r, and r_s the radius of the considered region.

By computing the Fourier transform of the second differences, we can check that the period of the peaks are twice the acoustic depth of the discontinuities in the sound speed. As discussed in Castro and Vauclair [11], the behavior of the second differences is very different if one takes helium diffusion into account or not. When helium diffusion is not computed, two peaks appear in the Fourier transform: a first one, close to the surface, is due to the helium ionization zone while a second peak due to the base of the convective zone can be identified. On the other hand, when helium diffusion is introduced, the peak due to the helium ionization zone disappears whereas a strong peak due to the diffusion-induced helium gradient develops. This is the peak visualized in Fig. 6.3.

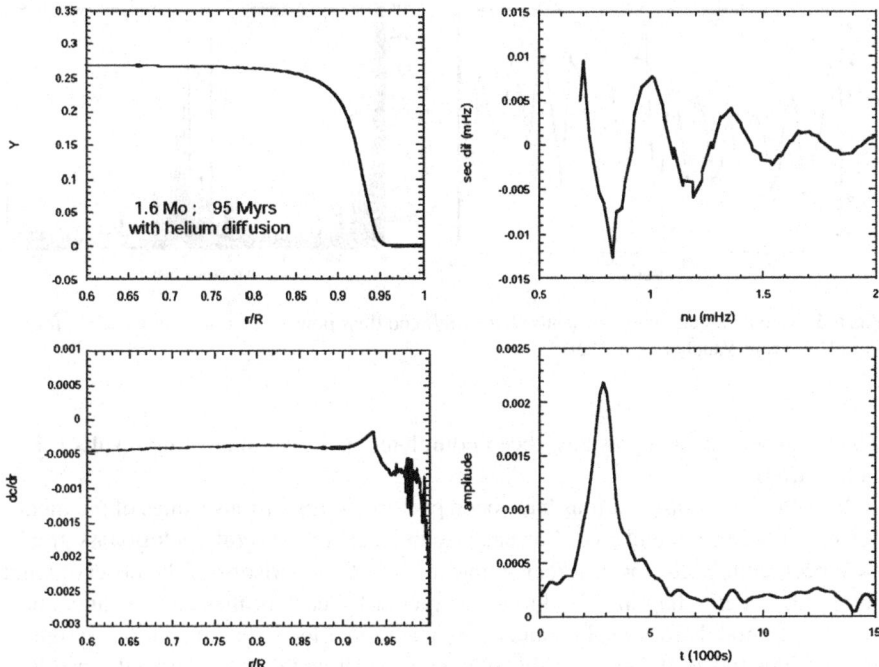

Fig. 6.3 Helium profile (*upper-left panel*), first derivative of the sound speed (*lower-left panel*), second differences (*upper-right panel*), and Fourier transform of the previous graph (*lower-left panel*) for a model of 1.6 M$_\odot$, 95 Myr, with helium diffusion (from Vauclair and Théado [10])

6.4 Two Examples of Seismic Analyses for Main Sequence Stars

As for exoplanet searches, the methods for stellar oscillations detection and observations are different from the ground and from space. On the ground, the radial velocity method is used, with spectrographs like HARPS (La Silla Observatory, Chile) or SOPHIE (Haute-Provence Observatory, France). In space, we rely on photometric methods. In the first case, we obtain radial velocity curves, in the second case light curves. Then we perform Fourier transforms, evaluate the mean large separation, draw the echelle diagram and compare the observed modes to the results of the model computations.

We recently focused on the seismology of exoplanet-host stars. All the exoplanet-host stars that we have observed, using the radial velocity method, proved to have oscillations. Four of these stars have been observed for a long enough time to be analyzed in detail (more than 1 week), i.e. μ Arae (HARPS, 2004), ι Hor (HARPS 2006), 51 Peg (SOPHIE 2007) and 94 Ceti (HARPS 2007). The first results obtained for the CoRoT main target HD52265, using the photometric method, also show evidences of oscillations. Here we discuss the cases of two exoplanet-host stars for

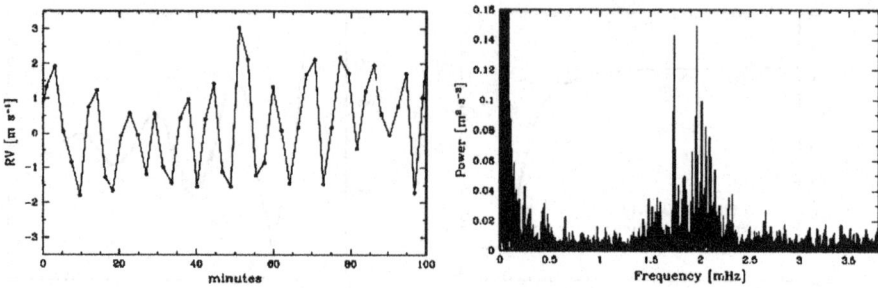

Fig. 6.4 Radial velocity measurements (*left panel*) and their power spectrum (*right panel*), for the star *ι* Hor (from Vauclair et al. [12])

which the seismic analysis have been completed and give interesting results (*ι* Hor and *μ* Arae).

In both cases, computations have been performed for various values of the metallicity and helium abundance. For each abundance set, several evolutionary tracks have been computed, for different stellar masses. Comparisons of the observed and computed echelle diagrams lead to a best model for each of these tracks, and a best of the best models for the abundance set. Then the effective temperatures, gravities and luminosities of the best models obtained for all metallicities and helium values are compared with the spectroscopic error boxes. With such a method, we were able to evaluate the helium content of these two overmetallic stars. We found evidence of a low helium abundance (lower than solar) for *ι* Hor while the helium abundance obtained for *μ* Arae is high, as expected from chemical evolution of galaxies.

6.4.1 The Star *ι* Hor

The case of the exoplanet-host star *ι* Hor was discussed by Vauclair et al. [12]. This star belongs to the Hyades stream: it has the same kinematical characteristics than the Hyades cluster in the Galaxy. *ι* Hor was observed in November 2006 with the HARPS spectrometer, in La Silla Observatory (Fig. 6.4). The analysis of the radial velocity time series led to the identification of 25 oscillation modes.

A seismic analysis of this star was carried out as described above, by computing models for various values of the metallicity and helium abundance, and comparing with the observations. The results are given in Fig. 6.5 and Table 6.1.

The values seismically obtained for *ι* Hor are also characteristic of the stars which belong to the Hyades cluster [13]. As *ι* Hor also behaves like the Hyades cluster in the Galaxy, although situated 40 pc away from it, we concluded that this exoplanet-host star was formed together with the cluster and evaporated. These results also lead to the conclusion that its overmetallicity was primordial and not due to accretion of planetesimals.

Fig. 6.5 Log g–Log T_{eff} diagram, for the star ι Hor. The spectroscopic error boxes are presented, as well as six different "best models" for various metallicity and helium content (from Vauclair et al. [12]). The models satisfying the spectroscopic observed parameters have a low helium abundance

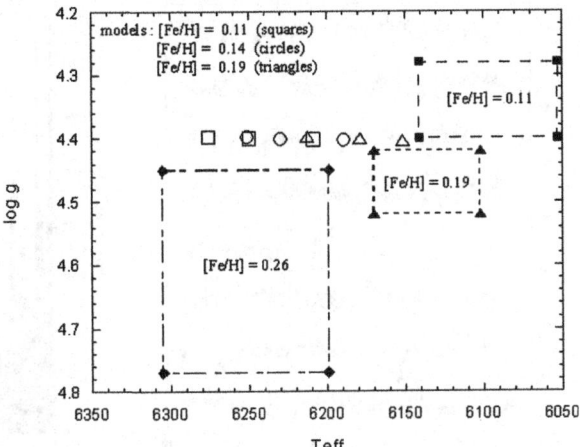

Table 6.1 iota Horologii: parameters of the three best models (cols. 2−4), compared to the Hyades values (col. 5, Ref.: Lebreton et al. [12]; from Vauclair et al. [13])

[Fe/H]	0.19	0.19	0.14	0.14 ± 0.05
Y	0.271	0.255	0.255	0.255 ± 0.013
Age (Myr)	620	627	627	625 ± 25
Mass ($M_\odot$)	1.24	1.26	1.25	
T_{eff} (K)	6179	6136	6189	
Log g	4.40	4.40	4.40	
Log $L/L_\odot$	0.245	0.237	0.250	

6.4.2 The Star μ Arae

The exoplanet-host star μ Arae (HD 160691) is a G3 IV–V type star, with a visual magnitude $V = 5.15$ (Simbad astronomical Database). A first value of the parallax was initially derived by Perryman et al. [14]: $\pi = 65.5\pm0.8$ mas. This value was used by Bazot et al. [15] to determine the parameters of this star. Recently, a new analysis of the Hipparcos data was achieved by Van Leeuwen [16], who found a new value for the parallax: $\pi = 64.48 \pm 0.31$ mas. This new value was used by Soriano and Vauclair [17] (herafter SV10) to derive an absolute magnitude of $M_V = 4.20\pm0.04$ and a luminosity of log(L/L$_\odot$) = 0.25 ± 0.03.

Five groups of observers (references in SV09) have derived different values of external parameters (T_{eff}, log g and metallicity).

This star was observed in 2004 with the HARPS spectrograph to obtain radial velocity time series (Fig. 6.4). These seismic observations led to the identification of 43 p-modes for degrees $\ell = 0$ to $\ell = 3$ [18].

SV09 performed a new seismic analysis of this star, using the same method as for ι Hor (Fig. 6.6). The results are given in Fig. 6.7 and Table 6.2.

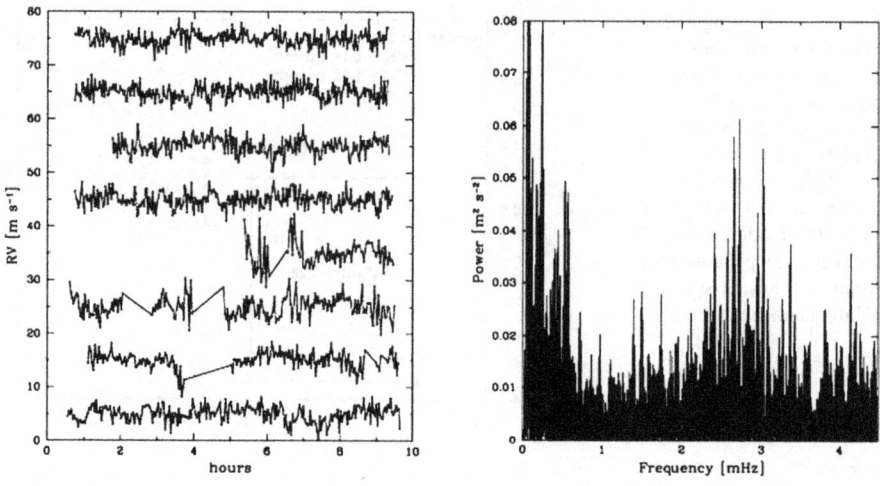

Fig. 6.6 Radial velocity measurements (*left panel*) and their power spectrum (*right panel*), for the star μ Arae (from Bouchy et al. [18])

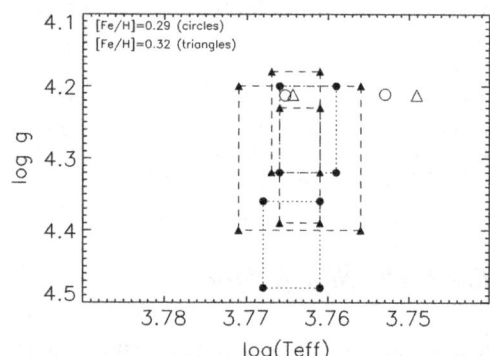

Fig. 6.7 Log g–Log T_{eff} diagram, for the star μ Arae. The spectroscopic error boxes are presented, as well as different "best models" obtained for various metallicity and helium content (from Soriano and Vauclair [17]). Contrary to the case of ι Hor, the models satisfying the spectroscopic observed parameters have a high helium abundance

Table 6.2 Parameters of μ *Arae* (from Soriano and Vauclair [17])

M/M$_\odot$	1.10 ± 0.01	T_{eff} (K)	5820 ± 40
R/R$_\odot$	1.36 ± 0.01	[Fe/H]	0.30 ± 0.01
Log g	4.215 ± 0.005	Y	0.301 ± 0.01
L/L$_\odot$	1.90 ± 0.10	Age (Gyr)	6.34 ± 0.40

Contrary to the case of ι Hor, the results for μ Arae give evidence of a high helium abundance in this star. SV10 also computed models with overshooting, treated as an extension of the convective core. They found that the overshooting extent cannot be more than 5% of the pressure height scale.

6.5 Conclusions

Asteroseismology is a blooming science, which brings a new dimension to the study of stellar structure and evolution. In this introductory paper, we focused on solar-type stars. We showed that precise parameters can be obtained, as well as constraints on the stellar interiors. In particular, crossed comparisons between seismic and spectroscopic observations can lead to the helium abundance, in stars for which it cannot be determined directly from spectroscopy alone. We have also access on precise determinations of the masses, gravities and therefore radii of these stars.

For the two stars that we have presented here, two very different results have been obtained for helium. One of these stars (ι Hor) has a low helium abundance (Y of order 0.255), in spite of its high metallicity (about a factor 2). This is difficult to explain in terms of the chemical evolution of the Galaxy, but it is also the case for all the stars of the Hyades cluster. We found that this star must have been formed with the cluster and evaporated. The second star, μ Arae, has a high helium abundance (Y of order 0.3), consistent with its surmetallicity in the framework of the chemical evolution law. These examples show how important maybe the consequences of seismic results for stars.

Precise determination of stellar parameters needs long time observations of these stars. More than one week is needed when using the radial velocity method (as with HARPS or SOPHIE). More than one month is better from space, using the photometric method. Statistical analysis methods have been proposed to rapidly derive parameters for a large number of stars [19]. These methods rely on general variations of frequency combinations with mass and age of the stars. They are not precise however, as real stars can deviate from these average relations. For the future, multisite campaigns would be helpful, like those proposed with the SONG network.

References

1. Leighton, R.B., Noyes, R.W., Simon, G.W.: ApJ **135**, 474 (1962)
2. Ulrich, R.K.: ApJ **162**, 993 (1970)
3. Leibacher, J.W., Stein, R.F.: ApL **7**, 191 (1971)
4. Tassoul, M.: ApJS **43**, 469 (1980)
5. Gough, D.O. In: Osaki, Y. (ed.) Hydrodynamic and Magnetohydrodynamic Problems in the Sun and Stars. University of Tokyo Press, Tokyo (1986)
6. Roxburgh, I.W., Vorontsov, S.V.: MNRAS **267**, 297 (1994)
7. Soriano, M., Vauclair, S., Vauclair, G., Laymand, M.: A&A **471**, 885 (2007)
8. Soriano, M., Vauclair, S.: A&A **488**, 975 (2008)
9. Gough, D.: Progress of seismology of the sun and stars. In: Osaki, H. Shibahashi (ed.) Oji International Seminar Hakone Lect Notes Phys, vol. 367, pp. 283. Springer, Japan (2009)
10. Vauclair, S., Théado, S.: A&A **425**, 179 (2004)
11. Castro, M., Vauclair, S.: A&A **456**, 611 (2006)
12. Vauclair, S., Laymand, M., Bouchy, F., et al.: A&A **482**, L5 (2008)
13. Lebreton, Y., Fernandes, J., Lejeune, T.: A&A **374**, 540 (2001)
14. Perryman, M.A.C., Lindegren, L., Kovalevsky, J., et al.: A&A **323**, L49 (1997)

15. Bazot, M., Vauclair, S., Bouchy, F., Santos, N.C: A&A **440**, 615 (2005)
16. van Leeuwen, F. (ed.): Astrophys. Space Sci. Libr., vol. 350, Hipparcos, the New reduction of the Raw Data (2007)
17. Soriano, M., Vauclair, S.: A & A **513**, 49 (2010)
18. Bouchy, F., Bazot, M., Santos, N.C., Vauclair, S., Sosnowska, D.: A&A **440**, 609 (2005)
19. Stello, D., Chaplin, W.J., Bruntt, H., et al.: ApJ **700**, 1589 (2009)

Part IV
Section 2: Stellar Pulsations

R. G. Deupree (ICA-Saint Mary's University, Canada)
M. G. Goupil (GEPI-Paris Observatory, France)
F. Lignières (LATT-Toulouse University, France)
S. Mathis (CEA-Paris-Diderot University, France)
R. Samadi (LESIA-Paris Observatory, France)

Chapter 7
Issues Relating to Observables of Rapidly Rotating Stars

Robert G. Deupree

Abstract I discuss several ways in which rapid rotation can make the interpretation of what is observed complex and outline computational tools which are being brought to bear on the problem. The first requirement is to develop the capability for computing the fully two dimensional stellar structure. Once this is available the pulsation periods and the spectral energy distribution of the models must be calculated without resorting to commonly used approximate methods which are adequate if the rotation is a relatively small perturbation to the nonrotating model. A family of computational tools which perform these tasks is presented along with results on the structure of two dimensional rotating stellar models, the effects of rapid rotation on the low order p modes, and on the spectral energy distribution.

7.1 Introduction

In this paper I wish to explore the requirements for a self consistent interpretation of the collection of observables related to rapidly rotating stars. If a star is rotating sufficiently rapidly, nearly every feature we can observe will be affected by the rotation. Assuming the star is otherwise static, rapid rotation through the centrifugal force will affect the force balance and hence the structure of the star. Just these changes will affect the quantities which depend on the stellar structure, for example the luminosity produced and any oscillation frequencies which might be observed. Rotation also changes the surface from a spherical to a spheroidal, and possibly in some cases an ellipsoidal, shape. This change in shape is accompanied by a variation in the emergent flux, at least in radiative regions, with latitude, as first realized by von Zeipel [1]. The net result is that more flux flows out of the surface at higher latitudes (which thus have higher effective temperatures). This asymmetry means

Robert G. Deupree (✉)
Institute for Computational Astrophysics (ICA), Saint Mary's University, Halifax NS Canada
e-mail: bdeupree@ap.smu.ca

J.-P. Rozelot and C. Neiner (eds.), *The Pulsations of the Sun and the Stars*,
Lecture Notes in Physics 832, DOI: 10.1007/978-3-642-19928-8_7,
© Springer-Verlag Berlin Heidelberg 2011

Fig. 7.1 Plot of rotation rate
versus surface equatorial
velocity. The two are linearly
proportional at slow rotation,
but the large increase in the
surface equatorial radius
with small changes in the
rotation rate changes the
relation at high rotation.

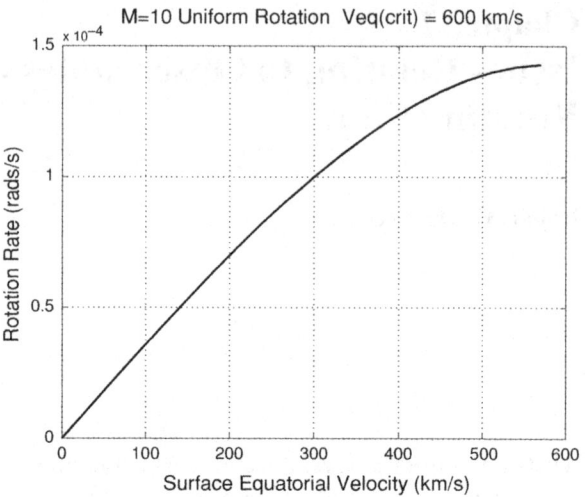

that everything we observe based on the light from a rapidly rotating star depends on
the inclination, i, between the rotation axis and the observer, something which must
itself be determined from the data. Included in these observables are the luminosity
and effective temperature we would deduce, and the spectral energy distributions,
including the line profiles. The one set of observables which are not affected by the
inclination is the collection of oscillation frequencies, although the deduction of what
mode a given frequency corresponds to might well be.

These various effects become important at different rotational velocities, leading
me to divide the effects of rotation into three categories. Throughout this work I will
use a collection of 10 and 12 $M_\odot$ ZAMS models at different rotation rates to quantify
these effects. Because most people think in terms of the surface equatorial velocity
(V_{eq}), I will use this parameter to quantify the amount of rotation. However, it is the
rotation rate (Ω) in terms of the critical rotation rate (Ω_c), that rate at which gravity
just balances the centrifugal force on the surface at the equator, which is the scaled
variable of importance. Where useful, I will give the appropriate surface equatorial
velocity with the ratio of the rotation rate to the critical rate in parenthesis.

The relationship between the surface equatorial velocity and the rotation rate is
shown in Fig. 7.1. At low rotation rates the surface equatorial velocity is nearly linearly proportional to the rotation rate, but near the critical rotation rate slight increases
in the rotation rate translate into sizeable increases in the surface equatorial velocity.
The reason is that the surface equatorial radius increases only slightly with increasing
rotation rate at low rotation but increases dramatically near critical rotation.

The three categories correspond to slow, moderate and rapid rotation.
For slow rotation, for which I will assign $V_{eq} \leq 100\,\text{km/s}$ ($\Omega \; \Omega_c \approx 0.25$ for my
$10\,M_\odot$ZAMS models), the effects of rotation are relatively marginal. The
surface asphericity is small, making the inclination effects small, and the effects of
rotation on the oscillation mode frequencies are small. Moderate rotation,

$100 \, \text{km/s} \leq V_{\text{eq}} \leq 300 \, \text{km/s}$ ($\Omega / \Omega_c \approx 0.69$), produces a noticeable inclination dependence and some of the oscillation frequencies require a more complex treatment than the usual methods for slow rotation. For example, the deduced $\log L$ varies by about 0.2 when seen pole on compared to when seen equator on for a surface equatorial rotational velocity of 300 km/s. This corresponds to a one solar mass uncertainty at ten solar masses if I assume the star being observed is not rotating. Rapid rotation is any rate higher than 300 km/s, for which rotation appreciable alters nearly all the observable properties of the star.

In order to model these properties, I need a variety of computational tools. I will first describe the modeling tools with which to deduce information about moderately and rapidly rotating stars and then discuss the effects of rotation on these properties.

7.2 Modelling Tools

What modeling tools are required to include the effects of rotation? First, I must be able to produce structural models of rotating stars. This can actually be done in a quasi one dimensional framework for conservative rotation laws (e.g. Monaghan and Roxburgh [2], Roxburgh et al. [3], Faulkner et al. [4], Kippenhahn and Thomas [5], Sackmann and Anand [6]) because the state variables are constant on equipotential surfaces. The solution is particularly easy if the distortion in the deep interior is sufficiently small that the gravitational potential can be treated as spherically symmetric. This is close to the truth for all uniformly rotating models, even those near critical rotation. The situation is somewhat more complicated for differential, but still conservative, rotation laws for which the gravitational potential must be computed in a two dimensional (2D) framework. This has been done for some time (e.g., Clement [7]; Jackson et al. [8]). These calculations tend to use the self consistent field approach [9] in which one iterates between solving Poisson's equation for the gravitational potential, given the current density distribution, and the stellar structure equations. However, such calculations do not really help solve the problems that make rotating stars different from nonrotating ones—the collection of hydrodynamic and secular (i.e., thermal time scale) instabilities (e.g., Endal and Sofia [10, 11]) which can redistribute the angular momentum and the composition within the star.

Two separate computational tools are required to deal with these instabilities. There is probably no complete substitute for a 3D hydrodynamics, and possibly magnetohydrodynamics, code to compute the hydrodynamic instabilities. Including the whole star in the calculation probably means that only an explicit hydrodynamics code is necessary, because there will probably be reasonably high speed (some reasonable fraction of the local sound speed) flows somewhere in the model. If one includes only a part of the star in the model, such as the convective core, one probably needs to perform an implicit calculation (to avoid the Courant condition) or at least impose the anelastic modal approximation to keep the time steps sufficiently large to compute the hydrodynamic effects in a reasonable amount of computer time. Even under these conditions it is unlikely that these calculations can be carried sufficiently far in time to determine the precise effects on the long term angular momentum and composition redistributions, so some extrapolation of the results will be required.

The second tool required is a 2.5D stellar evolution code capable of simulating the large scale flows associated with thermal instabilities. The extra half dimension means that there is an azimuthal momentum conservation equation, even though azimuthal symmetry is maintained. This is required to determine the effects of the instabilities on the angular momentum distribution. Because the motions associated with secular instabilities are expected to be well subsonic, this calculation needs to be implicit to avoid the time step constraints of the Courant condition. I have been working on such a code sporadically for a considerable period of time [12–14], and the major difficulty has been the treatment near the stellar surface. An early solution for meridional circulation currents found by Sweet [15] for uniform rotation revealed that the circulation velocity would become infinite at the surface. Baker and Kippenhahn [16] showed that the rate at which the velocity went to infinity was even worse for differential rotation. Considerable work by Smith [17, 18] showed that more realistic treatments of the atmosphere could make the velocity there finite, although it still remained large. Tassoul and Tassoul [19, 20] argued that high velocities near the surface would generate turbulence, and that the turbulent viscosity would limit the amplitude. It is not universally accepted that turbulent viscosity is the only or the complete answer [21, 22], and disagreements remain [23]. Large velocities near the surface limit the ability to perform time dependent calculations on an evolutionary time scale, and some methods have been developed which limit this problem, albeit with a reduction of realism in the treatment near the surface.

It is reasonable to ask why one does not perform a full 3D implicit stellar evolution calculation. The primary reason is numerical. The azimuthal motion in a 3D calculation will generate an accuracy constraint that the time step be less than the minimum over all the cells of the azimuthal zoning divided by the azimuthal velocity. This azimuthal velocity is just the rotational velocity, which will be large for sufficiently rapidly rotating models. One might think that this could be avoided by performing the calculation in a rotating coordinate system, but then the velocity in the constraint above would be just the difference between the local rotational velocity and the rotation velocity that corresponds to the rotational velocity of the coordinate system. Thus, once there is a noticeable departure from uniform rotation, the time step would become far too small to be able to perform even implicit calculations that cover secular and evolutionary time scales.

With the large scale flow version of my 2.5D evolution code still under development and testing, here I shall restrict my discussion to ZAMS models with an imposed rotation law and no motion other than the rotation itself.

Once a suitable model has been computed, several tools are required to compute information which can be compared with observational data. One such tool would compute oscillation frequencies for the model. Because my interest is in moderately and rapidly rotating stars, I want to avoid the usual assumptions of perturbation theory and a single spherical harmonic employed in these calculations for rotating stars. The tool I am using is a linear, adiabatic pulsation code developed by Clement [24] and updated to include differential rotation by Lovekin et al. [25]. This assumes that the latitudinal variation of the eigenfunctions can be written in terms of a finite number of spherical harmonics (for azimuthally symmetric rotating models the azimuthal quantum number, m, is still valid). The number of spherical harmonics included is user determined. More provide greater accuracy of the eigenfrequency but make it harder to determine a given mode's relation to modes in other models (for example, models of the same mass and evolutionary state but rotating slightly slower or faster).

The remaining observables come from the spectral energy distribution (SED). With rotation the SED observed is produced by a weighted integral of the emitted intensity in the direction of the observer from all the locations on the stellar surface visible to the observer. The effective temperatures and effective gravities at these locations differ, pronouncedly so if the rotation is sufficiently large. Any attempt to relate any component of the SED to the stellar properties must be interpreted through this distorting lens. In order to make this interpretation, two tools are required. The first is a model atmospheres code which provides the emergent intensity at each point on the stellar surface, and the second is a code to perform the integral to obtain the SED that would be observed as a function of inclination.

I have already made an implicit assumption that a 1D model atmosphere code (and plane parallel as well) will provide the emergent intensity. This will be a good approximation as long as the horizontal variation over a photon mean free path is small. In fact this holds quite well until the rotation rate is very close to critical and the surface of the model near the equator quite extended. Even here it may not be too bad an assumption because most of the light making up the SED will come from the much higher intensity areas at mid and high latitudes. The model atmosphere code chosen is the PHOENIX code [26], which is distinguished by the large number of energy levels allowed to be in non local thermodynamic equilibrium (NLTE). This certainly was a reason in the selection, as was the availability of a local user of PHOENIX at Saint Mary's. I have utilized PHOENIX to compute a grid of model atmospheres to produce the emergent intensity as a function of wavelength, angle from the local vertical, effective temperature, and effective gravity.

These PHOENIX results provide part of the integrand that goes into computing the observed SED. The remainder is merely geometry and the integral is given by

$$F_\lambda(i) = \int \int \frac{I_{\lambda_1}(\xi(\theta, \phi, i)) W(\lambda, \lambda_1)}{d^2} dA_{\text{proj}}(\theta, \phi) \qquad (7.1)$$

where F_λ is the flux observed at a given wavelength and I_λ is the emergent intensity emitted from the stellar surface at a location identified by the spherical polar coordinates (θ, ϕ) and emitted in the direction of the observer. This direction is given by the angle ξ, which is the angle between the surface normal and the direction to the observer and thus depends on the local surface shape, the specific location on the surface, and the direction to the observer. Of course, the integrals over θ and ϕ include only those combinations for which the local surface is visible to the observer. The Doppler shift is indicated by the function W, and dA_{proj} is the projected surface area. The distance between the star and the observer is d. Experience has shown that dividing the surface into about 80000 zones provides adequate resolution. The SED is just this flux distribution over all the wavelengths.

To obtain a SED in its most usable form, we require many wavelengths. The spectrum is generally computed over the wavelength region 300–10000 Å for our intermediate mass ZAMS models with a resolution of 0.02 Å. These are fortunately all independent if we are interested in broad band results so that we can ignore the Doppler shift, but including the Doppler shift introduces only a modest dependence among neighboring wavelengths. This modest dependence can be treated in such a way (see Lovekin and Deupree [27]) to keep the integration "embarrassingly parallel" so calculating the integral requires only a modest amount of computer time with a relatively small number of processors on a relatively slow interconnect computer cluster. This is a small penalty to pay for removing any reliance on poorly known limb and gravity darkening laws.

We may integrate the SED over wavelength to deduce a luminosity and use the shape of the spectrum to deduce an effective temperature. As has been long known (e.g., Collins [28], Collins and Harrington [29], Hardorp and Strittmatter [30], Maeder and Peytremann [31]) these deduced quantities will vary with inclination. In particular, the luminosity deduced may have only a modest relation to the luminosity actually emerging from the star.

I now turn to using some of these tools to determine properties related to rotating stellar models.

7.3 Results from Rotating Stellar Models

7.3.1 Structural Results

I shall consider only $10\,M_\odot$ ZAMS models with $Z=0.02$. The equation of state and opacity tables are taken from the OPAL tables [32, 33]) and the pp and CNO reaction rates are given by the composite rates of Fowler et al. [34]. The reason for using these composite rates is that the time dependent composition equations for those species important for energy production must be solved simultaneously with the other conservation laws for numerical stability and therefore the number of composition equations must be kept to a minimum. With these composite rates, I need to include only a hydrogen conservation equation for main sequence evolution. The ZAMS models are obtained from the imposed rotation law by solving the time independent conservation laws with the radial and latitudinal velocities set to zero. I shall include two rotation laws: uniform rotation and a particular conservative law in which the rotation rate decreases with increasing distance from the rotation axis. The differential rotation law is a generalization of one taken from Jackson et al. [8]:

$$\Omega(\varpi) = \frac{\Omega_0}{1 + (a\varpi)^\beta} \tag{7.2}$$

Fig. 7.2 Variation in the rotation rate as a function of distance from the rotation axis for the rotation law given in (7.2) with a = 2. The distance from the rotation axis is given in units of the surface equatorial radius. The rotation rate is scaled in units of the rotation rate for uniform rotation.

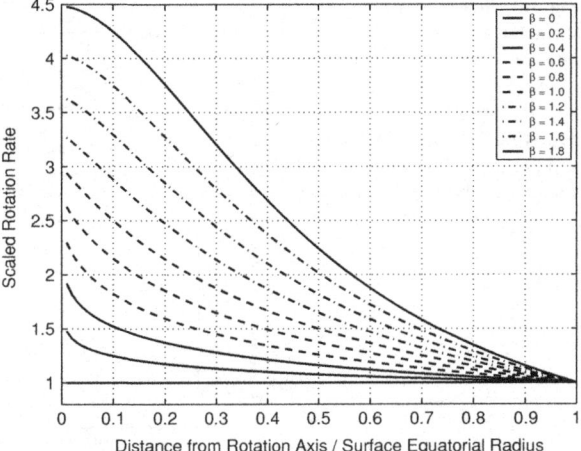

and is plotted in Fig. 7.2. This rotation law was studied because Jackson et al. [8] found that they could obtain the observed interferometric shape for Achernar deduced by Domiciano de Souza et al. [35] with models using it. While further research has indicated that the observed shape may be due to a circumstellar envelope (e.g., Vinicius et al. [36], Kanaan et al. [37], Kervella et al. [38]), it does raise the question of how one might be able to verify such a rotation law from other information.

From the resulting models I can obtain structural information, synthetic SEDs, and selected oscillation frequencies. The intent here is to examine how these quantities vary as the rotation law changes.

I show the surface shape for various surface equatorial rotation velocities for uniform rotation in Fig. 7.3. Note that the distortion of the surface is quite small for rotational velocities below approximately 250 km/s, changes only modestly until critical rotation is approached, at which point the distortion becomes very sensitive to the velocity. Even at the critical rotation velocity the oblateness is not that extreme, which led to the claim that the observed oblateness of Achernar could not be matched with uniformly rotating models. Note that the observed oblateness depends on the inclination, and the observed value of $v \sin(i)$ of 275 km/s for Achernar implies that the inclination should not be that large if it is truly near critical rotation. The limited oblateness for uniform rotation arises because the interior parts of the star where the mass is located never rotate very rapidly and thus do not modify the gravitational potential. Critical rotation occurs before significant interior rotation occurs.

The effect of this specific differential rotation law is to change the relationship between the surface equatorial velocity and the oblateness. This is shown in Fig. 7.4, where it is clear the oblateness, and indeed the surface shape, changes appreciably with the extent of the differential rotation. One feature that should be noticed for the highest value of β for $V_{eq} = 240$ km/s is that the distance of the surface from the equatorial plane decreases with decreasing distance from the rotation axis near the pole. This means that the incident radiation field at these locations will not be

Fig. 7.3 Shape of the surface for uniformly rotating stellar models. The rotational velocities are 0, 50, 90, 150, 210, 270, 330, 390, 450, 510, and 570 km/s, and the critical rotation velocity is about 600 km/s.

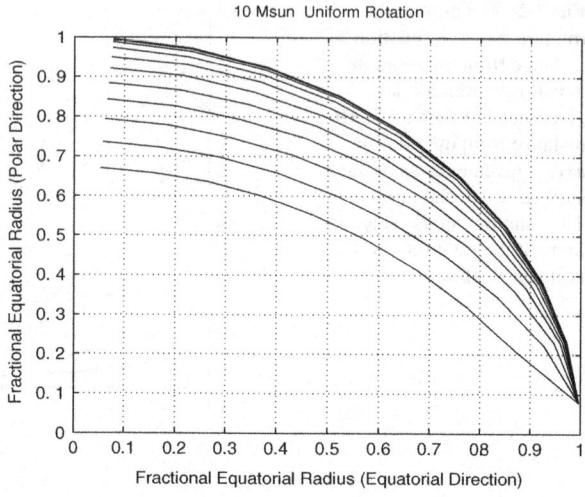

Fig. 7.4 Effects of the differential parameter β on the surface for stellar models with surface equatorial velocity of 120 km/s (*solid*) and 240 km/s (*dashed*). Values of β for $V_{eq} = 120$ km/s are 0, 0.2, 0.6, 1, 1.4, and 1.8. Values of β for $V_{eq} = 240$ km/s range from 0 to 2.0 in steps of 0.2. Increasing the rate of differential rotation inside the star makes the surface more oblate.

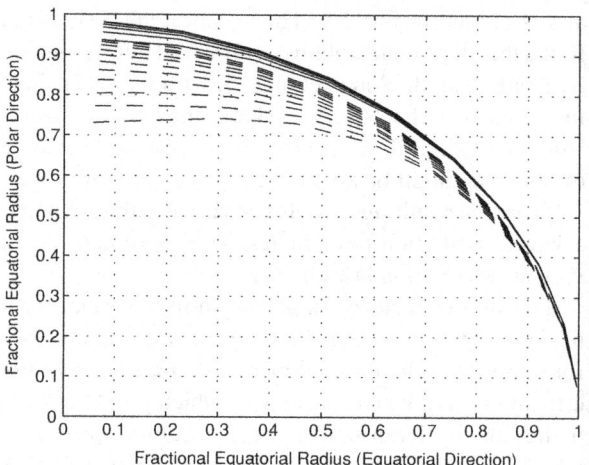

azimuthally symmetric about the normal to the surface here because light will be impinging from the direction toward the polar axis that will not be matched by light from the other directions. Thus, we cannot count on being able to use our 1D model atmosphere models for the SED integration for this case.

7.3.2 SEDs

I wish to utilize the rotating stellar structure models and the plane parallel model atmospheres as input into terms of (7.1) to obtain information about the basic prop-

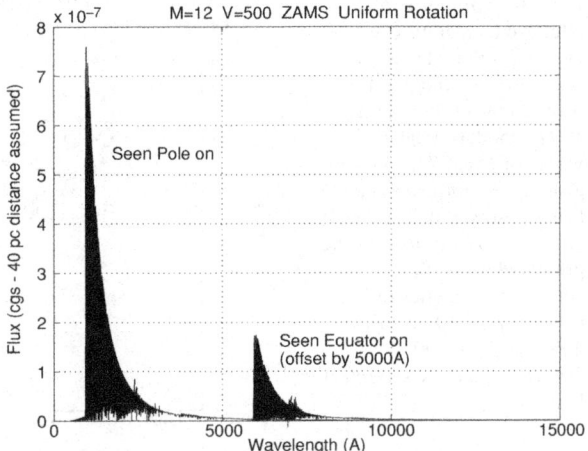

Fig. 7.5 The SED of a 12 $M_\odot$ ZAMS model uniformly rotating at 500 km/s when observed pole on (*left*) and equator on (*right*). The curve on the right has been offset in wavelength by 5000 Å

erties of the star. The emergent flux as a function of wavelength can be integrated over the wavelength to produce the luminosity divided by the surface area of a sphere of radius d. Because I imposed a value of d in (7.1), I can obtain the luminosity in a straightforward manner. Of course the result is not actually the luminosity, but what I would think the luminosity to be if I assumed the flux I observe arose from a spherically symmetric star. Because the observed flux depends on the inclination, this deduced luminosity may be decided different from the luminosity emitted by the actual model. The possible difference will be larger for more rapidly rotating models. I show in Fig. 7.5 how extreme this can get by comparing the SED one would observe when seen pole on with the SED seen equator on for the 10 $M_\odot$ ZAMS model rotating at 500 km/s. I have offset the two SEDs by 5000 Å to enable the visualization. The SEDs appear mostly filled because the stellar lines are so horizontally compressed. The most significant difference is the amplitude, although there are less obvious differences in the shape of the two curves. The difference between the two luminosities I would deduce based on my different orientations with respect to the rotation axis clearly indicates that I must obtain a reasonable estimate for the inclination if I am to be able to translate an observed magnitude into the amount of energy being emitted from the surface of the star per second.

The shape of the SED can be used to provide an estimate of the effective temperature. Because the shape of the curve also varies with inclination, the effective temperature I deduce will also depend on the inclination. Gillich [39] examined a number of wavelength bands which provided reasonable determinations of the effective temperature when applied to the collection of plane parallel model atmospheres. He settled on four, two in the ultraviolet and two in the visible. Four were needed because the atmospheres depend just enough on the gravity that the gravity variation must be considered to get a reasonable deduced effective temperature for the plane parallel models.

Fig. 7.6 Luminosity and effective temperature an observer would deduce as a function of inclination for several uniformly rotating 12 $M_\odot$ models. Higher values of L and T_{eff} on a given curve correspond to being observed progressively closer to pole on. Results are presented for surface equatorial velocities of 255 (*diamonds*), 310 (*pluses*), 405 (*squares*), 500 (*asterisks*), and 575 (*circles*) km/s.

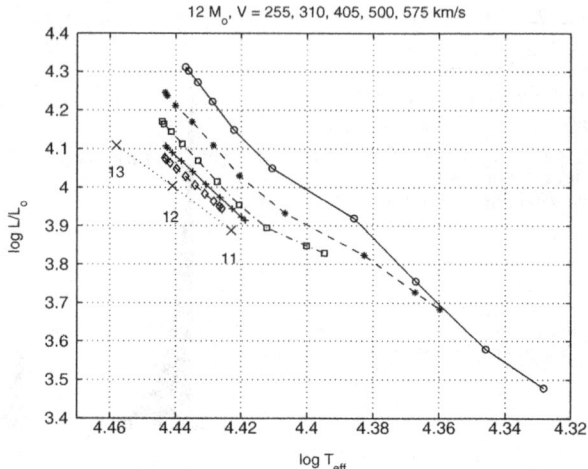

I show the results of this determination of deduced luminosity and effective temperature in Fig. 7.6 for some 12 $M_\odot$ ZAMS models. I have also included the ZAMS locations for 11 and 13 $M_\odot$ nonrotating models for reference. Data are presented for inclinations in 10° increments in the inclination between 0° and 90° I will define these curves as the inclination curves. The deduced luminosity and effective temperature I would find for a given model will be a point on the inclination curve, with the precise location given by the inclination. The highest effective temperatures and luminosities correspond to observations seen pole on. From this figure one can see that the total variation in deduced luminosity for a model rotating at about 300 km/s is equivalent to that between 11 and 13 $M_\odot$, something like 0.5 bolometric magnitudes. For the most rapidly rotating member of this sequence, the pole to equator variation in the deduced effective temperature is about 6100 K and in the luminosity is about 2.1 bolometric magnitudes. These general trends have been found by a number of authors (e.g., Linnell and Hubeny [40]; Frémat et al. [41]; Reiners [42]; Townsend et al. [43]) for many years using varying degrees of approximations related to the rotating model, the model atmospheres, or both.

The general trend in Fig. 7.6 is to move the curves to the right as the rotation rate is increased. The deduced effective temperature at the pole does not vary much with rotation rate until very close to critical rotation, while the deduced effective temperature seen equator on changes far more markedly. This is reflective of the change in the behavior of the effective temperature as a function of latitude as the rotation rate increases. As the rotation rate increases, the polar temperature does increase, but the equatorial temperature drops more dramatically. Gillich et al. [44] have shown that increasing the differential parameter β has the same effect as increasing the rotation rate as a whole: generally the inclination curves get longer and move to the right as β increases.

Fig. 7.7 Comparison of the effective temperature of uniformly rotating model as a function of colatitude (*solid*) with the deduced effective temperature as a function of inclination (*dash*).

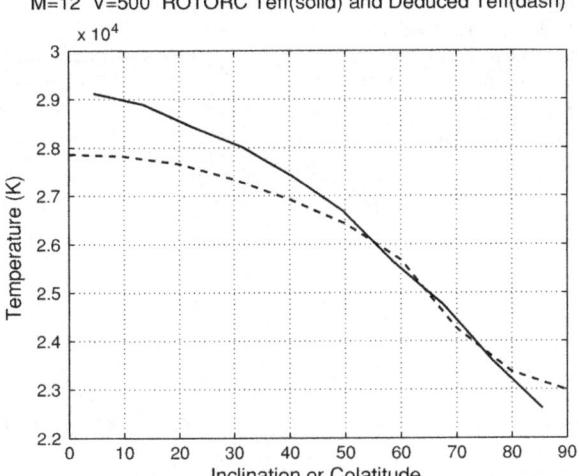

M=12 V=500 ROTORC Teff(solid) and Deduced Teff(dash)

I compare the latitudinal variation of the effective temperature in the rotating model with the deduced effective temperature as a function of inclination in Fig. 7.7. The deduced effective temperature variation is less than the actual variation on the model, which is reasonable given that the deduced effective temperature is based on a composite SED which is produced from a weighted integral of the light emitted for a range of effective temperatures.

7.3.3 Line Profiles

Spectral lines have the potential to supply much information about rotating stars, but like most other features of rapidly rotating stars, the information must be decomposed into its relevant parts. For example, line profiles contain information about elemental abundances, surface conditions, possible differential rotation, and inclination, but it is only useful when we can separate each of these from the others. From a modeling perspective, the process of computing a spectral line is the same as computing a SED: using (7.1) including the appropriate Doppler shift at each location on the surface. Thus, the observed spectral line profile will be a weighted integral of the Doppler shifted line intensity taken over the various conditions on the visible surface. At some level I assume that the shape of the line profile is dominated by the rotation rate, the rotation law, and the inclination, while the equivalent width is more indicative of the elemental abundances (as well as the atomic parameters for the line). This may get blurred if there is significant temperature variation from the pole to the equator and sufficient line sensitivity to the temperature and effective gravity.

One problem may be identified immediately by examining a line profile as a function of the surface equatorial velocity, as I show in Fig. 7.8. In this and subsequent

Fig. 7.8 Line profiles for
12 $M_\odot$ ZAMS models
rotating at $V_{eq} = 50$
(*deepest line core*), 100, 210,
350, and 500 km/s
(*shallowest line core*).

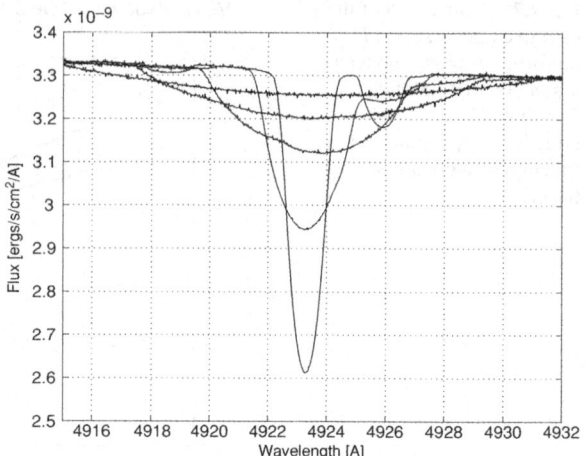

line profiles, all wavelengths are measured in a vacuum, not in air. This figure shows the change in the line profile (seen equator on) as a function of the surface equatorial velocity for a He I line with an O II line nearby. The rotation rate does not have to be too high before the two lines overlap. A further increase in rotation flattens the profile so much that determining the location of the continuum becomes difficult, and a small error in determining the location of the continuum can lead to a substantial error in the deduced equivalent width of the line. The blending of the lines because of the large Doppler shifts associated with the rapid rotation can also lead to significant errors in the equivalent width.

I show the dependence of the line profile on the differential rotation parameter β in (7.2) (Fig. 7.9). In common with previous work (e.g., Stoeckley [45]; Reiners and Schmitt [46]), this figure shows that increasing the rotation rate toward the pole broadens the wing of the line and decreases the depth of the core. The reason becomes clear in Fig. 7.10, a plot of the surface rotational velocity as a function of colatitude for selected values of β. As β increases, the rotational velocity at every point on the surface increases. The exception is the equator because the surface equatorial velocity is set to the same value there for all models. This increase in the surface rotation velocity with β means the Doppler shift is larger, and there is less of the surface contributing small Doppler shift absorption to the line profile. The observed line profile depends on the inclination as well as the rotation law, and it is not clear that these two can be disentangled in any obvious way (e.g., discussion in Reiners and Schmitt [46]). Spectro-interferometry appears to offer such a separation, as indicated by Domiciano de Souza et al. [47], but it can only applied to bright stars in the infrared. One final point I wish to make is that the effects of differential rotation, or at least my particular expression of it, appear to be most pronounced at mid latitudes. This can be seen by comparing the dependence of the line profile on β in Fig. 7.11 with that in Fig. 7.9.

Fig. 7.9 Three line profiles for a He I line observed equator on. The values of β indicate the degree of differential rotation. Differential rotation according to (7.2) broadens the wings of the line and makes the line core more shallow.

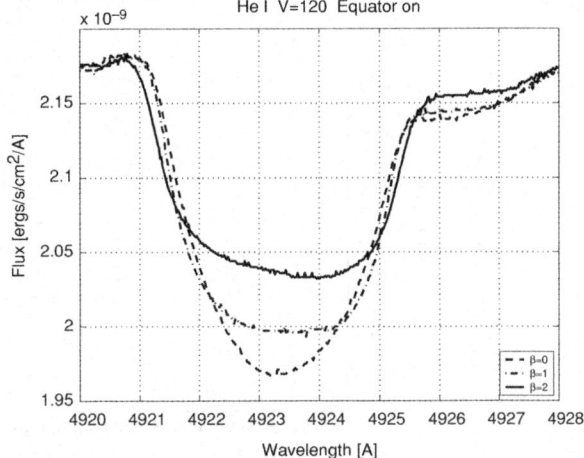

Fig. 7.10 Values of the surface rotational velocity as a function of colatitude. The lower curve is for uniform rotation, the middle curve for a value of the differential rotation parameter $\beta = 1$, and the upper curve for $\beta = 2$. The surface equatorial velocity in all three cases is 120 km/s.

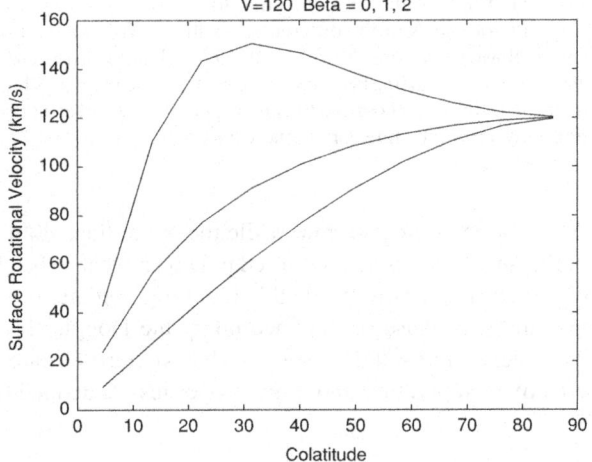

A question that needs some examination is whether the information derived from the SED as a whole, such as the effective temperature and luminosity, are consistent with information derived from lines. I have investigated this for a very rapidly rotating model by comparing the equivalent widths computed at various inclinations with those from the PHOENIX plane parallel atmospheres to determine an effective temperature for the line as a function of inclination. Doppler effects have been omitted so that I could compute reliable equivalent widths, but theoretically the equivalent width is independent of rotational broadening (e.g., Gray [48]). I have performed this calculation for several lines: He I 4471, He II 4686, C II 4267, N II 4631, O II 4642, Mg II 4481, Al II 1855, and Si II 4130. I compare the deduced effective temperatures at the pole and equator for each of these lines with the values deduced from the SED in Fig. 7.12. Interestingly, the two helium lines indicate temperatures consistent with

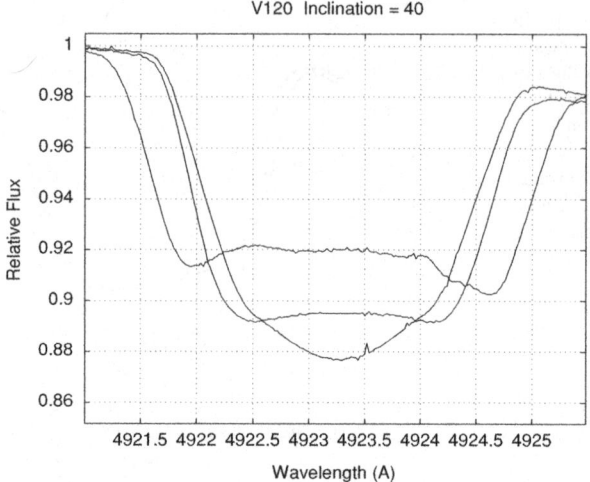

Fig. 7.11 Line profile for a He I line for a model rotating with a surface equatorial velocity of 120 km/s for values of the differential rotation parameter β of 0, 1 and 2 for the deepest, middle, and shallowest line cores, respectively. The inclination between the rotation axis and the observer is 40°. The core of the line becomes more shallow and the wings broader as β increases. For extremely high values of β it is possible to get an apparent core reversal because so much absorption is being transferred from the line core to the wings.

the broadband information, while the metal lines do not. The temperature variation of the metal lines when observed pole on appear to be closer to each other than when observed equator on. While this rapid rotation rate is an extreme case, and probably unrealistic because of the uncertainty the Doppler broadening would create in the equivalent widths, it does suggest that we must be careful in thinking about what the effective temperature and other properties we deduce from rotating stars mean.

7.3.4 Pulsation Frequencies

One property that does not depend on the inclination is the pulsation frequency of any given mode, although mode identification might be made more difficult because knowledge of the radius is much more uncertain for moderately and rapidly rotating stars. However, rotation does affect the pulsation frequencies because it changes the internal structure of the star. There are a number of approaches to computing pulsation frequencies for rotating stars, but most of them require the rotation to be small. One approach that does not has been developed by Clement [24], in which he numerically integrates the appropriate pulsation equations along several radial lines from the stellar center and stellar surface to some specific interior location. The oscillation frequencies of nonrotating stars can be written in terms of a single

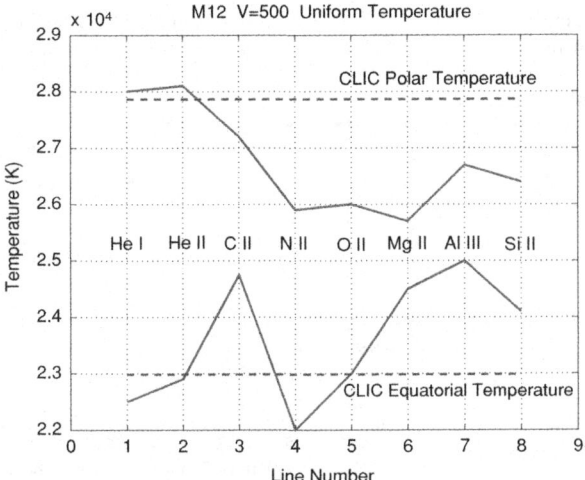

Fig. 7.12 Effective temperatures predicted for various lines (see text) by comparison with equivalent widths with nonrotating plane parallel model atmospheres for a rapidly rotating star seen pole on (*top curve*) and equator on (*bottom curve*). The upper dashed curve is the effective temperature deduced from the shape of the SED when seen pole on; the lower curve when seen equator on. We note that the two helium lines used tend to give the same temperature as the SEDs, while the metal lines typically do not.

spherical harmonic, but rotation brings in more spherical harmonics into the solution for any given mode. With the Clement approach, the coefficients of the spherical harmonics are determined by the results on the different radial lines. The code is adiabatic, but has been updated to include differential rotation by Lovekin et al. [25].

The usual terminology for a mode is to define it in terms of its radial, latitudinal, and azimuthal quantum numbers (n, ℓ, and m). With sufficient rotation, ℓ is no longer a valid quantum number, and we need a new way to describe the mode. One possible way is to use the value of ℓ, ℓ_0, which is the value of ℓ in the nonrotating star to which the given mode in the rotating star can be traced back. As the rotation rate gets higher, this becomes more difficult because the number of spherical harmonics one needs to include becomes larger (and hence the number of modes computed becomes larger) and the changes in the frequency from one rotation rate to the next are larger. Here I shall restrict my attention to low order radial ($n \leq 3$), low latitudinal ($\ell_0 \leq 3$), axisymmetric ($m = 0$) p modes. The linear pulsation calculations include six radial integrations, which corresponds to six spherical harmonics. Lovekin and Deupree [49] have found this to be satisfactory for 10 $M_\odot$ ZAMS models for rotation speeds up to about 360 km/s.

The frequencies for all of these modes decrease as the (uniform) rotation rate increases, with higher radial order modes decreasing more strongly. This can be seen in Fig. 7.13, a plot of the frequencies for several modes as a function of rotation rate. The frequencies are given in terms of the frequency for that mode in the nonrotating

Fig. 7.13 Pulsation frequencies as functions of the rotational velocity for several modes with $\ell_0 = 2$. The frequencies for each mode are scaled to the frequency for that mode in the nonrotating model.

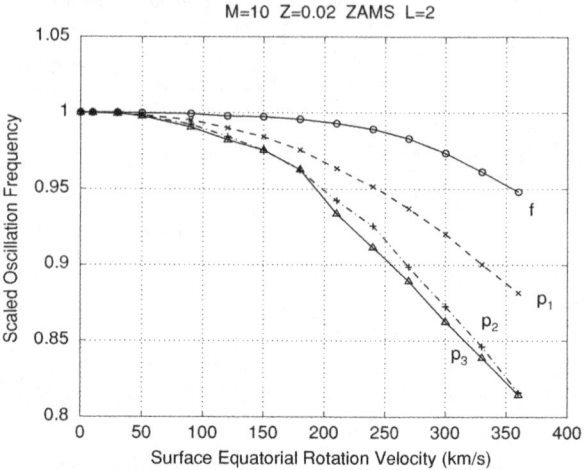

model. This decrease in frequency with increasing rotation rate is the trend that one would expect from the period mean density relation, although the pulsation constant is not constant as a function of rotation if one computes the mean density as the mass divided by the volume of the star. One might also expect higher order radial modes to be more affected than lower order ones because they sample the outer layers, where the effects of rotation are largest, more than lower radial order modes. The trends are such that the large separation, given by $v_{\ell,n+1} - v_{\ell,n}$, decreases and the small separation, given by $v_{\ell,n} - v_{\ell+2,n-1}$, increases with increasing rotation. When the rotation reaches about 300 km/s, both separations are about the same magnitude.

The effects of differential rotation on the pulsation frequencies are shown in Fig. 7.14 a plot of the frequency in terms of the frequency of the uniformly rotating model versus the differential rotation parameter β. The changes in frequency are small, although the change may either increase or decrease the frequency depending on the value of ℓ_0. The large separation is very modestly changed, whereas increasing β mimics a slight increase in the rotation rate for the small separation. From these results I conclude that it should be possible to deduce from a sufficient number of frequencies something about how much rotation there is, but it would be considerably more difficult to obtain information about how the angular momentum is distributed inside the star.

We might also note that the coupling of several spherical harmonics into the eigenfunction may make photometric mode identification (e.g., Stamford and Watson [50]; Watson [51]; Cugier et al. [52]; Heynderickx et al. [53]) more difficult for rapidly rotating stars.

7.4 Final Comments

I have presented a number of techniques which, when used together, might allow us to interpret observations of rapidly rotating stars to the extent that information

Fig. 7.14 Effects of the differential rotation parameter β on the pulsation frequency for several ℓ_0 modes. The surface equatorial velocity in all cases is 120 km/s. All frequencies for a given mode are scaled to the frequency of the uniformly rotating model for that mode. The changes are always small with variations of β and may be either positive or negative.

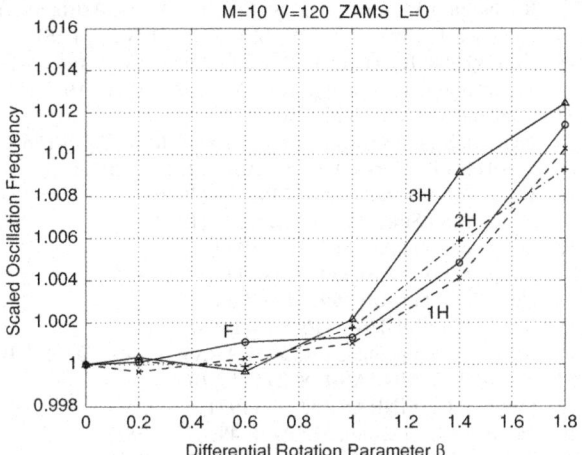

about the effects of rotation on the stellar structure might be deduced and possibly about the distribution of angular momentum inside the star. At the moment it is hard to say that any possible solutions for the type of information routinely available are even unique. Presumably more experience will determine what may and may not be possible.

It is a pleasure to thank a number of people who have contributed to this work. Maurice Clement modified his stellar oscillation code to be able to read models from my 2D structure code and to include differential rotation. Catherine Lovekin performed many of the calculations associated with the model atmospheres and the pulsation calculations, and Aaron Gillich also did much of the work on the generating the SEDs and the spectral lines. Ian Short provided much useful guidance and help in understanding and utilizing the model atmosphere code. Nathalie Toqué developed the scheme to compute the total potential on the model surface for differential rotation. Many of the calculations were performed on computational facilities provided by ACEnet, the high performance computing provider for Atlantic Canada. ACEnet is funded by the Canada Foundation for Innovation, the Atlantic Canada Opportunities Agency, the Nova Scotia Research Innovation Trust, Business New Brunswick, the New Brunswick Innovation Fund, and the Industrial Research and Innovation Fund of the province of Newfoundland and Labrador. All of this support is greatly appreciated.

References

1. von Zeipel, H.: MNRAS **84**, 665 (1924)
2. Monaghan, F.F., Roxburgh, I.W.: MNRAS **131**, 13 (1965)

3. Roxburgh, I.W., Griffiths, J.S., Sweet, P.A.: Zs. Astrophys. **61**, 203 (1965)
4. Faulkner, J., Roxburgh, I.W., Strittmatter, P.A.: ApJ **151**, 203 (1968)
5. Kippenhahn, R., Thomas, H.-C., In: Slettebak, A. (ed.) Stellar Rotation, p 20
6. Sackmann, I.-J., Anand, S.P.S.: ApJ **162**, 105 (1970)
7. Clement, M.J.: ApJ **230**, 230 (1979)
8. Jackson, S., MacGregor, K.B., Skumanich, A.: ApJS **156**, 245 (2005)
9. Ostriker, J.P., Marks, J.W.-K.: ApJ **151**, 1075 (1968)
10. Endal, A.S., Sofia, S.: ApJ **210**, 184 (1976)
11. Endal, A.S., Sofia, S.: ApJ **220**, 279 (1978)
12. Deupree, R.G.: ApJ **357**, 175 (1990)
13. Deupree, R.G.: ApJ **439**, 357 (1995)
14. Deupree, R.G.: ApJ **499**, 340 (1998)
15. Sweet, P.A.: MNRAS **110**, 548 (1950)
16. Baker, N., Kippenhahn, R., Zs. f Astrophys. **48**, 140 (1959)
17. Smith, R.C.: MNRAS **148**, 275 (1970)
18. Smith, R.C.: MNRAS **153**, 33 (1971)
19. Tassoul, J.L., Tassoul, M.: ApJS **49**, 317 (1982)
20. Tassoul, M., Tassoul, J.L.: ApJ **440**, 789 (1995)
21. Sakurai, T.: Geophys. Astrophys. Fluid Dyn. **36**, 257 (1986)
22. Zahn, J.P.: A&A **265**, 115 (1992)
23. Tassoul, M.: ApJ **427**, 388 (1994)
24. Clement, M.J.: ApJS **116**, 57 (1998)
25. Lovekin, C.C., Deupree, R.G., Clement, M.J.: APJ **693**, 677 (2009)
26. Hauschildt, P.H., Baron, E.: J. Comp. Appl. Math. **109**, 41 (1999)
27. Lovekin, C.C., Deupree, R.G.: Proc. High Perform. Comput. Symp. **2006**, 146 (2006)
28. Collins, G.W.: ApJ **146**, 914 (1966)
29. Collins, G.W., Harrington, J.P.: ApJ **146**, 152 (1966)
30. Hardorp, J., Strittmatter, P.A.: ApJ **151**, 1057 (1968)
31. Maeder, A., Peytremann, E.: A&A **7**, 120 (1970)
32. Iglesias, C.A., Rogers, F.J.: ApJ **464**, 943 (1996)
33. Rogers, F.J., Swenson, F.J., Iglesias, C.A.: ApJ **456**, 902 (1996)
34. Fowler, W.A., Caughlan, G.R., Zimmerman, B.A.: ARA&A **5**, 525 (1967)
35. Domiciano de Souza, A., Kervella, P., Jankov, S., Abe, L., Vakili, F., di Falco, E., Paresce, F.: A&A **407**, L47 (2003)
36. Vinicius, M.M.F., Zorec, J., Leister, N.V., Leverhagen, R.S.: A&A **446**, 643 (2006)
37. Kanaan, S., Meilland, A., Stee, P.h., Zorec, J., Domiciano de Souza, A., Frémat, Y., Briot, D.: A&A **486**, 785 (2008)
38. Kervella, P., Domiciano de Souza, A., Kanaan, S., Meilland, A., Spang, A., Stee, Ph.: A&A **493**, L53 (2009)
39. Gillich, A.: MSc. Thesis, Saint Mary's University (2007)
40. Linnell, A.P., Hubeny, I.: ApJ **434**, 738 (1994)
41. Frémat, Y., Zorec, J., Hubert, A.M., Floquet, M.: A&A **440**, 305 (2005)
42. Reiners, A.: A&A **408**, 707 (2003)
43. Townsend, R.H.D., Owocki, S.P., Howarth, I.D.: MNRAS **350**, 189 (2004)
44. Gillich, A., Deupree, R.G., Lovekin, C.C., Short, C.I., Toqué, N.: ApJ **683**, 441 (2008)
45. Stoeckley, T.R.: MNRAS **140**, 121 (1968)
46. Reiners, A., Schmitt, J.H.M.M.: A&A **384**, 155 (2002)
47. Domiciano de Souza, A., Zorec, J., Jankov, S., Vakili, F., Abe, L., Janot-Pacheco, E.: A&A **418**, 781 (2004)
48. Gray, D.F.: The Observation and Analysis of Stellar Photospheres, p. 375. Cambridge University Press, Cambridge (1992)
49. Lovekin, C.C., Deupree, R.G.: ApJ **679**, 1499 (2008)
50. Stamford, P., Watson, R.D.: Ap&SS **77**, 131 (1981)

51. Watson, R.D.: Ap&SS **140**, 255 (1988)
52. Cugier, H., Dziembowski, W.A., Pamyatnykh, A.A.: A&A **291**, 143 (1994)
53. Heynderickx, D., Waelkens, C., Smeyers, P.: A&AS **105**, 447 (1994)

Chapter 8
Seismic Diagnostics for Rotating Massive Main Sequence Stars

Mariejo Goupil

Abstract Effects of stellar rotation on adiabatic oscillation frequencies of β Cephei star are discussed. Methods to evaluate them are briefly described and some of the main results for four specific stars are presented.

8.1 Introduction

Main sequence (MS) massive stars are usually fast rotators and their fast rotation affects their internal structure as well as their evolution. The issue which is addressed here is what information can we obtain—about rotation—from the oscillations of these massive, main sequence stars?

The following seismic diagnostics for rotation using non axisymmetric modes will be discussed: (1) *rotational splittings* as direct probes of the rotation profile. More precisely, we study the effects of cubic order in the rotation rate compared to effects of a latitudinal dependence of the rotation on the splittings; (2) *splitting asymmetries* as a probe for centrifugal distortion. The case of (3) *axisymmetric modes* as indirect probes of rotation throughout effect of rotationally induced mixing on the structure will also be considered.

Results discussed here are obtained with perturbation methods. For nonperturbative methods and results, we refer the reader to Lignières et al. [1], Reese et al. [2] and references therein.

The paper is organized as follows: in Sect. 8.2, properties of pulsating B stars are recalled with emphasis on the uncertainties of their physical description that can be addressed by seismic analyses. Sect. 8.3 recalls the theoretical framework for seismic analyses of relevance here. In Sect. 8.4, seismic analyses of four β Cep are presented. In Sect. 8.5 a theoretical study compares the modifications of the rotation splittings

Mariejo Goupil (✉)
Observatoire de Paris, 5 place Jules Janssen 92190 Meudon principal Cecex France
e-mail: mariejo.goupil@obspm.fr

J.-P. Rozelot and C. Neiner (eds.), *The Pulsations of the Sun and the Stars*,
Lecture Notes in Physics 832, DOI: 10.1007/978-3-642-19928-8_8,
© Springer-Verlag Berlin Heidelberg 2011

due either to latitudinal dependence of the rotation rate, Ω, or to cubic order ($O(\Omega^3)$) frequency corrections. Some conclusions are given in Sect. 8.6.

8.2 Massive Main Sequence Stars

O-B stars are characterized by a convective core and an envelope which is essentially radiative apart a thin outer region related to the iron opacity bump. Important uncertainties regarding the structure and future evolution of these stars are:

- The extent of chemical element mixing beyond the central instable layers as defined by the Schwarzschild criterium
- Transport of angular momentum because the rotation can play a significant role in chemical element mixing.

Convective core overshoot. In 1D stellar models, the convective core is delimited by the radius r_{zc} according to the Schwarzschild criterium $\nabla_{ad} = \nabla_{rad}$. However this corresponds to a vanishing buoyancy force: the eddies are then strongly slowed down but still retain some velocity. Hence due to inertia, eddies move beyond the Schwarzschild radius till their velocity vanishes that is over a distance d_{ov} such that the effective convective core radius becomes $r_{ov} = r_{zc} + d_{ov}$. Despite theoretical investigations [3, 4], the overshooting distance computed in 1D stellar evolutionary models usually remains a rough prescription i.e. it is assumed that $d_{ov} = \alpha_{ov}\min(r_{zc}, H_p)$ with H_p the local pressure scale height and α_{ov} is a free parameter. Empirical determinations from observations ([5, 6, 7] and references therein) yield a wide range for α_{ov}, namely [0–0.5] H_p. The adopted value for this free parameter has important consequences for the evolution of a model with a given mass: with a higher luminosity, it is older at given central hydrogen content (X_c) on the MS and reaches the end of the MS with a larger mass core- total mass ratio. On a statistical point of view, the value of α_{ov} affects the thickness of the MS on a HR diagram as well as the isochrones. core overshoot has therefore an influence on stellar age determination [8, 9].

Rotationally induced mixing in radiative regions. Departure from thermal equilibrium generated by the oblateness of a rotating star causes large scale motions, the meridional circulation. As differential rotation also induces turbulence, competition of these two processes can result in (rotationally induced) diffusion of chemical elements (Zahn [10] and subsequent works). The evolution of a given chemical specie j with concentration c_j is governed by a diffusion equation (for a review [11, 12]):

$$\rho\frac{dc_j}{dt} = \rho\dot{c}_j + \frac{1}{r^2}\frac{\partial}{\partial r}[r^2\rho V_{ip}] + \frac{1}{r^2}\frac{\partial}{\partial r}\left[r^2\rho D_t\frac{\partial c_j}{\partial r}\right] \tag{8.1}$$

where the first term represents nuclear transformation and the second term atomic diffusion with $V_{j,p}$ the diffusion velocity of particles j with respect to protons and where the turbulent diffusion coefficient $D_t = D_{eff} + D_v$, D_{eff} comes from the meridional circulation and D_v from the turbulence. As D_{eff} depends on the vertical meridional velocity U_r, chemical and angular momentum evolutions must be solved together. Hence one also solves an (diffusion-advection) evolution equation for the angular momentum:

$$\rho \frac{dr^2 \Omega}{dt} = \frac{1}{5r^2} \frac{\partial}{\partial r} [r^4 \rho \Omega U_r] + \frac{1}{r^4} \frac{\partial}{\partial r} [r^4 \rho v_v \frac{\partial \Omega}{\partial r}] \tag{8.2}$$

where v_v is the vertical turbulent viscosity related to rotational instabilities.

The current picture is that the vigor of the meridional circulation is controlled by the magnitude of the surface losses of angular momentum. Hence for hot, high mass stars which lose mass but much less angular momentum, one expects no efficient angular momentum internal transport. The rotation profile then essentially results from expansion and contraction within the star during its evolution: i.e. high ratio of core rotation over surface rotation. This is well reproduced by rotationally induced mixing of type I [13]. On the other hand, for cool stars with extended convective outer layers, dynamo generates an efficient magnetic driven wind which is efficient to drive important angular momentum losses and internal transport. This mechanism however is not sufficient enough in the solar case to make the observed rigid rotation in the radiative solar interior and one must calls for to other mechanisms (waves, magnetic field) (see [14, 15] for reviews). This shows that many open questions related to stellar internal rotation and its gradients subsist. An important issue then is to locate regions of uniform rotation and regions of differential rotation (depth and/or latitude dependence) inside the star ($\Omega_{core}/\Omega_{surf}$). Another problem which must be solved is to disentangle effects of overshooting and rotation on mixed central regions and extension of convective cores. Indeed the rotationally induced chemical mixing affects the evolution of the star, its internal structure and oscillation frequencies as does core overshoot although in a different way [13, 16–19]). Figure 8.1 illustrates the respective effects of element mixing by core overshoot and rotation on the evolution of a 9 M$_\odot$ main sequence model in a HR diagram.

Seismology of O-B stars can bring some light about these processes. More specifically, β Cephei stars are good candidates for this purpose [16–18, 20–22]. Indeed, unlike δ Scuti stars, β Cephei stars do not present significant outer convective layers which makes the mode identification more trustworthy provided the star is slowly rotating or that its fast rotation is taken into account in the mode identification process [1, 2, 21, 23].

8.2.1 β Cephei Stars

β Cephei stars are main sequence stars with masses roughly larger than 5–7 M$_\odot$ (Fig. 8.2). They oscillate with a few low degree, low radial order modes around the fundamental radial mode i.e. with periods around 3–8 h. The modes are excited by the kappa mechanism due to the iron bump opacity. These pulsating stars are located

Fig. 8.1 Evolutionary tracks for 9 M$_\odot$ models with neither rotation, nor overshoot included (*dotted line*), with overshoot included but not rotation (*short dashed line*) and with rotation included but not overshoot (*long dashed* and *solid lines*) (from [13])

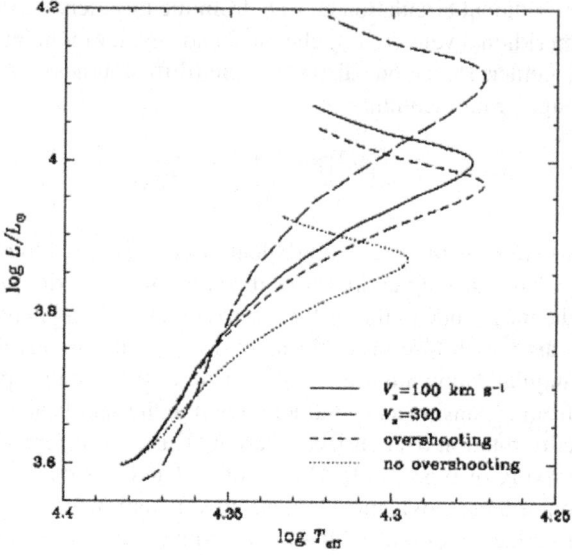

Fig. 8.2 *Left* HR diagram and instability strip for β Cephei stars. *Full dots* represent confirmed β Cephei stars and *open dots* candidates. The *dashed lines* delimitate the instability strip for the fundamental radial mode (adapted from [25])

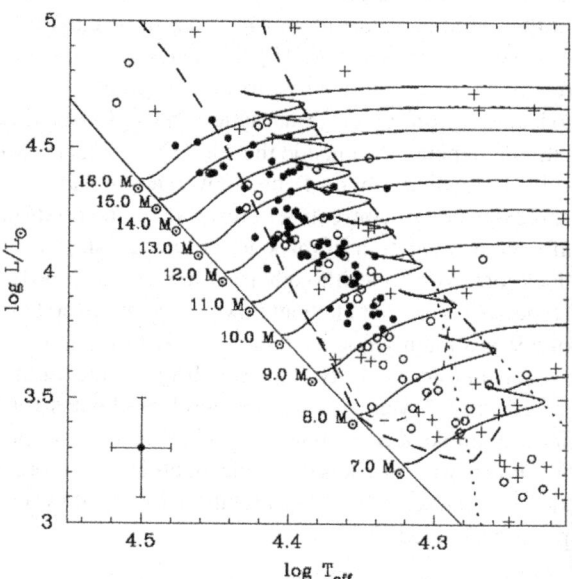

at the intersection of the main sequence and their instability strip in a HR diagram (Fig. 8.2). For more details about β Cephei stars, we refer to reviews by Handler [24], Stankov and Handler [25], Pigulski [26], and Aerts [27].

So far the observed modes have been identified as p_1, p_2, g_1 modes. We recall that p modes are propagative when $\omega^2 > N^2$ and $\omega^2 > S_l^2$ (for more details, see [28]). The squared Brünt–Väissälä (buoyancy) frequency is defined as

Fig. 8.3 Propagation diagram for model **A**: a 8.5 M$_\odot$ model with $T_{\text{eff}} = 22230$ K and initial $X = 0.7$, $Z = 0.019$ (no rotation, no overshoot). The lamb frequency (*dashed line*) is plotted for $\ell = 1$. Normalized squared frequencies discussed here are found in the range $\sigma^2 = 5$–15

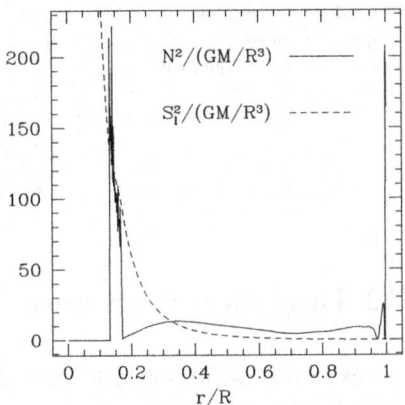

$$N^2 = \frac{g}{r}\left(\frac{1}{\Gamma_1}\frac{d\ln p}{d\ln r} - \frac{d\ln \rho}{d\ln r}\right) \tag{8.3}$$

with p, ρ, g respectively the pressure, density and gravity of the stellar medium and Γ_1 the adiabatic index. The squared lamb frequency is defined as

$$S_l^2 = (k_h c_s)^2 = \ell(\ell + 1)\frac{c_s^2}{r^2} \tag{8.4}$$

with k_h the horizontal wavenumber of the pulsation mode and ℓ the degree of the mode (when its surface distribution is described with a spherical harmonics $Y_\ell^m(\theta, \phi)$). The local sound speed c_s is given by:

$$c_s = \left(\frac{\Gamma_1 p}{\rho}\right)^{1/2} \tag{8.5}$$

For g-modes, the propagative region is delimitated by $\omega^2 < N^2$ and $\omega^2 < S_l^2$.

For β Cephei stars, mixed modes propagate as g mode in the inner part and as p mode in the outer part of the star. Depending on the evolutionary stage of the star, one expects some of the detected modes to be of mixed p and g nature. Modes with frequencies around that of the fundamental radial one (normalized frequency $\sigma = \omega/\Omega_K \sim 2$–$3$ with $\Omega_K = (GM/R^3)^{1/2}$, R the radius and M the mass of the star) can be mixed modes. This can be seen in Fig. 8.3 which shows a propagation diagram for a typical case, model **A**, a model with a mass 8.5 M$_\odot$ and an age $= 19.9$ Myr, a solar metallicity to hydrogen ratio $Z/X = 0.019$ with $X = 0.7$ and $\log T_{\text{eff}} = 4.347$ and $\log L/L_\odot = 3.723$ that therefore lies in the middle of the main sequence and instability strip for these stars.

Rotation of β Cephei stars ranges from slow (rotational velocity $v < 50$ km/s) to extremely rapid ($v > 250$ km/s) (Fig. 8.4). Effects of uniform rotation start to modify significantly the tracks in a HR diagram beyond $v = 100$ km/s for these masses [23]. For $v = 100$ km/s, with a stellar radius $R = 4.94$ R$_\odot$, model **A** is characterized by $\Omega/\Omega_K \sim 0.175$ where $\Omega_K = (GM/R^3)^{1/2}$ is the break up angular velocity.

Fig. 8.4 Histogram of projected rotational velocities for β Cephei stars (from [25])

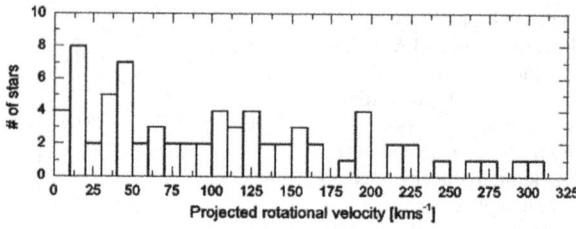

8.3 Theoretical Framework

In this section, we recall the theoretical framework within which seismic observations of these stars can be interpreted in terms of rotation (for more details, the reader is referred to Goupil [29] and references therein). For sake of notation, we recall first the non rotating case.

8.3.1 No Rotation

Adiabatic pulsation studies consider the linearized conservation equations for a compressible, stratified fluid about a *static equilibrium* stellar model characterized by $P_0, \rho_0, \Gamma_1, \phi_0$ respectively pressure, density, adiabatic index, gravitational potential profiles. The equation for hydrostatic equilibrium is:

$$\nabla p_0 = -\rho_0 \nabla \phi_0 \tag{8.6}$$

Assuming the fluid displacement $\delta \mathbf{r}(\mathbf{r}, t)$ of the form $\delta \mathbf{r}(\mathbf{r}, t) = \boldsymbol{\xi}(\mathbf{r}) \exp(i\omega_0 t)$, the linearized momentum equation then is:

$$\mathcal{L}_0(\boldsymbol{\xi}) - \rho_0 \omega_0^2 \boldsymbol{\xi} = 0 \tag{8.7}$$

with

$$\mathcal{L}_0(\boldsymbol{\xi}) \equiv \nabla p' + \rho_0 \nabla \phi' + \rho' \nabla \phi_0$$

where $\mathcal{L}_0$ is a differential operator acting on $\boldsymbol{\xi}$; p', ρ', ϕ' are the Eulerian perturbation for the pressure, density and gravitational potential respectively. One must add boundary conditions [30] and this gives rise to an eigenvalue problem where ω_0 is the eigenvalue for the nonrotating case and $\boldsymbol{\xi}$ is the eigenfunction for the fluid displacement. In the following, we will keep the notation: ν in μHz or c/d; ω in rad/s; $\sigma = \omega/(GM/R^3)^{1/2}$ the normalized frequency. One defines the scalar product:

$$\langle \mathbf{a} | \mathbf{b} \rangle \equiv \int_V \mathbf{a}^* \cdot \mathbf{b} \, d^3 \mathbf{r} \tag{8.8}$$

where $*$ means complex conjugate and where V is the stellar volume. The scalar product of $\boldsymbol{\xi}$ with (8.7) then yields:

$$\langle \boldsymbol{\xi} | \mathcal{L}_0(\boldsymbol{\xi}) - \rho_0 \omega_0^2 \boldsymbol{\xi} \rangle \equiv \int_V \boldsymbol{\xi}^* \cdot (\mathcal{L}_0(\boldsymbol{\xi}) - \rho_0 \omega_0^2 \boldsymbol{\xi}) d^3 \mathbf{r} = 0$$

The eigenfrequency can be obtained as an integral expression:

$$\omega_0^2 = \frac{1}{I} \langle \boldsymbol{\xi} | \mathcal{L}_0(\boldsymbol{\xi}) \rangle \tag{8.9}$$

or

$$\omega_0^2 = \frac{1}{I} \int_V \boldsymbol{\xi}^* \cdot (\nabla p' + \rho_0 \nabla \phi' + \rho' \nabla \phi_0) d^3 \mathbf{r} \tag{8.10}$$

with

$$I = \int_V (\boldsymbol{\xi}^* \cdot \boldsymbol{\xi}) \rho_0 \, d^3 \mathbf{r} \tag{8.11}$$

In absence of rotation, the eigenmode displacement is written in a spherical coordinate system with a single harmonics, $Y_\ell^m(\theta, \phi)$, with a spherical degree ℓ, an azimuthal number m being the number of nodes along the equator

$$\boldsymbol{\xi}(\mathbf{r}) = \xi_r(r) Y_\ell^m \mathbf{e}_r + \xi_h(r) \nabla_h Y_\ell^m \tag{8.12}$$

where the first part is the radial component and the second term the horizontal component of the fluid displacement. The horizontal divergence is

$$\nabla_h = \mathbf{e}_\theta \frac{\partial}{\partial \theta} + \mathbf{e}_\phi \frac{1}{\sin \theta} \frac{\partial}{\partial \phi}$$

The divergence of the fluid displacement is written as:

$$\nabla \cdot \boldsymbol{\xi} = \lambda Y_\ell^m \tag{8.13}$$

with

$$\lambda = \frac{1}{r^2} \frac{dr^2 \xi_r}{dr} - \frac{\Lambda}{r} \xi_h \tag{8.14}$$

and $\Lambda = \ell(\ell+1)$. The perturbed density $\rho'(\mathbf{r}) = \rho'(r) Y_\ell^m$ is given by the linearized continuity equation:

$$\rho'(r) = -\nabla \cdot (\rho_0 \boldsymbol{\xi}) = -\frac{d\rho_0}{dr} \xi_r - \rho_0 \lambda \tag{8.15}$$

The perturbed gravitational potential $\phi'(\mathbf{r}) = \phi'(r)Y_\ell^m$ is given by the perturbed Poisson equation:

$$\nabla^2 \phi' = 4\pi G \rho' \tag{8.16}$$

The pressure perturbation $p'(r)$ is related to the density perturbation $\rho'(r)$ by the adiabatic relation [30] where δ means here a Lagrangean variation:

$$\frac{\delta p}{p_0} = \Gamma_1 \frac{\delta p}{p_0}$$

8.3.2 Including Rotation

In presence of rotation the centrifugal and Coriolis accelerations come into play. The centrifugal force affects the structure of the star—the star is distorted—and causes a departure from thermal equilibrium which generates large scale meridional circulation and chemical mixing. Accordingly, the resonant cavities of the modes are modified. The static equilibrium (averaged over horizontal surfaces) 1D stellar model is modified and characterized by $P_{0,\Omega}(r)$, $\rho_{0,\Omega}(r)$, $\Gamma_{1,\Omega}(r)$, $\phi_{0,\Omega}(r)$ with $\Omega(r,\theta)$ the rotation rate. The Coriolis force enters the equation of motion and affects the motion of waves and frequencies of normal modes. The linearized equation of motion is modified. As rotation breaks the azimuthal symmetry, it lifts the frequency degeneracy: without rotation, $2\ell + 1$ modes with given n, ℓ, m ($m = -\ell, \ell$) have the same frequency ω_0 (omitting for shortness the subscripts n, ℓ). With rotation, the same $2\ell + 1$ modes have different frequencies ω_m and the rotational splitting is defined as: $S_m = (\omega_m - \omega_0)/(m)$. One also uses $S_m = \omega_m - \omega_{m-1}$ and the generalized rotational splitting:

$$S_m = \frac{\omega_m - \omega_{-m}}{2m} \tag{8.17}$$

where ω_m is the mode frequency. These various definitions are equivalent only at first perturbation order in the rotation rate Ω; the first two are used when only a few components are available.

8.3.3 Rotational Splittings

At first perturbation order in Ω, only the Coriolis acceleration plays a role. The linearized equation of motion including the effect of Coriolis acceleration ($2\Omega \times \mathbf{v}$) in a frame of inertia is

$$\mathcal{L}_0(\xi) - \rho_0(\omega_m + m\Omega)^2 \xi + 2\rho_0(\omega_m + m\Omega)\Omega \mathcal{K}\xi = 0 \tag{8.18}$$

with $\mathcal{K}\boldsymbol{\xi} = i\mathbf{e}_z \times \boldsymbol{\xi}$ and $\boldsymbol{\xi}$ is the displacement eigenvector in absence of rotation and $\mathbf{e}_z$ is the vertical unit vector in cylindrical coordinates. The nonrotating case is recovered by setting $\Omega = 0$. One then expands the displacement eigenfunction as $\boldsymbol{\xi} = \boldsymbol{\xi}_0 + \boldsymbol{\xi}_1$ and the eigenfrequency as $\omega_m = \omega_{\Omega=0} + \omega_{1,m}$ where $\omega_{\Omega=0}$, $\boldsymbol{\xi}_0$ correspond to the eigenfrequency and eigenfunction for a nonrotating star and $\omega_{1,m}$, $\boldsymbol{\xi}_1$ give the first order correction due to Coriolis acceleration. Keeping only terms up to $O(\Omega)$, one obtains:

$$\mathcal{L}_0(\boldsymbol{\xi}_1) - \rho_0\omega_{\Omega=0}^2\boldsymbol{\xi}_1 - 2\rho_0\omega_{\Omega=0}(\omega_{1,m} + m\Omega)\boldsymbol{\xi}_0 + 2\rho_0\omega_{\Omega=0}\Omega\mathcal{K}\boldsymbol{\xi}_0 = 0 \quad (8.19)$$

The correction to the eigenfunction $\boldsymbol{\xi}_1$ can be chosen so that $\langle\boldsymbol{\xi}_0|\boldsymbol{\xi}_1\rangle = 0$. Taking the scalar product (8.8) of $\boldsymbol{\xi}_0$ with (8.19) and keeping only terms up to $O(\Omega)$ yields:

$$\langle\boldsymbol{\xi}_0|\,[\,\mathcal{L}_0(\boldsymbol{\xi}_1) - \rho_0\omega_{\Omega=0}^2\boldsymbol{\xi}_1 - 2\rho_0\omega_{\Omega=0}(\omega_{1,m} + m\Omega)\boldsymbol{\xi}_0 + 2\rho_0\omega_{\Omega=0}\Omega\mathcal{K}\boldsymbol{\xi}_0\,]\rangle = 0 \tag{8.20}$$

from which one derives for a mode with given n, ℓ

$$\omega_{1,m}I_0 = \int_V \boldsymbol{\xi}_0^* \cdot (\Omega\mathcal{K} - m\Omega)\boldsymbol{\xi}_0\rho_0 d^3\mathbf{r} \tag{8.21}$$

which is rewritten as:

$$\omega_{1,m} = \int_0^R \int_0^\pi K_m(r,\theta)\Omega(r,\theta)d\theta dr \tag{8.22}$$

where the analytical expression for the kernels K_m is given in Appendix. At first order $O(\Omega)$, the generalized splitting (8.17) then is given by

$$S_m = \frac{\omega_{1,m} - \omega_{1,-m}}{2m} \tag{8.23}$$

Assuming a shelllular rotation $\Omega(r)$, the splitting S_m becomes m independent and one has:

$$S = \int_0^R K(r)\Omega(r)dr \tag{8.24}$$

with the rotational kernel

$$K(r) = -\frac{1}{I}\left(\xi_r^2 - 2\xi_r\xi_h + (\Lambda - 1)\xi_h^2\right)\rho_0 r^2 \tag{8.25}$$

and mode inertia (8.11):

$$I = \int_0^R (\xi_r^2 + \Lambda \xi_h^2) \rho_0 r^2 dr \tag{8.26}$$

with again $\Lambda = \ell(\ell+1)$ and R the stellar radius. For a uniform rotation, this further simplifies to

$$S = \Omega\beta; \quad \beta = \int_0^R K(r)dr \tag{8.27}$$

β is assumed to be known from an appropriate stellar model, S is measured and Ω is inferred. This will be used in Sect. 8.4 for β Cep stars.

When only a few measured splittings are available, information about the internal rotation is limited so one assumes for instance a uniform rotation for the convective core with the angular velocity $\Omega = \Omega_c$ (for $x = r/R \le x_c$) and a uniform rotation for the envelope $\Omega = \Omega_e$ for $x > x_c$. Both values are the unknowns. Inserting into (8.24),

$$S = \int_0^1 K(x)\Omega(x)dx = \Omega_c\beta_c + \Omega_e\beta_e$$

with

$$\beta_c = \int_0^{x_c} K(x)dx; \quad \beta_e = \int_{x_c}^1 K(x)dx$$

The detection of two triplets $\ell = 1$ for instance yields Ω_c, Ω_e and Ω_c/Ω_e provided β_c, β_e are given by a model close to the observed star. This type of approach was used to determine whether the star is in rigid rotation or not for a δ Scuti star [31]; for white dwarfs [32, 33] and recently for β Cephei stars (Sect. 8.4) and SdB stars [34].

8.3.4 Splitting Asymmetries: Distortion

At second order in the rotationrate, the centrifugal acceleration comes into play. This has several consequences on the oscillation frequencies (for a review [29]). One is that the split components are no longer equally spaced. It is then convenient to define A_m the splitting asymmetry as

$$A_m = \omega_0 - \frac{1}{2}(\omega_m + \omega_{-m}) \tag{8.28}$$

In order to interpret observed asymmetries, let consider a given multiplet of modes (i.e. with specified n, ℓ). Its oscillation frequencies, ω_m ($m = -\ell, \ldots, \ell$), are computed up to second order $O(\Omega^2)$ as:

$$\omega_m = \omega_{0,\Omega} + m S_{|m|} + \frac{\bar{\Omega}^2}{\omega_{0,\Omega}}(X_1 + m^2 X_2) \tag{8.29}$$

where $\omega_{0,\Omega}$ is the eigenfrequency for a static model including the horizontally averaged centrifugal acceleration. The second term is the splitting (8.23) due to Coriolis effect and $\bar{\Omega}$ is an averaged rotation rate. The last term is the asymmetry due to the non spherical part of the centrifugal distortion which dominates for high radial order modes. Expressions for X_1, X_2 can be found in Goupil [29], Saio [35], Dziembowski and Goode [36], Soufi et al. [37], and Suarez et al. [38]. For low radial modes such as those excited in β Cep stars, the second order Coriolis contributions to X_1, X_2 remain significant. According to (8.29), the asymmetry is then given by:

$$A_m = \left(\frac{\bar{\Omega}^2}{\omega_{0,\Omega}}\right) m^2 X_2 \tag{8.30}$$

Let consider again the linearized equation of motion including now the centrifugal acceleration:

$$\mathcal{L}_{0,\Omega}(\boldsymbol{\xi}) - \rho_{0,\Omega}\hat{\omega}^2\boldsymbol{\xi} + 2\rho_{0,\Omega}\hat{\omega}\Omega K(\boldsymbol{\xi}) + \mathcal{L}_2(\boldsymbol{\xi}) - \rho_2\hat{\omega}^2\boldsymbol{\xi} = 0 \tag{8.31}$$

where $\hat{\omega} = \omega_m + m\Omega$. The spherical part of the centrifugal acceleration is included in the spherical 1D model, therefore the linear operator depends on the rotation rate i.e.

$$\mathcal{L}_{0,\Omega}(\boldsymbol{\xi}) \equiv \nabla p' + \rho_{0,\Omega}\nabla\phi' + \rho'\nabla\phi_{0,\Omega} \tag{8.32}$$

and for the non spherical distortion

$$\mathcal{L}_2(\boldsymbol{\xi}) = \rho'\left(\frac{p_2}{\rho_{0,\Omega}}\nabla p_{0,\Omega} - \nabla p_2\right) + \rho_2\nabla\phi' + \rho_{0,\Omega}\mathbf{e}_s r \sin\theta\nabla\Omega^2 \cdot \boldsymbol{\xi} \tag{8.33}$$

where $\mathbf{e}_s = \sin\theta\mathbf{e}_r + \cos\theta\mathbf{e}_\theta$ in a spherical coordinate system $(\mathbf{e}_r, \mathbf{e}_\theta, \mathbf{e}_\phi)$. The subscript 2 indicates departure from sphericity p_2, ρ_2, ϕ_2 for the pressure, density and graviational potential respectively. Again using the scalar product (8.8), one writes

$$\langle\boldsymbol{\xi}_0|\mathcal{L}_{0,\Omega}(\boldsymbol{\xi}) - \rho_{0,\Omega}\hat{\omega}^2\boldsymbol{\xi} + 2\rho_{0,\Omega}\hat{\omega}\Omega K(\boldsymbol{\xi})\rangle + \langle\boldsymbol{\xi}|(\mathcal{L}_2(\boldsymbol{\xi}) - \rho_2\hat{\omega}^2\boldsymbol{\xi})\rangle = 0 \tag{8.34}$$

One then assumes an eigenfunction of the form $\boldsymbol{\xi} = \boldsymbol{\xi}_0 + \boldsymbol{\xi}_1 + \boldsymbol{\xi}_2$ and the eigenfrequency as $\omega_m = \omega_{0,\Omega} + \omega_{1m} + \omega_2$ where the unknown now is ω_2. Solving (8.34) for ω_{2m} leads to an integral expression for X_1, X_2 and therefore an expression for A_m of the form:

$$A_m = m^2 \int_0^1 \Omega^2(x) K_2(x) dx \qquad (8.35)$$

where $K_2(x)$ include effects of distortion of the structure throughout p_2, ρ_2 and depend on the eigenfunction $\boldsymbol{\xi}$. Detailed expression for $K_2^{(j)}(x)$ can be found in Goupil [29], Dziembowski and Goode [36], Soufi et al. [37], Suarez et al. [38], and Karami [39]. An example for $K_2(x)$ is shown in Fig. 8.7 and discussed in Sect. 8.4.2.

Splitting asymmetries can provide probes of the internal structure which differ from those derived with the splittings S_m as the corresponding kernels are different. When only a few observed asymmetries are available, one can proceed as for the splittings (Sect. 8.3 above). Assuming a rotation profile of the simplified form:

$$\Omega^2(x) = \Omega_c^2 \quad \text{for} \quad x_c < x$$
$$\Omega^2(x) = \Omega_c^2 + 2(x - x_c)\Omega'\Omega_c + (x - x_c)^2 \Omega'^2 \quad \text{for} \quad x_c \le x \le x_e \qquad (8.36)$$
$$\Omega^2(x) = \Omega_e^2 \quad \text{for} \quad x_e < x$$

with

$$\Omega' = \frac{\Omega_e - \Omega_c}{x_e - x_c} \qquad (8.37)$$

then

$$A_m = m^2 \left(\Omega_c^2 \beta_{2,0} + 2\Omega'\Omega_c \beta_{2,1} + \beta_{2,2}\Omega'^2 + \beta_{2,e}\Omega_e^2 \right) \qquad (8.38)$$

where Ω_c, Ω' are assumed known from the splittings (Sect. 8.3.3) and

$$\beta_{2,q} = \int_{x_c}^{x_e} (x - x_c)^q K_2(x) dx \qquad (8.39)$$

$$\beta_{2,e} = \int_{x_e}^1 K_2(x) dx \qquad (8.40)$$

Determination of the β_2 coefficients then brings some information on the kernels $K_2(x)$ with the promising prospect of deriving constrains on the rotationally distorted part of the stellar structure.

8.3.5 Axisymmetric Modes: Rotationally Induced Mixing

Centrifugal departure from spherical symmetry has important effects on all modes including the axisymmetric modes. Indeed the values of the $m = 0$ mode frequencies are shifted when compared to those of non rotating models. Hence the differences

$$\delta\omega = \omega_0 - \omega(\Omega = 0) = \left(\frac{\bar{\Omega}^2}{\omega_{0,\Omega}}\right) m^2 X_1 \qquad (8.41)$$

from (8.29) between frequencies of a given mode from a model including rotation and a non rotating model can be an efficient diagnostic for rotation effects although some care must be taken when defining the $\Omega = 0$ stellar model for comparison. This has been extensively discussed in past publications ([28, 35, 36, 40, 41], for a review, see [29]).

Another (indirect) effect of the star oblateness on frequencies, as already mentioned in Sect. 8.2, is due to the departure from radiative equilibrium which generates large scale motions (meridional circulation), differential rotation and consequently shear turbulence. All this concurs to affect the rotation profile. It also causes mixing of chemical elements which affects the prior evolution of the observed star and therefore its structure. These structure changes must be computed by coupling both evolutions of the angular momentum and the chemicals, as already mentioned in Sect. 8.2. These equilibrium structure modifications affect all modes as compared to those of a nonrotating star, including the axisymmetric modes. The effect on the frequencies can be quite significant as was discussed by Goupil and Talon [16] and quantified by Montalban et al. [17], Miglio et al. [18], and Goupil and Talon [42] (see Sect. 8.4.3).

We consider here only the effect of the structure modifications due to rotationally induced mixing on the axisymmetric mode frequencies. The Coriolis or the centrifugal accelerations then are not included in the linearized equation of motion. Hence the linearized equation of motion including rotationally induced mixing yields the usual integral expression for the eigenfrequency of a nonrotating model, (8.10), except for the structure quantities such as the density, the pressure, the gravity (ρ, p, g resp.) etc. which are modified by the rotationally induced mixing. As they now depend on the rotation rate, we write them as ρ_Ω, p_Ω, g_Ω ... The linearized equation of motion including rotationally induced mixing in a 1D spherically symmetric stellar model then is given by:

$$\omega_{0,\Omega}^2 = \frac{1}{I_\Omega} \int_V \boldsymbol{\xi}_\Omega^* \cdot (\nabla p' + \rho_\Omega \nabla \phi' + \rho' \nabla \phi_\Omega) d^3\mathbf{r} \qquad (8.42)$$

with the mode inertia:

$$I_\Omega = \int_V (\boldsymbol{\xi}_\Omega^* \cdot \boldsymbol{\xi}_\Omega) \rho_\Omega d^3\mathbf{r}$$

From now on for sake of shortness, we omit the subscript Ω for the eigenfunctions. We define the dimensionless variables according to Dziembowski [43] (see also [30]):

$$y_1 = \frac{\xi_r}{r}; \quad y_2 = \frac{1}{g\Omega r}\left(\phi' + \frac{p'}{\rho\Omega}\right); \quad y_3 = \frac{1}{g\Omega r}\phi' \qquad (8.43)$$

Starting with (8.42), integrations over surface angles and a few integrations by part for the radial part yield:

$$\omega_{0,\Omega}^2 = \frac{1}{I_\Omega}\int_R \left(-\lambda(y_1 + y_2) - \frac{d\ln \rho\Omega}{d\ln r}y_1(y_1 + y_3)\right) g\Omega\rho\Omega r^3 dr \qquad (8.44)$$

where we have assumed that the surface integrals vanish. From its definition (8.14),

$$\lambda = V_{g,\Omega}(y_1 - y_2 + y_3)$$

with

$$V_{g,\Omega} = -\frac{1}{\Gamma_{1,\Omega}}\frac{d\ln \rho\Omega}{d\ln r}$$

Note that there are several alternative equivalent expressions for $\omega_{0,\Omega}^2$.

Differences between the structure of a model including rotationally induced mixing and that of a model which does not result in differences in the eigenfrequencies which we note $\delta\omega = \omega_{0,\Omega} - \omega_{0,\Omega=0}$. We will see in Sect. 8.4.3 that the structures of the models indicate that $\rho\Omega$ and its derivative with respect to the radius, the gravity $g\Omega$, the density $\rho\Omega$ are not significantly modified compared to the derivative of the density with respect to the radius. Accordingly using (8.44) and keeping only the first order terms, one obtains:

$$\delta\omega \approx -\frac{1}{2\omega_{\Omega=0}I_{\Omega=0}}$$
$$\times \int_R \delta\left(\frac{d\ln \rho\Omega}{d\ln r}\right) y_1(y_1 + y_3)g_{\Omega=0}\rho_{\Omega=0}r^3 dr \qquad (8.45)$$

where we have also assumed that the perturbations of the eigenfunctions $y_{j,\Omega} - y_{j,\Omega=0}$ ($j = 1, 3$) are negligible at first order. For massive main sequence stars the largest difference $\delta(d\ln \rho\Omega/d\ln r)$ arises near the convective core (Sect. 8.4.3). Largest frequency differences therefore are expected for mixed modes compared to p-modes. Note that the same interpretation can be obtained with differences in the Brünt–Väissälä approximation, one has from (8.3)

$$\delta\left(\frac{rN^2}{g}\right) = -\delta\left(\frac{d\ln \rho\Omega}{d\ln r}\right) \qquad (8.46)$$

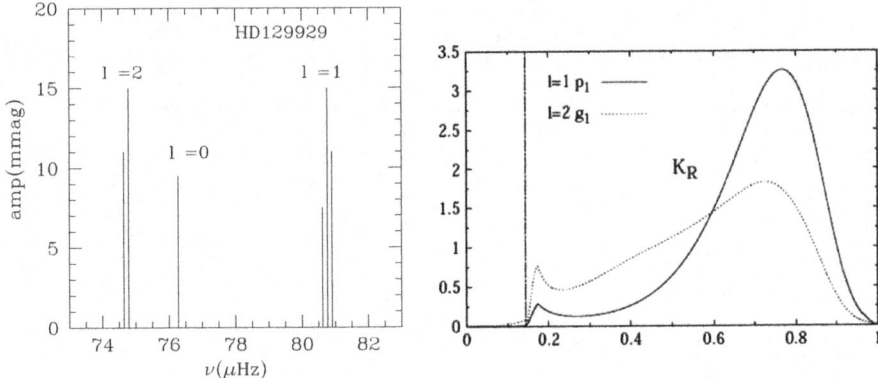

Fig. 8.5 *Left* Schematic representation of the frequency spectrum of HD129929 (data from [44]). *Right* Rotational kernels for the excited p_1 and g_1 modes of HD129929 in function of the radius r/R normalized to the stellar radius (from [45]).

For high frequency (i.e. pure) p-mode which propagate significantly above the $\nabla\mu$ region, the difference $\delta\,(d\ln\rho_\Omega/d\ln r)$ is essentially negative. In addition $|y_3| \ll |y_1|$ so that we obtain

$$\delta\omega \approx -\frac{1}{2\omega_{\Omega=0}I_{\Omega=0}}$$
$$\times \int_R \delta\left(\frac{d\ln\rho_\Omega}{d\ln r}\right) y_1^2 g_{\Omega=0}\rho_{\Omega=0} r^3 dr > 0$$

$$(8.47)$$

which is small and positive. For mixed modes having high amplitude in the $\nabla\mu$ region, $\delta\,(d\ln\rho_\Omega/d\ln r)$ can be positive and the frequency difference can be large and negative as illustrated in Sect. 8.4.3. The difference $\delta\omega$ is quantified and discussed in the case of a β Cephei model in Sect. 8.4.3.

8.4 Seismic Analyses of Four β Cephei Stars

We discuss four β Cephei stars which have been the subject of seismic analyses and for which information about rotation and core overshoot has been inferred: V836 Cen (HD 129929); ν Eridani, θ Ophiuchi and 12 Lacertae (see also [19]). Schematic representations of the frequency spectra for the first three stars are displayed in Figs. 8.5 and 8.6. These four stars are relatively slow rotators (with surface rotational velocities smaller than $\approx$70 km/s). Determination of the luminosity, effective temperature and location in the HR diagram for these slow rotators are not significantly affected by rotation.

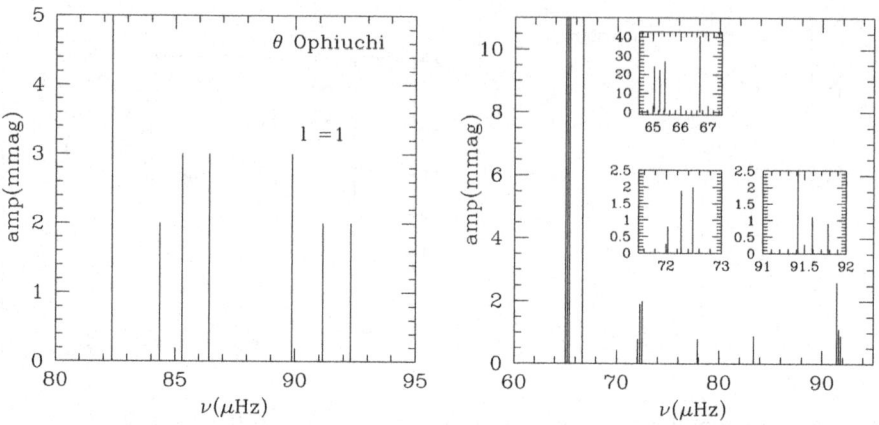

Fig. 8.6 Schematic representation of the power spectrum of *left* θ Ophiuchi (data from [46]) and *right* ν Eri (data from [67])

8.4.1 Rotational Splittings

HD129929 is a main sequence $\sim 9\,M_\odot$ star for which one $\ell = 1$, p_1 triplet has been detected and identified as well as one radial mode and 2 successive components of the $\ell = 2$, g_1 mode as represented in Fig. 8.5 [44, 45]. From the triplet and assuming a solid body rotation, one uses $S = \Omega\beta$ (8.54) as explained in Sect. 8.3. With β known from an appropriate stellar model, the measured splitting for the $\ell = 1$, $p = 1$ triplet S gives $v_{\text{rot}} = 3.61$ km/s but from the two successive components of the $\ell = 2$ multiplet, one obtains $v_{\text{rot}} = 4.21$ km/s, clearly indicating a nonuniform rotation [45]. Assuming therefore a uniform rotation for the convective core with angular velocity $\Omega = \Omega_c$ and a uniform rotation $\Omega = \Omega_e$ for the envelope of the star, the splittings then obey $S = \Omega_c\beta_c + \Omega_e\beta_e$ where β_j are the integral for the core or the envelope (Sect. 8.3). It is found that $|\beta_c| \ll |\beta_e|$ that is actually the detected modes do not efficiently probe the convective core. This can be seen with the associated rotational kernels in Fig. 8.5 which have no amplitude in the core. Therefore Ω_c is taken as the rotation rate of the radiative region in the μ-gradient region above the convective core (with μ the mean molecular weight). Assuming a linear depth variation of the angular velocity in the envelope $\Omega(x) = \Omega_0 + (x - x_0)\Omega_1$, the splittings must obey $S = \Omega_0\beta_0 + \Omega_1\beta_1$ where again β_0 and β_1 are known from the stellar model;

$$\beta_0 = \int\limits_0^{x_c} K(x)dx; \quad \beta_e = \int\limits_{x_c}^{x_e} K(x)(x - x_c)dx$$

The knowledge of S_1 and S_2 then yields Ω_0 and Ω_1. A rotation gradient in the envelope with $\Omega_c/\Omega_e = 3.6$ is obtained.

In addition, the seismic modelling of the detected axisymmetric modes favors a core overshooting distance of $\sim$0.1 pressure scale height (H_p) rather than 0 while an overshoot of $0.2H_p$ is rejected.

θ *Ophiuchi* is also a main sequence $\sim$9 $M_\odot$ star with an effective temperature $T_{\rm eff} \sim 22900$ K. Three multisite campains seismic observations and data analyses reveal seven identified frequencies: the radial fundamental $\ell = 0\,(p_1)$; one triplet $\ell = 1\,(p_1)$ and 3 components $(m = -1, 1, 2)$ of a quintuplet $\ell = 2\,(g_1)$ [46]. A seismic analysis led Briquet et al. [47] to conclude that the case of θ Ophiuchi is similar to HD129929. The detected modes do not provide strong constraint about the rotation of the convective core. On the other hand, unlike HD129929, the data for θ Ophiuchi are compatible with a uniform or a quite slowly varying rotation of the envelope. The convective core overshoot distance is found to be $(0.44 \pm 0.07)H_p$ This is a much larger amount than found for HD129929. Whether this difference must be related to the fact that θ Ophiuchi seems to rotate more than 10 times faster than HD129929 remains an open issue.

ν *Eri* is a very interesting case as it oscillates with three triplets $\ell = 1\,(g_1, p_1, p_2)$, one radial mode p_1 and one $\ell = 2$ component. Seismic studies show that the detected modes are able to probe the rotation of the core, which is rotating faster than the envelope [49–51]. Dziembowski and Pamyatnykh [50] further assumed a linear gradient as a transition (in the μ-gradient zone) between the uniform fast rotation $\Omega = \Omega_c$ of the core and the uniform slow rotation of the envelope $\Omega = \Omega_e$ above the μ-gradient region. They find a ratio $\Omega_c/\Omega_e = 5.3$–5.8. Model fitting based on the 3 axisymmetric $\ell = 1$ modes yield an extension of the mixed central region of 0.1–0.28 H_p above the convective core radius depending on the adopted chemical mixture and metallicity value [50, 51].

12 Lac Several frequencies have been detected for this star [52] but only four of the detected frequencies correspond to identified (ℓ, m) modes [53]. Only two successive components of one $\ell = 1$ triplet are known which is not enough to provide information on the inner/surface rotation ratio. One can use as an additional information the equatorial surface value, $v_{\rm eq} = 49 \pm 3$ km/s as derived by Desmet et al. [53]. One needs the stellar radius which is derived from a seismic modelling of the star. The resulting seismic model and its radius depend on the radial orders identified for the modes [50, 53]. Second order (centrifugal) effects on the frequencies must also be taken into account as the rotation for 12 Lac seems to be fast enough as recognized by Dziembowski and Pamyatnykh [50]. Taking then a value for the stellar radius in the broad range $R = 7$–9$R_\odot$, the equatorial surface value, $v_{\rm eq} = 49 \pm 3$ km/s and the observed splitting of 1.3032 μHz yields a ratio $\Omega_{\rm inner}/\Omega_{\rm surf}$ in the range [1.8–5] definitely indicating a non rigid rotation. There is not yet an agreement concerning the radial order of the identified modes but the triplet seems in any case to be of mixed nature and therefore able to probe the core rotation. Dziembowski and Pamyatnykh [50] did not consider overshoot and Desmet et al. [53] found that core overshoot must be smaller than 0.4 Hp.

Summary These studies lead to the conclusion that a few rotationally split modes can provide important information about internal rotation and core overshoot of β Cephei stars if the modes are identified, enough precise measurements are obtained

Table 8.1 Overshoot versus rotation rate for several stars from seismic analysis

β Cep	V_{eq} (km/s)	α_{ov}	$\Omega_{inner}/\Omega_{env}$	Z	ref
HD129929	~ 2	0.1 ± 0.05	$\Omega_{0.2}/\Omega_{surf} \sim 3.1$	0.019 ± 0.003	Dupret et al. [45]
θ Ophiuchi	29 ± 7	$0.44^* \pm 0.07$	env. unif. rotation	0.012 ± 0.003	Briquet et al. [47]
ν Eri	~ 6	0.15 ± 0.05	$\Omega_c/\Omega_{env} \sim 5.5$–$5.8$	0.0172 ± 0.0013	Pamyathnyck et al. [48]
12 Lac	~ 49	<0.4	$\Omega_c/\Omega_{env} \sim 1.8$–$5$	0.01–0.015	(Dziembowski and Pamyatnykh [50], Desmet et al. [53]

V_{eq} the derived the equatorial velocity, α_{ov} the overshoot parameter, $\Omega_{inner}/\Omega_{env}$ the ratio of the the rotation rate in the inner layers to that of the surface, Z the metallicity. The modellings assume a Grevesse–Noels mixture except for 12 Lac. [a]Asplund mixture

and the age of the star is such that excited modes have mixed g, p nature. Trying to disentangle overshoot and rotation effects on core element mixing is only starting with a measure of their relative magnitude as is illustrated in Table 8.1. As emphasized by Dziembowski and Pamyatnykh [50], in that respect, seismic modelling of fast rotators are needed. Once the size of the mixed core and the ratio of core to surface rotation are reliably determined, the next issue is to estimate ,what part in the seismically measured extension of the core, d_{ov}, comes from convective eddy overshooting and what part comes from other transport processes such as those induced by rotation.

8.4.2 Splitting Asymmetries: Distortion

The splitting asymmetry, A_m (8.35), for acoustic modes is mainly due to the oblateness of the star caused by the centrifugal force although for low radial order modes, the Coriolis contribution is also significant. Figure 8.7 represents the normalized splitting asymmetries:

$$\mathcal{R}_m \equiv A_m/v_{0,1,0} \tag{8.48}$$

for the $\ell = 1$, p_1 and g_1 modes in function of the scaled frequency $y = v_{\ell,n,0}/v_{0,1,0}$ where $v_{0,1,0}$ is the frequency of the radial fundamental mode. $\mathcal{R}_m$ is plotted for θ Ophiuchi, HD129929 and ν Eri. The same quantities for 8.2 $M_\odot$ stellar models are also represented. The models have been computed with CESAM2k code [54] assuming standard physics [42, 55] including a core overshooting distance of 0.1 H_p and an initial hydrogen abundance $X = 0.71$ and metal abundance $Z = 0.014$. The evolution of the selected models is represented by the central hydrogen content X_c from 0.5 to 0.2. The frequencies have been computed using a second order perturbation method and an adiabatic oscillation code WAR(saw)M(eudon) adapted from the Warsaw's oscillation code [56]. For each model, two sets of fre-

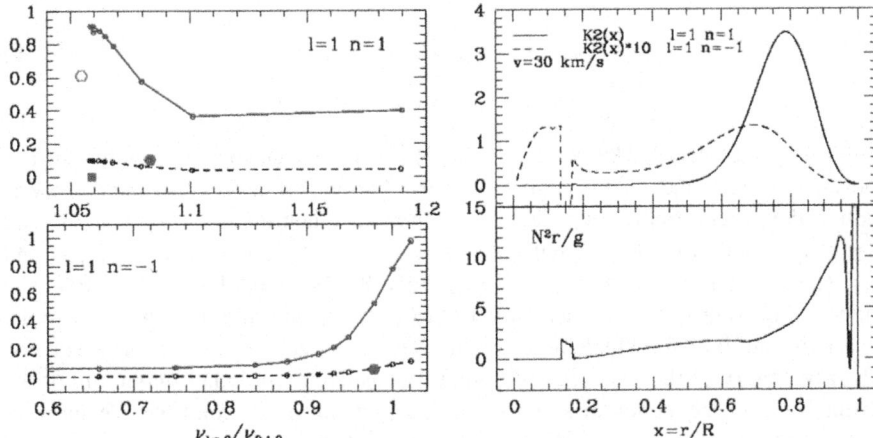

Fig. 8.7 *Left* Scaled asymmetries $\mathcal{R}_m \, 10^3$ for $\ell = 1 \; n = 1$, (*top*) and $n = -1$ (*bottom*) modes in function of the $m = 0$ frequency scaled by the radial fundamental mode frequency. The open dot (resp. *full dot, full square*) represents the observed asymmetry for θ Oph, (resp. for V386 Cen, ν Eri). The solid (resp. *dashed*) line corresponds to $\nu = 30$ km/s (resp. 10 km/s) $8.2 \, M_\odot$ models. The central hydrogen content X_c is decreasing toward the right. *Top right* Kernels $K_2(x)$ for splitting asymmetries of $\ell = 1, n = 1$ (p_1) mode (*solid line*) and $\ell = 1, n = -1$ (g_1) mode (*dashed line*) for model with $Xc = 0.35$. The abscissae is the normalized radius. *Bottom right*: Run of the normalized Brünt–Väissälä profile $N^2 r/g$ for the corresponding model with r/R. (from [59])

quencies are computed assuming a uniform rotation corresponding to $v = 30$ km/s and $v = 10$ km/s respectively. These sequences of models do not represent true evolutionary sequences as in realistic conditions, the rotation changes with time and can be non uniform. They however illustrate the evolution of the asymmetry when a mode changes its nature during evolution, from pure p mode to mixed p and g mode for instance. Indeed pure g modes have small asymmetries compared with pure p modes because they have much smaller amplitude in the outer envelope where distortion has its most significant effect. This is illustrated in Fig. 8.7. In a perturbation description, one finds that $\mathcal{R}_m$ is a second order effect proportional to Ω^2 ([29, 36, 57] and references therein). The variation of $\mathcal{R}_m$ with the scaled frequency y (i.e. with stellar evolution) is similar for the $v = 30$ and $v = 10$ km/s sequences of models but $\mathcal{R}_m$ is roughly nine times (ie ratio of Ω^2) larger for $v = 30$ km/s models than $v = 10$ km/s models. For pure p modes, the asymmetry amounts to $\mathcal{R}_m \sim 0.8 \times 10^{-3}$ whereas for pure g modes it almost vanishes. $\mathcal{R}_m$ for the $\ell = 1, n = 1$ mode decreases for older models (larger y). The reverse happens for the $\ell = 1, n = -1$ mode. The reason is that for young models, $\ell = 1, n = 1$ and $n = -1$ modes are pure p and g modes respectively. When the model is more evolved, these two modes experience an avoided crossing and exchange their nature. From a perturbative approach, one derives:

$$A_m = v_{\ell,n,0} \int_0^1 \hat{\Omega}^2(x) K_2(x) dx \qquad (8.49)$$

where $\hat{\Omega}^2 = \Omega^2/(GM/R^3)$ and $x = r/R$ the radius normalized to the surface radius. $K_2(x)$ depends on the centrifugal perturbation part of pressure and density as well as the differential rotation $\Omega(x)$ and the mode eigenfunction. Figure 8.7 shows $K_2(x)$ in function of the normalized radius $x = r/R$ for $\ell = 1, n = 1 (p_1)$ and $\ell = 1, n = -1 (g_1)$ modes for the $v = 30$ km/s, 8.2 $M_\odot$ model with $X_c = 0.5$. The inner layers contribute to the asymmetry of $\ell = 1, g_1$ multiplet in contrast with the $\ell = 1, p_1$ multiplet for which the kernel K_2 is concentrated toward the surface layers. The asymmetry of the $\ell = 1, g_1$ multiplet is sensitive to the inner maximum of the Brünt–Väissälä frequency, arising from the μ-gradient, which contributes negatively to K_2. As the negative contribution is very localized, it decreases the asymmetry only slightly compared to a pure p mode for a uniform rotation. However, one can expect a larger decrease in case of a rotation faster in the inner regions than the surface.

Theoretical estimates seem to disagree with observed asymmetries deduced from $\ell = 2$ modes for θ Ophiuchi [47] and v Eridani for $\ell = 1, p_2$ [48, 51, 58]. Is the disagreement real? The question has some relevance as the asymmetry values are only marginally above the observation uncertainties. Or can it be that the observed frequencies do not belong to the same multiplet as suggested by Dziembowski and Pamyatnykh [50] for v Eri?

8.4.3 Axisymmetric Modes: Mixing

Rotationally induced mixing of chemical elements changes the structure and in particular affects the Brünt–Väissälä frequency N at the border of the convective core. As a consequence, at a given location in a HR diagram corresponding to an observed star, one can find several models with different structures and therefore likely different values of the mode frequencies including axisymmetric modes which can then be used as diagnostics for mixing.

Uniform and constant diffusion coefficient D_t. Montalban et al. [17] and Miglio et al. [18] investigated the effect of turbulent mixing on a g-mode frequency spectrum and the ability of such modes to probe the size of stellar convective cores. They assumed a constant in time and uniform in space global diffusion coefficient $D_t = D_{\text{eff}} + D_v$ in (8.1). The constant value for D_t is chosen so as to correspond to the value near the convective core provided by a Geneva stellar model including rotationally induced mixing. This is valid for g-modes which have most of their amplitude there (see [18]). The model is a mid main sequence ($X_c = 0.3$) 10 $M_\odot$ with $D_t = 710^4$ cm^2/s chosen to correspond to a rotational velocity $v = 50$ km/s.

Figure 8.8 shows the Brünt–Väissälä frequency (N) profile for a model with turbulent chemical element mixing and a model with no turbulent chemical element

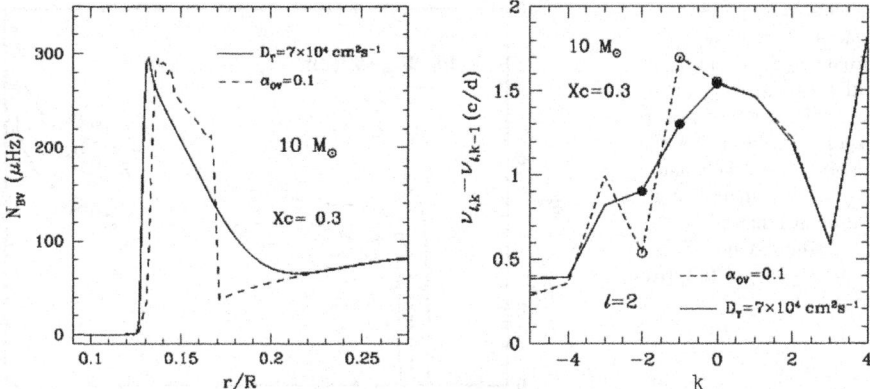

Fig. 8.8 *Left* Brünt–Väissälä profile in the central region of a 10 M$_\odot$ model with $X_c = 0.3$ and an initial velocity of 50 km/s. *Right* the large separation $v_{\ell,n,0} - v_{\ell,n-1,0}$ in function of the radial order n for $\ell = 2$ modes for a model including turbulent mixing (*solid*) line and a model including a $0.1 H_p$ overshoot instead (*dashed line*) (from [17])

mixing but including instead core overshoot assuming an overshoot distance of 0.1 H_p. Differences can be seen at the edge of the convective core. The Brünt–Väissälä frequency of the model with turbulent mixing behaves more smoothly in the μ-gradient region above the convective core than for the model computed with no turbulent mixing but with an overshoot distance of $0.1 H_p$. From Geneva code calculations, the evolution of the rotation profile leads to a core to envelope ratio of 1.6. The differences between the two profiles arising at the edge of the convective core cause significant changes on frequencies of g-modes and mixed modes. The frequency separations $\Delta_{n,\ell} = v_{\ell,n,0} - v_{\ell,n-1,0}$ differ by a few μHz for radial order $n = -1$ and $n = -2$, $\ell = 2$ modes between the model with overshoot $0.1 H_p$ and the model with turbulent mixing (Fig. 8.8). At higher frequencies for pure p-modes, no differences in $\Delta_{n,\ell}$ are seen when adding turbulent mixing or not.

Rotationally Induced Diffusion Coefficient. In this section, we consider stellar models which are computed with the Toulouse–Geneva evolutionary code which includes the coupling between rotationally induced mixing and momentum transport (8.1) and (8.2) as described by Talon [11]. The rotational evolution of the star begins from solid body when the core is still radiative, shortly after the star leaves the Hayashi track. A 8.5 M$_\odot$ mass has been chosen so that the models evolve through the HR diagram to a location where the star θ Ophiuchi is expected (log $L/L_\odot =$ 3.73, $T_{\text{eff}} = 4.35$). This corresponds to a mid main sequence model, $\mathbf{V_{15}}$, with a central hydrogen content $X_c = 0.3$. The evolution has been initiated with a uniform rotational velocity $v = 15$ km/s on the pms; the rotation profile then evolves to strongly differential rotation so that $\mathbf{V_{15}}$ has a surface velocity of $v = 48.2$ km/s and a ratio $\Omega_{\text{core}}/\Omega_{\text{surf}} = 1.6$ when crossing the θ Ophiuchi location in the HR diagram at an age of 19.65 Myr.

Fig. 8.9 Run of the
rotationally induced
turbulent coefficient, D_t,
with the relative shell mass at
three different evolutionary
stages with ages 0.5, 1 and
1.5 Myr respectively and
labelled with their central
hydrogen content
X_c- leading to the stellar
model V_{15} ($X_c = 0.3$) (from
[59])

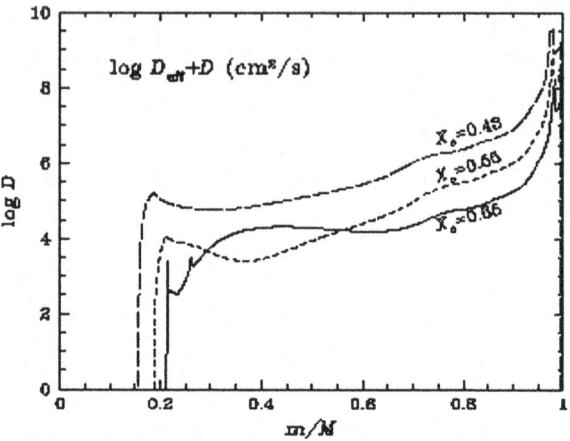

The diffusion coefficient, D_t, depends on the meridional circulation velocity and the local turbulence strength. It varies with depth and evolves with time as illustrated in Fig. 8.9. The D_t profile is represented for 3 models with ages 0.5, 1 and 1.5 Myr built assuming an initial 15 km/s velocity on the pms. The rotation evolving from uniform to strongly differential rotation causes a relaxation toward a stationary profile which persists with only an adjustment due to expansion and contraction with evolution [59].

Effect of rotationally induced mixing on the structure is significant at the edge of the convective core as emphasized in Fig. 8.9 where we compare the squared Brünt–Väissälä profile, N^2, in the vicinity of the edge of convective core for model V_{15} and a model V_0 which includes neither rotationally induced mixing Fig. 8.10 nor overshoot. Inclusion of rotationally induced mixing leads to the model V_{15} which shows a narrower maximum of Brünt–Väissälä profile at the edge of the convective core compared with that of V_0.

To illustrate the impact of such a difference on the oscillation frequencies, we compare low radial order frequencies of the models V_{15}, and V_0. Modes p_1, p_2, g_1 for these models have amplitudes near the edge of the convective core. Figure 8.11 shows that this can result in significant frequency differences for the same mode easily detectable with CoRoT observations. The frequencies of these modes are quite sensitive to the detail of the Brünt–Väissälä profile in this region. This means that some care must be taken when computing these frequencies and drawing conclusions. The frequencies of these modes are indeed sensitive not only to the physics but unfortunately also to the numerics which can be quite inaccurate in this region of the star.

The sign and magnitude of $\delta\omega = \omega_{V15} - \omega_{V0}$ are dependent on the mode when it has amplitude in the regions where the nonrotating model and the model with rotationally induced mixing differ. We consider here, as in Sect. 8.3.5, only the effect of rotationally induced mixing on the spherically symmetric structure. Differences in the structure of the model V_{15} which includes rotationally induced mixing and

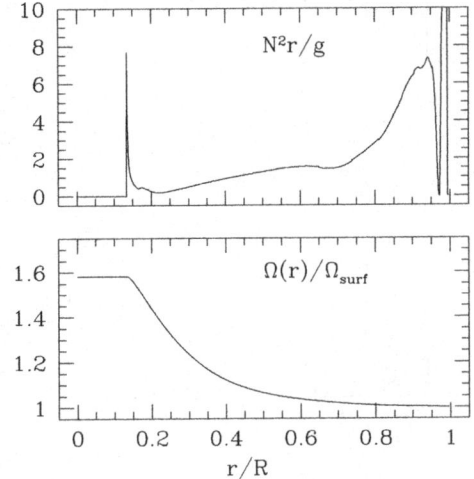

Fig. 8.10 Model V_{15} with a surface rotational velocity $v = 48.2$ km/s. *Top:* Profile of of the normalized Brünt–Väisälä frequency as defined by $N^2 r/g$ in function of the normalized radius r/R. *Bottom:* rotation profile normalized to its surface value. The core to surface ratio for the rotation rate then is 1.6 [60]

the model V_0 which does not result in differences in the eigenfrequencies which we note $\delta\omega = \omega_{0,\Omega} - \omega_{0,\Omega=0}$.

The structure of the models V_{15} and V_0 indicates that p_Ω and its derivative, the gravity g_Ω, the density ρ_Ω are not significantly modified compared to the derivative of the density. Figure 8.11 shows that the largest difference $\delta(d \log \rho_\Omega/d \log r)$ arises near the convective core. Accordingly from (8.47), one expect larger frequency differences $\delta\omega$ for mixed modes compared to p-modes. This is what is observed in Fig. 8.11. As explained in Sect. 8.3.5, with the help of the integral relation for $\delta\omega$, the frequency differences for high frequency (i.e. pure) p-mode is small and positive. For lower frequency mixed modes, $\delta\left(\frac{d \ln \rho_\Omega}{d \ln r}\right)$ can be positive and the frequency difference can be large and negative as illustrated in Fig. 8.11.

8.5 Cubic Order Versus Latitudinal Dependence

It has been known for a long time that latitudinal variations of the rotation rate generate departures from linear splitting. On the other hand, a fast uniform rotation can generate cubic order corrections to the frequency of non axisymmetric modes which also cause departure from linear splitting. The latitudinal correction to the linear splitting is proportional to the Ω gradient whereas cubic order effects, as their name indicate, are proportional to Ω^3. It is expected that the dependence of these corrections with the frequency differs when it is due to latitudinal differential rotation or to cubic effects.

Low mass stars are known to be slow rotators. Indeed due to their outer convection zone, they undergo magnetic braking. Due again to their outer convection zone, observational evidences exist for surface latitudinal differential rotation. Hence

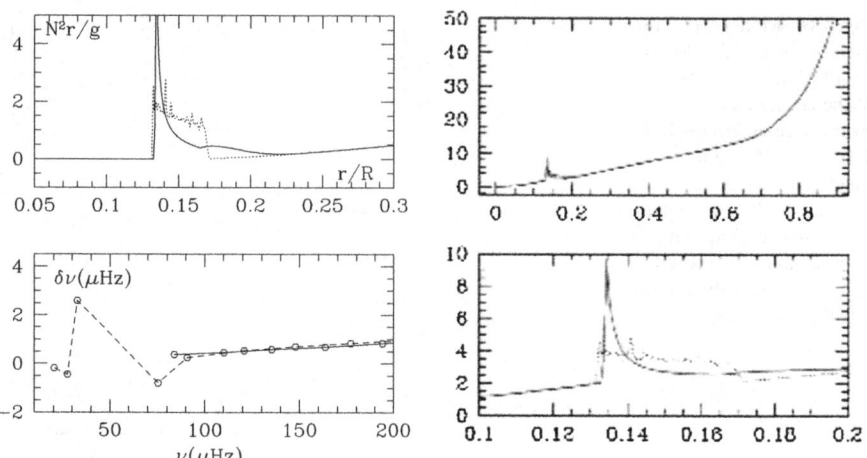

Fig. 8.11 *Left top:* Zoom of Brünt–Väissälä frequency profile in the vicinity of the edge of convective core in function of the normalized radius r/R for model V_{15} (*dashed line*) and model V_0 (*solid line*). The local maximum of N^2r/g corresponds to a nonzero μ-gradient. It decreases more sharply in presence of rotationally induced mixing because mixing results in smoothing the μ-gradient. *Left bottom:* Differences $\delta\nu = \delta\omega/(2\pi)$ between frequencies computed from model V_0 (no rotationally induced mixing included) and model V_{15} for $\ell = 0$ (*solid line*) and $\ell = 1$ (*dashed line*) $m = 0$ modes in μHz (from Goupil and Talon [28]). *Right d* $\ln \rho/d \ln r$ in function of the normalized radius r for model V_{15} (*dashed line*) and model V_0 (*solid line*), *right top:* from center to surface, *right bottom:* in the central region [60]

for these stars, the averaged rotation rate Ω is small and $\Delta\Omega = \Omega_{\text{equa}} - \Omega_{\text{pole}}$, the difference between the rotation rates at the equator and the poles, can be large (25–30% for the Sun, between 1% and 45% for a star like Procyon, Bonanno et al. [61]). One therefore expects that latitudinal corrections to the splittings dominate over cubic order ones which are negligible. On the other hand, more massive stars on the main sequence have shallower convection zones which even disappear above $\sim$3–5 $M_\odot$. These stars usually are fast rotators with a radiative envelope which may or may not be in latitudinal differential rotation. For these fast rotators, one can wonder what is the minimal latitudinal shear which dominates over cubic order effects and can therefore be detectable. Here we quantify this issue with the help of a polytropic model with index 3. The constants characterizing the polytrope are taken to correspond to model **A** considered in Sect. 9. We establish first the splitting correction due to latitudinal differential rotation. This is then compared with the splitting correction arising from cubic order effects as derived by previous works. We assume a rotation velocity of 100 km/s.

8.5.1 Latitudinal Dependence

Hansen et al. [62] derived the expression for the rotational splitting of adiabatic non-radial oscillations for slow differential (steady, axially symmetric) rotation $\Omega(r, \theta)$ and applied it to numerical models of white dwarfs and of massive main sequence stars assuming a cylindrically symmetric rotation law. In the solar case, the effects of latitudinal differential rotation on theoretical frequencies were investigated by Dziembowski and Goode [36], Gough and Thompson [41]; Dziembowski and Goode [63] who also considered the case of δ Scuti stars.

In order to be able to compute the splittings from (8.17) and (8.22), one must specify a rotation law. It is convenient to assume a rotation of the type:

$$\Omega(r, \theta) = \sum_{s=0}^{s_{\max}} \Omega_{2s}(r)(\cos \theta)^{2s} \qquad (8.50)$$

where θ is the colatitude and we take $s_{\max} = 2$. The surface rotation at the equator is $\Omega(r = R, \theta = \pi/2) = \Omega_0(r = R)$. Note that in the solar case, Ω_2, Ω_4 are negative and the equator rotates faster than the poles [63, 64]. As shown in Appendix, inserting (8.50) into (8.22) yields the following expression for the generalized splitting ((8.86) in Appendix):

$$S_m = \int_0^R \Omega_0(r) K(r) dr + \sum_{s=0}^{s=2} m^{2s} H_s(\Omega) \qquad (8.51)$$

with $K(r)$ defined in (8.25) and

$$H_s(\Omega) = -\frac{1}{I} \int_0^R \Omega_0(r) \left[R_s \left(\xi_r^2 - 2\xi_r\xi_h + \xi_h^2 ((\Lambda - 1)) + Q_s\xi_h^2 \right) \right] \rho_0 r^2 dr \qquad (8.52)$$

where R_s and Q_s depend on Ω_2, Ω_4 and $\Lambda = \ell(\ell + 1)$ and are given by (8.84) and (8.86) (Appendix) respectively.

(a) *Uniform rotation.* In that case, $\Omega(r, \theta) = \Omega_0$, $\forall r$, θ; Ω_2, $\Omega_4 = 0$ i.e. R_j, $Q_j = 0$ for $j = 0, 2$ hence $H_{m,j} = 0$. One recovers the well known expression:

$$S_m = \Omega_0 \beta \qquad (8.53)$$

where, for later purpose, we have defined

$$\beta = \int_0^R K(r) dr = -\frac{1}{I} \int_0^R [\, \xi_r^2 - 2\xi_r\xi_h + (\Lambda - 1)\xi_h^2 \,] \rho_0 r^2 dr \qquad (8.54)$$

This is usually rewritten as:

$$S_m = \Omega_0(C_L - 1)$$

where C_L is the [65] constant

$$C_L = \frac{1}{I} \int\limits_0^R \left[2\xi_r\xi_h + \xi_h^2 \right] \rho_0 r^2 dr$$

(b) *Shellular rotation* then $\Omega(r, \theta) = \Omega_0(r)$ and $s_{max} = 0$; again here: $\Omega_2, \Omega_4 = 0$ ie $R_j = Q_j = 0$ for $j = 0, 2$ and

$$S_m = -\frac{1}{I} \int\limits_0^R \Omega_0(r) [\, \xi_r^2 - 2\xi_r\xi_h + \Lambda\xi_h^2 \,] \rho_0 r^2 dr \qquad (8.55)$$

(c) *Latitudinally differential rotation only.* In that case, Ω_{2j}, $j = 0, 2$ are depth independent and R_s and Q_s are constant and

$$S_m = \Omega_0\beta + \Omega_0 \sum_{s=0}^{s=2} m^{2s}(R_s(\Omega)\beta + Q_s(\Omega)\gamma) \qquad (8.56)$$

with β defined in (8.54) and

$$\gamma = -\frac{1}{I} \int\limits_0^R \xi_h^2 \rho_0 r^2 dr$$

For a triplet $\ell = 1$, $m = 1$ ($\Lambda = 2$) then

$$S_1 = \Omega_0\beta + \Omega_0(R(\Omega)\beta + Q(\Omega)\gamma) \qquad (8.57)$$

with (using (8.84) and (8.86)):

$$R(\Omega) = \sum_{s=0}^{s=2} R_s(\Omega) = \frac{1}{5}\frac{\Omega_2}{\Omega_0} + \frac{3}{7}\frac{\Omega_4}{\Omega_0} \qquad (8.58)$$

$$Q(\Omega) = \sum_{s=0}^{s=2} Q_s(\Omega) = -\frac{24}{5}\frac{\Omega_4}{\Omega_0} \qquad (8.59)$$

In the solar case, $\beta \sim -1$ and $|\beta| \gg |\gamma|$ for the excited high frequency p-modes.

$$S_1 \approx -\Omega_0 \left(1 + \frac{1}{5} \frac{\Omega_2}{\Omega_0} + \frac{3}{7} \frac{\Omega_4}{\Omega_0} \right) \qquad (8.60)$$

With $\Omega_2/\Omega_0 = -0.127$, $\Omega_4/\Omega_0 = -0.159$, one obtains a departure from linear splitting of $|S_1/\Omega_0 + 1| = 0.093$ i.e. a $\approx 10\%$ change in the solar case. For upper main sequence stars, excited modes are around the fundamental radial mode and may be mixed modes with $|\beta| \sim |\gamma| \sim 1/2$. This leads for instance to $|S_1/\Omega_0 + 1/2| \approx 5\%$ for Ω_2/Ω_0 and Ω_4/Ω_0 equal to 1/5 of the solar values.

8.5.2 Latitudinal Dependence Versus Cubic Order Effects

Let assume on one side a pulsating star uniformly rotating with a rate Ω_0 high enough that cubic order $(O(\Omega_0^3))$ contributions are significant. On the other side, one also considers a model rotating with a latitudinally differential rotation (uniform in radius). One issue then is which one of these two effects dominate over the other one since the cubic one is $O(\Omega^3)$ whereas the other one is $O(\Delta\Omega)$?

For stars other than the Sun, one can simply assume the rotational latitudinal shear $\Delta\Omega = \Omega_2$ with $\Omega_4 = 0$ and $\Omega(\theta) = \Omega_0 + \Delta\Omega \cos^2 \theta$.

For $\ell = 1$ modes, (8.57) becomes

$$S_1(lat) = \Omega_0 \beta \left(1 + \frac{1}{5} \frac{\Delta\Omega}{\Omega_0} \right) \qquad (8.61)$$

Expressions for the frequency correction (in rad/s) for cubic order effects assuming a uniform rotation has been derived by Soufi et al. [37]. Part of the cubic order effect is included in the eigenfrequency $\omega_{0,\Omega}$ and therefore is also included in second order coefficients which indeed involve $\omega_{0,\Omega}$. Another part of the cubic order effects is included as an additive correction to the frequency.

Frequency up to 3rd order were computed for models of δ Scuti stars by Goupil et al. [66], Goupil and Talon [16], Pamyatnykh [67], Goupil et al. [68] and Karami [39] rederived the cubic order effects following Soufi et al.'s [38] approach and Karami [39, 69] applied it to a ZAMS model of a 12 $M_\odot$ β Cephei star. He found that cubic order effects are of the order of 0.01% for a $l = 2, n = 2$ and 0.5% for a $n = 14$ mode for a 100 km/s rotational velocity. Values of the third order additive correction to the frequency were listed for $\ell = 1$ p-modes of a polytrope of index 3 by Goupil [29].

Here we write the splitting under the form:

$$S_m(\text{cubic}) = \Omega_0 \beta + \Omega_0 \left(\frac{\hat{\Omega}_0}{\sigma_0} \right)^2 T_{|m|} \qquad (8.62)$$

where the last term represents the *full* cubic order contribution with σ_0 is the normalized frequency of the nonrotating polytrope and $\hat{\Omega}_0 = \Omega_0/\Omega_K$.

Table 8.2 Coefficients assuming a uniform rotation for a polytrope with polytropic index 3 and adiabatic index $\gamma = 5/3$

n	σ_0^2	C_{L0}	X_1	X_2	Y_1	Y_2	T_1/σ_0^2	$-\beta$	$-\gamma$
				$\ell = 1$					
-7	0.22	0.479	0.417	0.008	0.012	-0.018	0.592	0.521	0.456
-6	0.28	0.476	0.419	0.005	0.015	-0.023	0.462	0.524	0.450
-5	0.37	0.473	0.422	0.001	0.020	-0.029	0.351	0.527	0.441
-4	0.52	0.469	0.425	-0.004	0.026	-0.039	0.254	0.531	0.431
-3	0.78	0.466	0.427	-0.013	0.038	-0.056	0.164	0.534	0.410
-2	1.28	0.466	0.428	-0.024	0.059	-0.089	0.073	0.535	0.386
-1	2.51	0.473	0.422	-0.035	0.106	-0.159	-0.025	0.528	0.269
1	11.37	0.029	0.777	0.877	2.890	-4.335	0.024	0.970	0.025
2	21.49	0.034	0.773	0.864	5.802	-8.703	-0.034	0.966	0.028
3	34.83	0.033	0.773	0.851	9.624	-14.436	-0.063	0.966	0.027
4	51.39	0.031	0.776	0.840	14.340	-21.511	-0.077	0.969	0.026
5	71.15	0.027	0.778	0.832	19.940	-29.909	-0.084	0.973	0.025
6	94.09	0.024	0.781	0.826	26.414	-39.621	-0.088	0.976	0.023
7	120.19	0.021	0.783	0.821	33.757	-50.635	-0.089	0.979	0.022
8	149.43	0.019	0.785	0.817	41.964	-62.946	-0.089	0.981	0.020
9	181.81	0.017	0.787	0.814	51.032	-76.548	-0.089	0.984	0.019
10	217.32	0.015	0.788	0.811	60.958	-91.437	-0.089	0.985	0.018
11	255.94	0.013	0.789	0.809	71.739	-107.609	-0.088	0.987	0.017
12	297.67	0.012	0.790	0.807	83.375	-125.062	-0.087	0.988	0.017
13	342.51	0.011	0.791	0.805	95.862	-143.793	-0.087	0.989	0.016
14	390.44	0.010	0.792	0.804	109.201	-163.802	-0.086	0.990	0.015
15	441.47	0.009	0.793	0.803	123.392	-185.087	-0.085	0.991	0.014
16	495.59	0.008	0.793	0.802	138.432	-207.648	-0.085	0.992	0.014
17	552.80	0.008	0.794	0.801	154.323	-231.484	-0.084	0.993	0.013
18	613.09	0.007	0.794	0.800	171.064	-256.595	-0.084	0.993	0.013
19	676.47	0.006	0.795	0.799	188.655	-282.982	-0.083	0.994	0.012
20	742.93	0.006	0.795	0.798	207.097	-310.645	-0.083	0.994	0.012
21	812.46	0.006	0.796	0.798	226.389	-339.584	-0.082	0.995	0.012
22	885.08	0.005	0.796	0.797	246.534	-369.800	-0.082	0.995	0.011
23	960.78	0.005	0.796	0.797	267.530	-401.295	-0.082	0.995	0.011

The squared frequency σ_0^2 is the dimensionless squared frequency $\omega^2/(GM/R^3)$. Spherical centrifugal distortion of the polytrope has not been included

Table. 8.2 lists the value of the dimensionless coefficients T_1/σ_0^2 and $-\beta$, $-\gamma$ for $\ell = 1$ modes for a polytrope with a polytropic index 3.

The coefficient T_1/σ_0^2 remains nearly constant with increasing frequency for frequencies above $\sigma_0 > 10$ i.e. for p modes For $\sigma_0 > 10$ (p-modes), $-\beta \approx 1$ and $T_1/\sigma_0^2 \approx -0.09$. The splitting is decreased by a latitudinal dependence with $\Delta\Omega < 0$ whereas it is increased by cubic order effects $T_1/\beta > 0$. In absolute values, the effect of latitudinal differential rotation on the splittings then dominates over cubic order effects whenever:

$$\left|\frac{\Delta\Omega}{\Omega_0}\right| > 0.45\,\hat{\Omega}_0^2$$

For model **A** and a rotational velocity $100\,\text{km/s}$, $\hat{\Omega}_0 = 0.174$ then $\left|\frac{\Delta\Omega}{\Omega_0}\right| > 1.36\%$ For a faster rotator with for instance $200\,\text{km/s}$, the latitudinal shear must be larger i.e. $\left|\frac{\Delta\Omega}{\Omega_0}\right| > 5.45\%$.

For the slowly rotating β Cep stars considered in Sect. 8.4 above ($v < 50\,\text{km/s}$), cubic order effects in the splittings can be neglected in front of latitudinal effects equal or larger than 0.34%. At this low level, both effects are comparable to the observational uncertainties (0.1%).

8.6 Conclusions

We have seen along this review that several efficient seismic tools can be designed to obtain valuable information on the internal structure and dynamics of main sequence massive stars which oscillate with a few identified modes. Identification of the detected modes requires a high signal to noise which is made available due to the large amplitudes of these opacity-driven modes. On the other hand, these stars oscillate with low frequencies lying near/in the dense part of the spectrum where p modes, mixed modes and g modes can be encountered. While this is a great advantage in order to probe the inner layers of the star, resolution and precise measurement of quite close frequencies in a Fourier spectrum requires very long time series. This explains the yet still small number of β Cephei stars for which a successful seismic analysis has been obtained, despite the appealing prospects that a better knowledge of their structure bring up valuable constrains on their still poorly understood life end as supernovae. It is expected that the space experiments CoRoT (Michel et al. 1995) and Kepler [70] will increase the number of O-B stars for which fruitful seismic analyses can be carried out as well as possibly enlarge the sample to fast rotators. Mode identification can be at first difficult to perform for fast rotators but some of these fast rotating stars might also show oscillations of solar-like type which characteristics could help the mode identification. This interesting perspective has recently emerged with the discovery of the first chimera star with the CoRoT mission [71].

8.7 Appendix: Differential Rotation

The expression for the mode splitting of adiabatic nonradial oscillations due to a differential rotation $\Omega(r, \theta)$ can be put into the compact form [29, 36, 62, 63, 72], [73, 74]:

$$\delta\omega_m = m \int_0^R \int_0^\pi \mathcal{K}_m(r,\theta)\Omega(r,\theta)d\theta dr \tag{8.63}$$

where $\mathcal{K}_m$ is called *rotational kernel*:

$$\mathcal{K}_m(r,\theta) = -\frac{\rho_0 r^2}{I}\frac{\sin\theta}{2}\Omega(r,\theta)$$

$$\times \int \frac{d\phi}{2\pi}\left(\left(|\xi_r|^2 - (\xi_r^*\xi_h + cc)\right)|Y_\ell^m|^2 + |\xi_h|^2\left(\nabla_H Y_\ell^{m*}\cdot\nabla_H Y_\ell^m - \frac{\partial|Y_\ell^m|^2}{\partial\theta}\frac{\cos\theta}{\sin\theta}\right)\right)$$

$$\tag{8.64}$$

where the spherical harmonics Y_ℓ^m are normalized such that

$$\int (Y_\ell^{m'})^*(\theta,\phi)Y_\ell^m(\theta,\phi)\frac{d\Omega}{4\pi} = \delta_{\ell,\ell'}\delta_{m,m'}$$

where $d\Omega = \sin\theta d\theta d\phi$ is the solid angle elemental variation and $\delta_{\ell,\ell'}$ is the Kroenecker symbol. Mode inertia I is given by

$$I = \int_0^R \left(|\xi_r|^2 + \Lambda|\xi_h|^2\right)\rho_0 r^2 dr \tag{8.65}$$

with $\Lambda = \ell(\ell+1)$.

It is convenient to assume a rotation of the type:

$$\Omega(r,\theta) = \sum_{s=0}^{s_{max}} \Omega_{2s}(r)(\cos\theta)^{2s} \tag{8.66}$$

where θ is the colatitude. (8.63) becomes:

$$\delta\omega_m = -\frac{m}{I}\sum_{s=0}^{s_{max}}\int_0^R \Omega_{2s}(r)$$

$$\times \left[(|\xi_r|^2 - (\xi_r^*\xi_h + cc))S_s + |\xi_h|^2(B_1 + B_2)\right]\rho_0 r^2 dr \tag{8.67}$$

where we have defined

$$S_s \equiv \int |Y_\ell^m|^2(\cos\theta)^{2s}\frac{d\Omega}{4\pi} = \int_0^1 \mu^{2s}|Y_\ell^m(\theta,\phi)|^2 d\mu \tag{8.68}$$

with $\mu = \cos\theta$ and

$$B_1 = \int \left(\nabla_H Y_\ell^{m*}\cdot\nabla_H Y_\ell^m\right)(\cos\theta)^{2s}\frac{d\Omega}{4\pi} \tag{8.69}$$

$$B_2 = -\int \left(\frac{\partial |Y_\ell^m|^2}{\partial \theta} \frac{\cos \theta}{\sin \theta} \right) (\cos \theta)^{2s} \frac{d\Omega}{4\pi} \qquad (8.70)$$

The term in $|\xi_h|^2$ requires a little care. Consider first $B1$. Integration by part leads to

$$B1 = -\int \frac{d\Omega}{4\pi} Y_\ell^{m*}$$
$$\times [(\nabla_H^2 Y_\ell^m)(\cos \theta)^{2s} + (\nabla_H Y_\ell^m) \cdot \nabla_H ((\cos \theta)^{2s})] \qquad (8.71)$$

Recalling that $\nabla_H^2 Y_\ell^m = -\Lambda Y_\ell^m$, one gets

$$B1 = \Lambda \mathcal{S}_s - \frac{1}{2} \int \left[Y_\ell^{m*} \frac{\partial Y_\ell^m}{\partial \theta} \frac{d(\cos \theta)^{2s})}{d\theta} + cc \right] \frac{d\Omega}{4\pi} \qquad (8.72)$$

where cc means complex conjugate. Again an integration by part yields

$$B1 = \Lambda \mathcal{S}_s + \frac{1}{2} \int |Y_\ell^m|^2 \frac{d}{d\theta} \left[\sin \theta \frac{d(\cos \theta)^{2s})}{d\theta} \right] \frac{d\theta}{2} \qquad (8.73)$$

One finally obtains

$$B1 = \Lambda \mathcal{S}_s + s \left[(2s-1)\mathcal{S}_{s-1} - (2s+1)\mathcal{S}_s \right] \qquad (8.74)$$

Turning to the second term B_2 in (8.68), an integration by part yields

$$B_2 = -(2s+1)\mathcal{S}_s \qquad (8.75)$$

Inserting expressions (8.74) and (8.75) into (8.68), one obtains

$$\delta \omega_m = m \sum_{s=0}^{s_{max}} \int_0^R \Omega_{2s}(r) K_{m,s}(r) dr \qquad (8.76)$$

with

$$K_{m,s}(r) = K(r)\mathcal{S}_s - \frac{1}{I} \rho_0 r^2 |\xi_h|^2 s \left[(2s-1)\mathcal{S}_{s-1} - (2s+3)\mathcal{S}_s \right] \qquad (8.77)$$

and

$$K(r) = -\frac{1}{I} \left[|\xi_r|^2 - (\xi_r^* \xi_h + cc) + |\xi_h|^2 (\Lambda - 1) \right] \rho_0 r^2 \qquad (8.78)$$

Expression (8.77) is equivalent to (8.25) in [36]. For any s, $\mathcal{S}_s$ is given by a recurrent relation (8.31) in Dziembowski and Goode [36]). Note that $\delta\omega_m = \delta\omega_{-m}$. Let define the generalized splitting

$$S_m = \frac{\omega_m - \omega_{-m}}{2m} = \frac{\delta\omega_m - \delta\omega_{-m}}{2m} = \frac{\delta\omega_m}{m}$$

We limit the expression for the rotation to $s_{max} = 2$ i.e.:

$$\Omega(r,\theta) = \Omega_0(r) + \Omega_2(r)\cos^2\theta + \Omega_4(r)\cos^4\theta \qquad (8.79)$$

then for adiabatic oscillations ($\xi_r(r)$ and $\xi_h(r)$ are real):

$$K_{m,0}(r) = K(r)$$

$$K_{m,1}(r) = K(r)\mathcal{S}_1 - \frac{1}{I}\xi_h^2[1 - 5\mathcal{S}_1]\rho_0 r^2 \qquad (8.80)$$

$$K_{m,2}(r) = K(r)\mathcal{S}_2 - \frac{1}{I}\xi_h^2 2[3\mathcal{S}_1 - 7\mathcal{S}_2]\rho_0 r^2$$

where we have used $\mathcal{S}_{-1} = 0$; $\mathcal{S}_0 = 1$.

We obtain a formulation for the generalized splittings with a m dependence of the form:

$$S_m = \int_0^R (\Omega_0 + \Omega_2\mathcal{S}_1 + \Omega_4\mathcal{S}_2)\, K(r)dr$$

$$- \frac{1}{I}\int_0^R (\Omega_2(1 - 5\mathcal{S}_1) + \Omega_4 2(3\mathcal{S}_1 - 7\mathcal{S}_2))\,\xi_h^2\rho_0 r^2 dr \qquad (8.81)$$

One needs S_1 and S_2 (computed from (8.31) in [35]):

$$S_1 = \frac{1}{4\Lambda - 3}(-2m^2 + 2\Lambda - 1) = \frac{2\Lambda - 1}{4\Lambda - 3} - m^2\frac{2}{4\Lambda - 3}$$

$$S_2 = \frac{1}{4\Lambda - 15}\frac{3}{2}\left[S_1(-2m^2 + 2\Lambda - 9) + 1\right]$$

The first term in brackets in (8.81) becomes

$$(\Omega_0 + \Omega_2\mathcal{S}_1 + \Omega_4\mathcal{S}_2) = \Omega_0(1 + R_0 + m^2 R_1 + m^4 R_2) \qquad (8.82)$$

where

$$R_0 = \frac{\Omega_2}{\Omega_0}\frac{2\Lambda - 1}{4\Lambda - 3} + 3\frac{\Omega_4}{\Omega_0}\frac{[(2\Lambda^2 - 8\Lambda + 3)]}{(4\Lambda - 15)(4\Lambda - 3)}$$

$$R_1 = -\frac{2}{4\Lambda - 3}\left[\frac{\Omega_2}{\Omega_0} + 3\frac{\Omega_4}{\Omega_0}\frac{(2\Lambda - 5)}{(4\Lambda - 15)}\right] \qquad (8.83)$$

$$R_2 = \frac{\Omega_4}{\Omega_0}\frac{6}{(4\Lambda - 15)(4\Lambda - 3)}$$

For the second term in (8.81), one has:

$$\Omega_2(1 - 5S_1) + \Omega_4 2(3S_1 - 7S_2) = \Omega_0(Q_0 + m^2 Q_2 + m^4 Q_2) \tag{8.84}$$

where

$$
\begin{aligned}
Q_0 &= \frac{2}{4\Lambda - 3}\left[\frac{\Omega_2}{\Omega_0}(1 - 3\Lambda) - 6\frac{\Omega_4}{\Omega_0}\frac{(3\Lambda^2 - 11\Lambda + 3)}{4\Lambda - 15}\right] \\
Q_1 &= \frac{10}{4\Lambda - 3}\left[\frac{\Omega_2}{\Omega_0} + 12\frac{\Omega_4}{\Omega_0}\frac{(\Lambda - 2)}{(4\Lambda - 15)}\right] \\
Q_2 &= -\frac{4}{(4\Lambda - 3)}\frac{21}{(4\Lambda - 15)}\frac{\Omega_4}{\Omega_0}
\end{aligned}
\tag{8.85}
$$

Collecting terms from (8.82) and (8.84), the generalized splitting (8.81) takes the expression:

$$S_m = \int_0^R \Omega_0(r)K(r)dr + \sum_{s=0}^{s=2} m^{2s} H_s(\Omega) \tag{8.86}$$

with

$$H_s(\Omega) = \int_0^R \Omega_0(r)\left[R_s K(r) - Q_s\frac{1}{I}\xi_h^2\right]\rho_0 r^2 dr \tag{8.87}$$

For a depth independent rotation law, $\Omega(\theta)$, Ω_{2j}, $j = 0, 2$ are depth independent and R_s and Q_s are constant. then for a triplet $\ell = 1$ ($\Lambda = 2$):

$$S_1 = \Omega_0\beta + \Omega_0\left(\sum_{s=0}^{s=2} R_s(\Omega)\right)\beta + \left(\sum_{s=0}^{s=2} Q_s(\Omega)\right)\gamma \tag{8.88}$$

with

$$\sum_{s=0}^{s=2} R_s = \frac{1}{5}\frac{\Omega_2}{\Omega_0} + \frac{3}{7}\frac{\Omega_4}{\Omega_0} \tag{8.89}$$

$$\sum_{s=0}^{s=2} Q_s = -\frac{24}{5}\frac{\Omega_4}{\Omega_0} \tag{8.90}$$

and

$$\beta = \int_0^R K(r)dr \tag{8.91}$$

$$\gamma = -\frac{1}{I}\int_0^R \xi_h^2\rho_0 r^2 dr \tag{8.92}$$

References

1. Lovekin, C.C., Deupree, R.G., Clement, M.J.: ApJ **693**, 677 (2009)
2. Reese, D.R., MacGregor, K.B., Jackson, S., et al.: In: Dikpati, M., Arentoft, T., González Hernández, I., Lindsey, C., Hill, F. (eds.) Solar-Stellar Dynamos as Revealed by Helio- and Asteroseismology: GONG 2008/SOHO ASPC 416, 395 (2009).
3. Zahn, J-P.: A&A **252**, 179 (1991)
4. Roxburgh, I.W.: A&A **266**, 291 (1992)
5. Schaller, G., Schaerer, D., Meynet, G., Maeder, A.: A&AS **96**, 269 (1992)
6. Cordier, D., Lebreton, Y., Goupil, M.J., et al.: A&A **392**, 169 (2002)
7. Claret, A.: A&A 475, 1019 (2007).
8. Lebreton, Y.: IAU Symp. **248**, 411 (2008)
9. Lebreton, Y., Michel, E., Goupil, M.J., et al.: IAU Symp. **166**, 135 (1995)
10. Zahn, J-P.: A&A **265**, 115 (1992)
11. Talon, S.: In: Charbonnel, C., Zahn, J.-P. (eds.) Transport Processes in Stars: Diffusion, Rotation, Magnetic fields and Internal Waves. EAS Publications Series 32, 81 (2008)
12. Decressin, T., Mathis, S., Palacios, A., et al.: A&A **495**, 271 (2009)
13. Talon, S., Zahn, J-P., Maeder, A., Meynet, G.: A&A **322**, 209 (1997)
14. Talon, S.: In: Proceedings of the Aussois School "Stellar Nucleosynthesis: 50 Years After B2FH" ArXiv e-prints, 708 (2007)
15. Rieutord, M.: sf2a Conf., 501 (2006)
16. Goupil, M.J., Talon, S.: ASPC **259**, 306 (2002)
17. Montalbán, J., Miglio, A., Eggenberger, P., et al.: AN **329**, 535 (2008)
18. Miglio, A., Montalbán, J., Eggenberger, P., Noels, A.: CoAst **158**, 233 (2009)
19. Thoul, A.: CoAst.159, 35 (2009)
20. Miglio, A., Montalbán, J., Noels, A., et al.: MNRAS **386**, 1487 (2008)
21. Lignières, F., Rieutord, M., Reese, D.: A&A **455**, 607 (2006)
22. Lovekin, C.C., Goupil, M.J.: A&A **515**, 58 (2010)
23. Lovekin, C.C., Deupree, R.G.: ApJ **679**, 1499 (2008)
24. Handler, G.: CoAst.147...31 (2006)
25. Stankov, A., Handler, G.: ApJS **158**, 193 (2005)
26. Pigulski, A.: CoAst **150**, 159 (2007)
27. Aerts, C., Waelkens, C., Daszyńska-Daszkiewicz, J., et al.: A&A **415**, 241 (2004)
28. Christensen-Dalsgaard, J.: Lecture Notes on Stellar Oscillation, 5th edn. Mai (2003)
29. Goupil, M.J., Michel, E., Lebreton, Y., Baglin, A.: AA **268**, 546 (1993)
30. Unno, W., Osaki, Y., Ando, H., et al.: Nonradial Oscillations of Stars, 2nd edn. University of Tokyo Press, Tokyo (1989)
31. Goupil, M.J.: Ap&SS **316**, 251 (2008)
32. Winget, D.E., Nather, R.E., Clemens, J.C., et al.: ApJ **430**, 839 (1994)
33. Kawaler, S.D., Sekii, T., Gough, D.: ApJ **516**, 349 (1999)
34. Charpinet, S., van Grootel, V., Reese, D., et al.: A&A **489**, 377 (2008)
35. Saio, H.: ApJ **244**, 299 (1981)
36. Dziembowski, W.A., Goode, P.R.: ApJ **394**, 670 (1992)
37. Soufi, F., Goupil, M.J., Dziembowski, W.A.: A&A **334**, 911 (1998)
38. Suárez, J.C., Goupil, M.J., Morel, P.: AA **449**, 673 (2006)
39. Karami, K.: ChJAA **8**, 285 (2008)
40. Chandrasekhar, S., Lebovitz, N.R.: ApJ **136**, 1105 (1962)
41. Gough, D.O., Thompson, M.J.: MNRAS **242**, 25 (1990)
42. Goupil, M.J., Talon, S.:CoAst **158**, 220 (2009)
43. Dziembowski, W.A.: Acta Astron. **21**, 289 (1971)
44. Aerts, C.: IAUS **250**, 237 (2008)
45. Dupret, M.A., Thoul, A., Scuflaire, R., et al.: A&A **415**, 251 (2004)
46. Handler, G., Shobbrook, R.R., Mokgwetsi, T.: MNRAS **362**, 612 (2005)

47. Briquet, M., Morel, T., Thoul, A., et al.: MNRAS **381**, 1482 (2007)
48. Pamyatnykh, A.A., Handler, G., Dziembowski, W.A.: MNRAS **350**, 1022 (2004)
49. Ausseloos, M., Scuflaire, R., Thoul, A., et al.: MNRAS **355**, 352 (2004)
50. Dziembowski, W.A., Pamyatnykh, A.A.: MNRAS **385**, 2061 (2008)
51. Suárez, J.C., Moya, A., Amado, P.J., et al.: ApJ **690**, 1401 (2009)
52. Handler, G., Jerzykiewicz, M., Rodríguez, E., et al.: MNRAS **365**, 327 (2006)
53. Desmet, M., Briquet, M., Thoul, A., et al.: MNRAS **396**, 1460 (2009)
54. Morel, P.: A&AS **124**, 597 (1997)
55. Lebreton, Y., Montalbán, J., Christensen-Dalsgaard, J., Roxburgh, I.W., Weiss, A.: Ap&SS 316, 187L (2008)
56. Daszynska-Daszkiewicz Dziembowski, W.A., Pamyatnykh, A.A., Goupil, M.J.: AA **392**, 151 (2002)
57. Goupil, M.J.: LNP **765**, 45 (2009)
58. Dziembowski, W.A., Jerzykiewicz, M.: A&A **341**, 480 (1999)
59. Goupil, M.J., Dziembowski, W.A , Pamyatnykh, A.A., Talon, S.: ASPC **210**, 267 (2000)
60. Goupil, M.J., Talon, S.: CoAst 158, 220 (2009)
61. Bonanno, A., Küker, M., Paterno, L.: A&A **462**, 1031 (2007)
62. Hansen, C.J., Cox, J.P., van Horn, H.M.: ApJ **217**, 151 (1977)
63. Dziembowski, W.A., Goode, P.R.: In: Cox, A.N., Livingston, W.C., Matthews, M.S. (eds.) Solar Interior and Atmosphere (A92-36201 14-92), vol. 501, pp. 501–518. University of Arizona Press, Tucson (1991)
64. Schou, J., Antia, H.M., Basu, S., et al.: ApJ **505**, 390 (1998)
65. Ledoux, P.: ApJ 114, 373 (1951)
66. Goupil, M.-J., Dziembowski, W.A., Pamyatnykh, A.A., Talon, S.: ASPC 210, 267 (2000)
67. Pamyatnykh, A.A.: Ap&SS 284, 97 (2003)
68. Goupil, M.J., Samadi, R., Lochard, J. et al.: In: Favata, F., Aigrain, S., Wilson A. (eds.) Stellar Structure and Habitable Planet Finding (ESASP 538), 133 ESA Special Publication 538, p. 133 (2004)
69. Karami, K.: Ap&SS **319**, 37 (2009)
70. Christensen-Dalsgaard, J., Arentoft, T., Brown, T.M., et al.: J Phys Conf Ser 118 Claret, A., 2007, A&A **475**, 1019 (2008)
71. Belkacem, K., Samadi, R., Goupil, M.J., et al.: Science **324**, 1540 (2009)
72. Pijpers, F.P.: MNRAS **326**, 1235 (1997)
73. Schou J., Tomczyk, S., Thompson, M.J.: AAS 185, 4401 (1994a)
74. Schou J., Christensen-Dalsgaard, J., Thompson, M.J.: ApJ 433, 389 (1994b)
75. Jerzykiewicz, M., Handler, G., Shobbrook, R.R., et al.: MNRAS **360**, 619 (2005)

Chapter 9
Asymptotic Theory of Stellar Oscillations Based on Ray Dynamics

F. Lignières

Abstract This chapter is concerned with the extension of the asymptotic theory of stellar oscillations beyond the case of a non-rotating, non-magnetic spherically symmetric star. It is shown that ray models that describe propagating waves in the short-wavelength limit provide a natural framework for this extension. The basic tools to construct an asymptotic theory from a ray model and some general results obtained in the context of quantum physics are first described. Then, a recent application to the high-frequency acoustic modes of rapidly rotating stars is presented.

9.1 Introduction

The asymptotic theory of stellar oscillations has played a major role in the development of helio and asteroseismology. By providing analytical formulas for the modes and the frequencies [1–5], the theory allows a deep understanding of the oscillation properties that, in turn, enables to construct identification and inversion tools for seismology [4, 6]. Although the theory is asymptotic (that is formally valid in the limit of modes of vanishing wavelength) and assumes linear adiabatic oscillations, it proved sufficiently accurate to describe observed modes like the high frequency p-modes of the Sun and the low frequency g-modes in white dwarfs.

The asymptotic theory of stellar oscillations has been however restricted to situations where the eigenvalue problem is fully separable. For a non-magnetic non-rotating spherically symmetric star, the modes are indeed separable in the three spherical coordinates. Spherical harmonics form the angular part of the mode while its radial part verifies a one-dimensional boundary value problem. The asymptotic theory then consists in applying a short-wavelength approximation to the radial eigen-

F. Lignières (✉)
Laboratoire d'Astrophysique de Toulouse-Tarbes (LATT), Université de Toulouse, UPS, Toulouse 31400 , France
Laboratoire d'Astrophysique de Toulouse-Tarbes (LATT), CNRS, Toulouse 31400 , France
e-mail: francois.lignieres@ast.obs-mip.fr

J.-P. Rozelot and C. Neiner (eds.), *The Pulsations of the Sun and the Stars*,
Lecture Notes in Physics 832, DOI: 10.1007/978-3-642-19928-8_9,
© Springer-Verlag Berlin Heidelberg 2011

value problem to obtain an analytical solution for the radial part of the mode and the eigenfrequencies (see the references above for the details). Such a simplification is however not possible in many cases of practical interest for stellar seismology. The rapid rotation of most upper-main-sequence pulsating stars (δ Scuti stars, γ Doradus stars, Be stars, pulsating B stars) destroys the mode separability due both to the centrifugal flattening of the star [7] and to the angular coupling induced by the Coriolis force [8]. Strong magnetic fields also prevent mode separability like in roAp stars [9]. An asymptotic theory for these stars would be of great importance to interpret their frequency spectra, notably as high quality data are acquired by spatial missions (MOST, CoRoT, Kepler).

In this lecture, I shall be concerned by the extension of the asymptotic theory of stellar oscillations to non-separable situations. Such a theory has been recently proposed for acoustic modes in rapidly rotating stars and has been successfully confronted with numerical computed high frequency p-modes of uniformly rotating polytropic stars [10]. I will thus mainly consider acoustic stellar waves in the following although many aspects of the construction of the asymptotic theory are general and should be also relevant for other stellar waves. In particular I will show that ray models of stellar waves provide a natural framework to extend the asymptotic theory of stellar oscillations to non-separable problems. Much as optical rays describe short-wavelength traveling electromagnetic waves in the geometrical optics limit, it is indeed possible to construct a ray model that describes traveling stellar waves in a short-wavelength asymptotic limit. But as we are interested in modes, that is in standing waves, the central issue for an asymptotic theory based on a ray model is to construct modes from positively interfering traveling waves. This is not an easy task in the general case. Fortunately, as commented in the next paragraph, this ray model route has already been taken in quantum physics to describe short-wavelength quantum waves and we can benefit from the results obtained in this field. Another objective of this lecture is to present cases where the asymptotic organization of the oscillation frequency spectrum is significantly more complex than in non-rotating, non magnetic, spherically symmetric stars. For example, the frequency spectrum of high-frequency acoustic modes in rapidly rotating stars can be described as a superposition of independent frequency subsets that are either regular (with different type of frequency patterns) or irregular but with generic statistical properties.

In quantum physics, the ray model of the quantum waves corresponds to the classical limit of the quantum system. Since Bohr's model of the Hydrogen atom, numerous efforts have been made to relate the classical and the quantum properties of quantum systems and in particular to compute the eigenstates and the energy levels from the classical trajectories. This is exactly the same issue as constructing stellar oscillation modes from the ray model of stellar waves. Early works in quantum physics have concentrated on the case where the Hamiltonian that describes the classical dynamics is integrable. In this case, a general procedure has been found that enables to construct the eigenstates and the energy levels from the classical dynamics. This procedure is known as the EBK semiclassical quantization after the name of its main contributors Einstein, Brillouin and Keller [11–13]. More recently, in the last 30 years, the issue of relating the properties of the quantum system to those of its classical limit has been considered in the wider context of non-integrable Hamiltonian dynamics. In particular, a basic issue has been to determine how the chaotic dynamics of a classical system manifests itself in the properties of the eigenstates and energy levels of the associated quantum system. This field of research has been called quantum chaos and it produced a number of important results which have since been applied to other wave phenomena, such as those observed in e.g. microwave resonators [14], lasing cavities [15], quartz blocks [16], and underwater waves [17].

Is it possible to use quantum chaos theory to construct an asymptotic theory of stellar oscillations from a ray model? Since quantum chaos results are based on the Hamiltonian character of the classical dynamics, they are in principle applicable to any wave problem whose ray model is governed by Hamiltonian dynamics. As we shall see in the following, this is indeed the case for many type of waves as long as the dissipative effects can be neglected and the boundary conditions do not destroy the Hamiltonian character of their ray dynamics.

The document is organized as follows. In Sect. 9.2, wave equations occurring in quantum physics, optics and acoustic stellar oscillations are written down to emphasize their similarities. The ray models of these three types of waves together with their Hamiltonian formulations are derived in a unified way. Some results of quantum chaos studies are presented in Sect. 9.3. In particular, features of the energy level spectra that are sensitive to the integrable or chaotic nature of the classical dynamics are described. In Sect. 9.4, a recent asymptotic analysis of acoustic modes in rapidly rotating stars based on acoustic ray dynamics is briefly presented.

9.2 Wave Equations and Ray Models

In this section, I first emphasize the similarities between three wave equations respectively governing the quantum eigenstates of a single particle in a potential, the monochromatic electromagnetic waves in a linear, isotropic, transparent medium and the adiabatic high-frequency acoustic waves in stars. This allows to describe in a unified way the short-wavelength approximation of these wave equations which then leads to the eikonal equation and the ray model. The Hamiltonian formulation of the equations governing the rays is then explicited.

In quantum physics, the eigenstates of a single non-relativistic particle in a potential V are solutions of the time-independent Schrödinger equation:

$$\Delta \Psi + \frac{2m}{\hbar^2}[E - V(x)]\Psi = 0 \tag{9.1}$$

where the wavefunction associated with the particle is $\psi(x, t) = \Psi(x)e^{(-iEt/\hbar)}$, $\Psi(x)$ is the eigenstate, E is the energy, m is the mass and $\hbar$ is the reduced Planck constant.

In optics, monochromatic electromagnetic waves in a linear, isotropic, transparent medium of refractive index n verify:

$$\Delta \hat{\mathbf{E}} + \left[\frac{\omega}{c} n(x) \right]^2 \hat{\mathbf{E}} = 0 \qquad (9.2)$$

where $\hat{\mathbf{E}}(x)$ is the complex amplitude of the electric field $\mathbf{E} = \Re\{\hat{\mathbf{E}} e^{(-i\omega t)}\}$, ω is the wave pulsation and c is the speed of light.

The wave equation that governs monochromatic high-frequency adiabatic linear acoustic waves in uniformly rotating stars can be written in the following form:

$$\Delta \hat{\Psi} + \frac{\omega^2 - \omega_c^2}{c_s^2} \hat{\Psi} = 0 \qquad (9.3)$$

where $\hat{\Psi}$ is complex amplitude of the full wave solution $\Psi = \Re\{\hat{\Psi} e^{(-i\omega t)}\}$, ω is the pulsation, c_s is the sound speed and $\omega_c(x, \omega)$ is the cut-off frequency which provokes the wave reflection at the star surface. The high-frequency hypothesis enables to neglect the gravity waves, the effect of the Coriolis force and the perturbation of the gravitational potential. Furthermore, while non-adiabatic effects are known to be important near the surface of stars, the adiabaticity hypothesis is generally good enough to compute accurate oscillation frequencies.

Note that different forms of (9.3) have been proposed in the literature, the expression of ω_c and the relation between Ψ and physical quantities such as the pressure perturbation or the Lagrangian displacement ξ depend on the choice of the dependent variable and on the assumptions made. For example, if the variation of the background gravity is neglected in the perturbation equation (see [18] p. 493), (9.3) is obtained with $\Psi = \rho_0^{1/2} c_s^2 \nabla \cdot \xi$ and $\omega_c = \frac{c_s}{2H_\rho}(1 - 2\mathbf{n_0} \cdot \nabla H_\rho)^{1/2}$ where ρ_0 is the background density, ξ is the Lagrangian displacement, H_ρ is the background density scaleheight and $\mathbf{n_0}$ a unit vector opposite to the gravity direction. If this approximation is not made, the expressions of ω_c and Ψ are more complex and ω_c generally depends on ω ([18], p. 439). It can also be shown that in centrifugally distorted stars high-frequency adiabatic acoustic waves are also governed by an equation of the form (9.3), the expressions of ω_c and Ψ being given in [10].

The three wave equations (9.1–9.3), have a similar form:

$$\Delta \Psi + K^2(x)\Psi = 0 \qquad (9.4)$$

$K(x)$ being equal to $\sqrt{2m(E - V)}/\hbar$ in the quantum case, $\omega n/c$ in the optical case and $\sqrt{\omega^2 - \omega_c^2}/c_s$ in the acoustic case. The solutions of the eigenvalue problem will thus only depend on the specific form of $K^2(x)$ and on the boundary conditions.

There is a particular situation where these three problems are identical. It occurs when K is constant and the domain of propagation is bounded by a closed curve where a given boundary condition is applied on Ψ. The two-dimensional version of this problem is called a quantum billiard because in the short wavelength limit the rays are straight lines and the reflections on the boundary are specular. Quantum billiards play an important role in quantum chaos theory (an example is shown in the next section). In quantum physics, this corresponds to the idealized situation where the potential V vanishes inside the domain and goes to infinity outside the domain.

K being equal to $\sqrt{2mE}/\hbar$. In optics, a quantum billiard is obtained with linearly polarized electromagnetic waves propagating in an homogeneous two-dimensional cavity bounded by a perfectly conducting medium, K is then equal to $\omega n/c$. For acoustic waves, the sound speed has to be uniform inside the domain ($K = \omega/c_s$) and the surface has to behave as a reflecting wall. This is not, however, a realistic model for the stellar acoustic waves since we know that c_s is strongly inhomogeneous in stars.

The short-wavelength approximation (also known as WKB, WKBJ, or JWKB approximation) of the wave equation (9.4) consists in looking for wave-like solutions of the form $\Psi = A(x) \exp[i\Phi(x)]$ under the assumption that their wavelength is much shorter than the typical lengthscale of variation of the background medium. The amplitude term $A(x)$ is assumed to vary on the background lengthscale L while the oscillating term $\exp[i\Phi(x)]$ varies much more rapidly. This suggests to expand the solution as

$$\Phi = \Lambda(\Phi_0 + \frac{1}{\Lambda}\Phi_1 \cdots) \quad \text{and} \quad A = A_0 + \frac{1}{\Lambda}A_1 \cdots \tag{9.5}$$

where $1/\Lambda$ is the ratio between the wavelength of the solution and the background lengthscale. When this expansion is introduced into (9.4), the dominant $O(\Lambda^2)$ term yields the so-called eikonal equation:

$$K(x)^2 = \Lambda^2(\nabla\Phi_0)^2. \tag{9.6}$$

This implies that $K(x)$ must be of the order of Λ which, according to the expressions of K, indicates that the small wavelength limit corresponds to high-energy levels for the quantum system and to high frequencies for the optical and acoustic systems. In the eikonal equation describing stellar acoustic waves, the ω_c term must be retained since its increase near the star surface is responsible for the back-reflection of the waves and thus eventually for the formation of the modes through constructive interferences. The next order of the expansion (9.5) enables to relate A_0 to Φ_0.

The eikonal equation can be viewed as a local dispersion relation. Indeed, a local wavevector k can be defined from the spatial phase term $\Phi(x)$ by the relation $k = \nabla\Phi$ (recall that, in an homogeneous medium, the spatial phase would be $\Phi(x) = k_0 \cdot x$ with k_0 a uniform wavevector) so that the eikonal equation reads:

$$D(k, \omega, x) = K^2 - k^2 = 0. \tag{9.7}$$

Instead of trying to solve directly the eikonal equation as a PDE (Partial Differential Equation) verified by the function $\Phi(x)$, the ray model consists in searching solutions for the phase along a given path $x(s)$. To find these solutions, one has to solve the coupled differential equations that determine the ray path and the evolution of $k(s)$ along it (and then integrate $k = \nabla\Phi$ along the ray).

We now demonstrate that, for a general eikonal equation $D(k, \omega, x) = 0$, these coupled equations can be written in a Hamiltonian form. Let's consider a general coordinate system $[x_1, x_2, x_3]$ and compute the partial derivative of D with respect to each coordinate x_i:

$$\frac{\partial D}{\partial x_i} + \sum_{j=1}^{N=3} \frac{\partial D}{\partial k_j} \frac{\partial k_j}{\partial x_i} = 0 \quad i = 1, \dots, 3 \tag{9.8}$$

where k_j is defined as $k_j = \frac{\partial \Phi}{\partial x_j}$. From the definition of k_j, we have that $\frac{\partial k_j}{\partial x_i} = \frac{\partial^2 \Phi}{\partial x_i \partial x_j} = \frac{\partial^2 \Phi}{\partial x_j \partial x_i} = \frac{\partial k_i}{\partial x_j}$. Thus, (9.8) can be written as:

$$\frac{\partial D}{\partial x_i} + \sum_{j=1}^{N=3} \frac{\partial D}{\partial k_j} \frac{\partial k_i}{\partial x_j} = 0 \quad i = 1, \dots, 3 \tag{9.9}$$

If we consider a path $x(s)$ defined by $\frac{dx_i}{ds} = \frac{\partial D}{\partial k_i}$, the derivative of k_i following this path is given by:

$$\frac{dk_i}{ds} = \sum_{j=1}^{N=3} \frac{\partial k_i}{\partial x_j} \frac{dx_j}{ds} = \sum_{j=1}^{N=3} \frac{\partial k_i}{\partial x_j} \frac{\partial D}{\partial k_j} \quad i = 1, \dots, 3 \tag{9.10}$$

As $\frac{dk_i}{ds}$ corresponds to the second term on the left hand side of (9.9), the equations defining the ray model are

$$\frac{dx_i}{ds} = \frac{\partial D}{\partial k_i} \quad i = 1, \dots, 3 \tag{9.11}$$

$$\frac{dk_i}{ds} = -\frac{\partial D}{\partial x_i} \quad i = 1, \dots, 3 \tag{9.12}$$

These are Hamilton's equations where D is the Hamiltonian and x_i and k_i are the conjugate variables, x_i the position variables and k_i the momentum variables (see [19] p. 317 for a similar demonstration). The above derivation is valid for any coordinate system $[x_i]$. The momentum variables k_i are the covariant component of the wave vector k in the natural basis associated with $[x_i]$, the definition of the natural basis being $e_i = \partial x / \partial x^i$.

This Hamiltonian formulation can be simplified in two special cases that are relevant for the there wave equations considered in this section. First, when D can be written as $D(k, \omega, x) = H(k, x) - \omega$, the above equations become:

$$\frac{dx_i}{ds} = \frac{\partial H}{\partial k_i} \quad i = 1, \dots, 3 \tag{9.13}$$

$$\frac{dk_i}{ds} = -\frac{\partial H}{\partial x_i} \quad i = 1, \dots, 3 \tag{9.14}$$

where the ray path now moves at the group velocity $\frac{\partial H}{\partial k_i}$. According to the expression of $K(x)$ for the three wave equations considered, this formulation holds with

$H = ck/n$ for the optical rays and $H = \sqrt{c_s^2 k^2 + \omega_c^2}$ for stellar acoustic rays. It is also the case for the quantum system with $D = H(\boldsymbol{p}, \boldsymbol{x}) - E$ where $H = \boldsymbol{p}^2/2m + V(\boldsymbol{x})$ is the Hamiltonian and $\boldsymbol{p} = \hbar \boldsymbol{k}$ is the momentum vector.

Second, when the Hamiltonian can take the form $D = \boldsymbol{p}^2/2m + V(\boldsymbol{x})$, Hamilton's equations reduce to the classical vectorial form:

$$\frac{d\boldsymbol{x}}{dt} = \frac{\boldsymbol{p}}{m} \tag{9.15}$$

$$\frac{d\boldsymbol{p}}{dt} = -\boldsymbol{\nabla} V \tag{9.16}$$

where the second equation is simply the Newton's second law for the conservative force associated with the potential V. The classical limit of the quantum system can obviously be written in this form. It is also possible in the other cases since the eikonal equations of the electromagnetic and acoustic waves can be written $0 = \boldsymbol{p}^2/2m + V(\boldsymbol{x})$ where $m = 1$, $\boldsymbol{p} = \boldsymbol{k}$ and the potential V is respectively $V = -\frac{1}{2}\left[\frac{\omega n}{c}\right]^2$ in the optical case and $V = -\frac{1}{2}\frac{\omega^2 - \omega_c^2}{c_s^2}$ in the acoustic case. The total energy D is thus fixed to zero but the frequency ω acts as a parameter that modifies the potential.

In this section, we have derived the ray models of three similar wave equations and have shown that they can be described by Hamiltonian dynamics. A direct consequence is that the bulk of knowledge accumulated on Hamiltonian dynamics is available to characterize the ray properties. For example, as shown in Sect. 9.4, acoustic rays become more and more chaotic as the rotation of the star increases [10]. The deep understanding of the transitions from integrability to chaos in Hamiltonian dynamics is extremely useful to characterize such an evolution. But what is still more important in the context of this lecture is that the special properties of the Hamiltonian systems can be used to construct an asymptotic theory of stellar oscillation based on the ray model. This will be considered in the next section.

Before concluding this section, it must be reminded that some effects which have not been considered in the present analysis would modify the Hamiltonian character of the ray equations. Dissipative effects produce a concentration of phase space volume that can not be described by Hamiltonian dynamics. Thus the acoustic ray model does not take into account non-adiabatic effects in stars. Another non-Hamiltonian effect can be induced by the presence of a sharp boundary between two media (like the strong gradients at the upper limit of a core convective zone) since an incident ray divides into a reflected ray and a transmitted ray. There have been however attempts to extend the ray dynamics approach to account for the splitting of rays at such discontinuities [20]. Finally, in some circumstances, the reflection of waves at a wall can lead to a focusing or defocusing effect that destroys the Hamiltonian character of the ray dynamics. This is for example the case for the reflection of gravity waves if the wall is inclined with respect to the direction of the gravity [21].

9.3 Regular Versus Irregular Energy Level Spectra
in Quantum Systems

As acoustic stellar waves of short-wavelength can be described by rays and as the
ray dynamics is Hamiltonian, one wonders whether the oscillation modes formed by
these waves are sensitive to the nature, chaotic or integrable, of the Hamiltonian ray
dynamics. This question has been considered in the context of quantum physics an
overview of the results being available in classical textbooks [22–24]. Here, I shall
focus on the results that concerns the organization of the energy level spectra for
quantum systems that are either classically integrable or completely chaotic. The
situation where the dynamics is mixed in the sense that regular and chaotic motions
coexist in phase space is mentioned in the next section, in the context of acoustic
rays in rapidly rotating stars. We also restrict ourselves to bounded systems where
the energy spectrum is known to be discrete.

9.3.1 Regular Spectrum

Energy level spectra of quantum system whose classical limit is integrable are said
to be regular in the sense that they can be described by a smooth function of N
integers $(n_1, n_2, n_3 \ldots, n_N)$, where N is the number of degree of freedom of the
Hamiltonian:

$$E_i = f(n_1, n_2, n_3, \ldots) \tag{9.17}$$

This remarkable property results from the fact that the phase space of integrable systems is entirely structured by N-dimensional invariant surfaces (also called invariant
tori because these surfaces have the topology of a N-torus). These surfaces are said
to be invariant because any trajectories starting on the surface remains on the surface
as time goes on.

Let us first come back to the construction of a solution $\Psi = A(x) \exp[i\Phi(x)]$
from a ray solution $[x(s), k(s)]$. To obtain the spatial phase $\Phi(x)$, the expression
$k = \nabla\Phi$ is integrated along the ray:

$$\Phi(x) = \Phi(x_0) + \int_{x_0}^{x} k(s) \cdot dx(s) \tag{9.18}$$

If a phase space trajectory crosses its starting position x_0 at a later time, the phase
function $\Phi(x)$ will be multivalued on that position. Thus, the necessary condition
that the function $\Psi(x) = A(x) \exp[i\Phi(x)]$ is single-valued on the position space
requires that the variation of Φ between these two phase space points $[x_0, k_0]$ and
$[x_0, k_1]$ is a multiple of 2π (provided the phase of A does not change which is true
outside the caustic). More generally, trajectories that crosses a surface $\Phi = const$
must also verify such a condition.

To implement this condition of positive interference is not easy in the general case, notably when the trajectories are chaotic. But, in integrable systems, trajectories stay on a well-defined structure of phase space and the condition of positive interferences can be shown to apply to any closed contour C on the torus (and not necessarily to a contour that follows a phase space trajectory). Furthermore, the fact that the action integral $\int_C \mathbf{k} \cdot d\mathbf{x}$ is identical for any contours C' obtained by continuously deforming C on the torus (known as the Poincaré–Cartan theorem) reduces the condition to N independent conditions:

$$\int_{C_i} \mathbf{k} \cdot d\mathbf{x} = 2\pi \left(n_i + \frac{\beta_i}{4} \right) \tag{9.19}$$

where C_i are N topologically independent closed paths on the N-dimensional torus. The integer β_i called the Maslov index is introduced to account for a $\pi/2$ phase lag that must be added each time the contour crosses a caustic. Indeed, the caustic corresponds to the boundary of the torus projection onto position space; the amplitude A taken in the position space is discontinuous there, leading to the $\pi/2$ phase loss (see [13] for details). Equation 9.19 is the EBK semiclassical quantization condition mentioned before. In practice, the usual way to apply it is to choose contours C_i for which the formulas (9.19) are simple to compute. Gough [18] applied the EBK quantization to acoustic rays in a non-rotating spherically symmetric star and found that the result is practically identical to the usual asymptotic theory that uses of the separability of the wave equations.

The existence of the function (9.17) defining the energy level spectrum then follows from the expression of the Hamiltonian in the action-angle coordinates $[\mathbf{I}, \boldsymbol{\theta}]$. This a particular coordinate system of integrable systems such that the momentum coordinates $I_1, I_2, I_3, \ldots$ are defined by the action integrals (9.19) (divided by 2π) and are constant of motions. The Hamiltonian is thus a function of the N actions only (since $dI_i/dt = \partial H/\partial \theta_i = 0$), $H(I_1, I_2, I_3 \ldots)$. Consequently, the EBK formulas (9.19) appears as quantization formulas for the actions, $I_i = n_i + \frac{\beta_i}{4}$, and the energy level spectrum is simply determined by $E_i = H(I_1, I_2, I_3 \ldots) = f(n_1, n_2, n_3, \ldots)$.

An important remark about the EBK quantization is that it essentially requires the presence of an invariant torus in phase space. Thus, as we shall see in the next section, it can be also applied to non-integrable systems if invariant tori are present in phase space.

9.3.2 Irregular Spectrum

When the classical dynamics is chaotic, a smooth function like (9.17) can not be found and the energy spectrum is said to be irregular. Instead, the spectrum of a classically chaotic quantum system is best characterized by its statistical properties.

To show this, it is first necessary to define the fluctuations of the density of energy level, $d^{\text{fluct}}(E)$. The total density $d(E)$ is such that the number of energy level comprised between E_a and E_b is equal to $\int_{E_a}^{E_b} d(E)\mathrm{d}E$. This quantity can then be split into a mean density of level $d^{\mathrm{av}}(E)$ and the deviation from the mean $d^{\mathrm{fluct}}(E)$:

$$d(E) = d^{\mathrm{av}}(E) + d^{\mathrm{fluct}}(E) \quad \text{where} \quad d^{\mathrm{av}}(E) = \frac{1}{2\Delta} \int\limits_{E-\Delta}^{E+\Delta} d(E)\mathrm{d}E \qquad (9.20)$$

and Δ is the averaging scale. The interest of this expression is that the mean density $d^{\mathrm{av}}(E)$ does not depend on the chaotic or integrable nature of the dynamics, while the fluctuations about this mean $d^{\mathrm{fluct}}(E)$ do.

The mean density $d^{\mathrm{av}}(E)$ depends on the global properties of the system considered. This has been shown by Weyl [25] who provided an analytical estimate of $d^{\mathrm{av}}(E)$ in the high-energy limit. Accordingly, the mean number of modes whose energy level is below E, $N(E) = \int_{-\infty}^{E} d^{\mathrm{av}}(E')\mathrm{d}E'$, is approximatively equal to the volume of phase space available (that is the volume of the $H < E$ region) divided by the mean phase space volume occupied by an individual mode, that is $(2\pi\hbar)^N$. The mean density $d^{\mathrm{av}}(E)$ is then obtained by derivating this quantity with respect to E. As an example, for a two-dimensional quantum billiard, the phase space volume such that $H < E$ is simply $\int_{H(p,x)<E} d^2p\,d^2x = 2\pi m E A$ where A is the area of the billiard, thus $d^{\mathrm{av}}(E) = mA/(2\pi\hbar^2)$. As expected, the mean level density does not depend on the nature of dynamics inside the billiard but only on its area (see [10] for an application of the Weyl's formula to acoustic stellar oscillations).

A simple way to characterize the fluctuations of the level density $d^{\mathrm{fluct}}(E)$ is to consider the statistical distribution of the spacing between consecutive energy levels $S_i = E_{i+1} - E_i$ (the energy levels E_i are labeled in ascending order). The mean level difference ΔE (computed over the averaging scale Δ) is the inverse of the mean level density $d^{\mathrm{av}}(E)$. Thus, to characterize the deviations from the mean, the energy differences are scaled by ΔE. Statistical distributions of $s_i = (E_{i+1} - E_i)/\Delta E$ have been determined for different systems, either experimentally or through the numerical computations of theoretical problem.

The first experimental evidence of a universal distribution for classically chaotic systems has been obtained from nuclear energy levels. Fig. 9.1 shows an histogram of s_i for 1726 consecutive energy level spacings which has been determined from the analysis of 27 different nuclei [26]. Also shown on this figure is a distribution $P(s) = \pi s/2 \exp(-\pi s^2/4)$ called the Wigner's surmise that fits closely the data. This distribution corresponds to an heuristic model that was proposed long before by Wigner. Confronted to the difficulty of defining an Hamiltonian for the nucleus, Wigner assumed that the statistical properties of nuclear spectra are similar to that of Hamiltonians taken at random. As the Hamiltonian operator projected on a basis of eigenstates is represented by infinite matrices, this idea can be pursued by looking at the eigenvalue spectra of random matrices. Basic requirements on the matrices, namely that the results should not depend on the choice of the eigenstate basis and that the matrix elements are independent random variables, enables to specify the matrix ensemble. For time-reversible problems, this ensemble called GOE for Gaussian

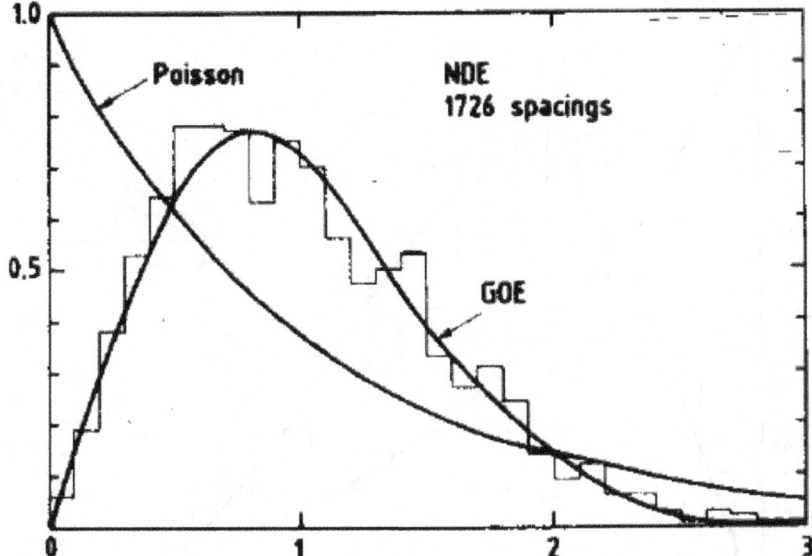

Fig. 9.1 Statistical distribution of the spacings between consecutive nuclear energy levels, the GOE distribution and the Poisson distribution. From Bohigas et al. [26]

Orthogonal Ensemble corresponds to real, symmetric matrices where each matrix element follows a Gaussian distribution, the width of the distribution of off-diagonal elements being twice that of diagonal elements. The Wigner distribution $P(s) = \pi s/2 \exp(-\pi s^2/4)$ provides a good approximation to the statistical distribution of their eigenvalue consecutive spacings. Thus, the experimental evidence shown in Fig. 9.1 provided the first striking agreement between real data and the prediction of the random matrix theory.

The level spacing distribution has been also determined for numerically computed spectra of quantum billiards. Fig. 9.2 presents the result obtained by Bohigas et al. [27] for a chaotic billiard, namely the Sinai billiard, showing again a good agreement with the Wigner's surmise. Since then, similar evidences have been obtained in quantum systems (from the atomic level of rare-earth atoms) and in other wave systems whose ray dynamics is chaotic (with dedicated experiments using microwave resonators [14], or quartz blocks [16]). This led to the conjecture that the distribution of consecutive level spacing is indeed universal in classically completely chaotic systems and corresponds to the prediction of the random matrix theory.

Conversely, the spectra of integrable systems are predicted to be uncorrelated, and in general this leads to fluctuations given by the Poisson distribution $P(s) = \exp(-s)$ if $N > 1$ [28]. As shown in Figs. 9.1 and 9.2, the prediction is strikingly different from the chaotic case. In particular, a distinctive property of classical chaotic system is that $P(s) = 0$ at $s \to 0$. This level repulsion effect can be interpreted as the consequence of avoiding crossing effects between coupled modes in non-integrable systems.

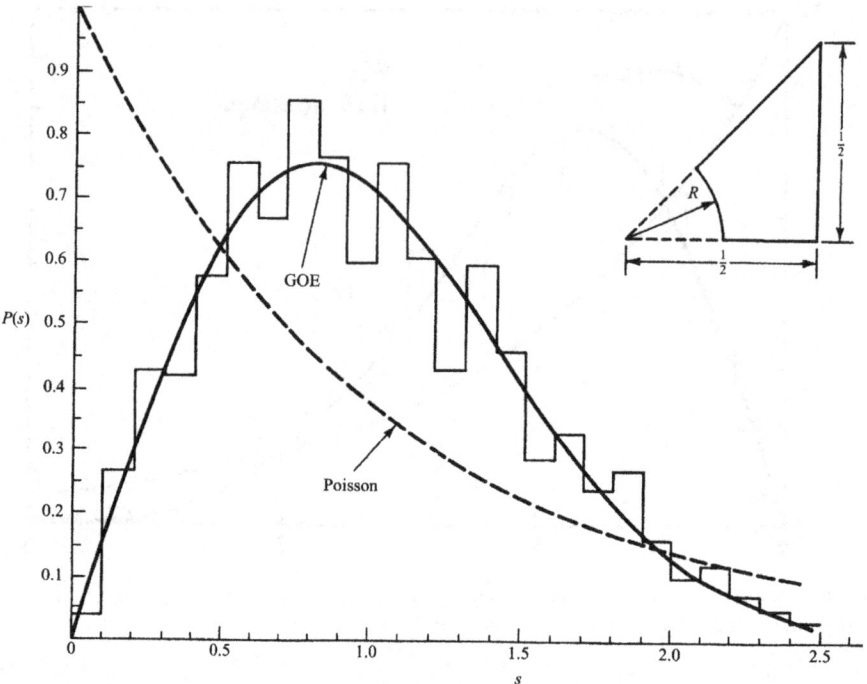

Fig. 9.2 Statistical distribution of the spacings between consecutive energy levels of the Sinaï billiard, the GOE distribution and the Poisson distribution. The Sinaï billiard is shown in the insert. From Bohigas et al. [27]

Other statistical properties of the energy level spectrum, the level clustering and the spectral rigidity, have been shown to be sensitive to the nature of the dynamics (see for example [22] for a brief description of these properties).

9.4 Application to the Asymptotic Theory of Acoustic Modes in Rapidly Rotating Stars

The basic tools to construct an asymptotic theory from a ray model and some general results obtained in the context of quantum physics have been presented in the previous sections. They have been used recently to propose an asymptotic theory of the high-frequency acoustic modes in rapidly rotating stars [10]. In this section, we give a brief description of this theory with emphasis on the type of predictions that can be made and on the confrontation of these predictions with the numerically computed modes.

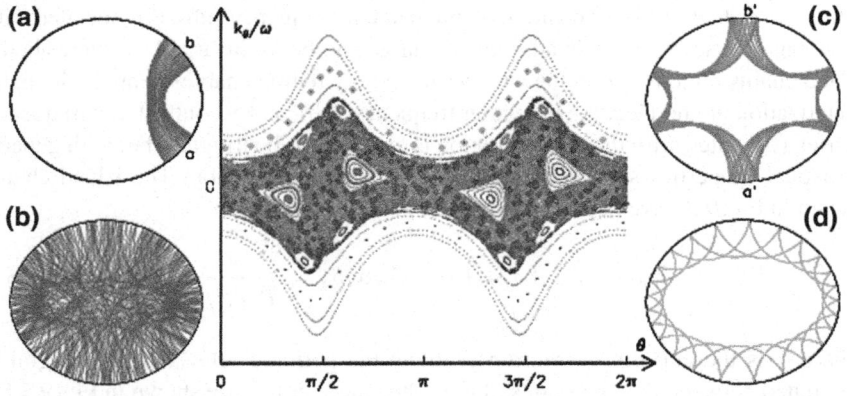

Fig. 9.3 PSS at $\Omega = 0.59(GM/R_e^3)^{1/2}$ and typical acoustic rays associated with the four main phase space structures: **a** a two-period island ray (*blue/dark grey*) and the associated periodic orbit with endpoints a and b (*orange/light grey*), **b** a chaotic ray (*red/grey*), **c** a six-period island ray (*magenta/light grey*) and **d** a whispering gallery ray (*green/light grey*). On the PSS, (*colored/grey*) symbols (diamonds for the chaotic and whispering gallery rays, crosses for the two-period and six-period island rays) specify the points where these trajectories cross the PSS. M denotes the mass of the star and R_e its equatorial radius

The acoustic ray dynamics has been studied in polytropic models of star whose rotation has been progressively increased. For each rotation rate, the Hamiltonian equations governing the ray dynamics are integrated numerically for many different initial conditions. Then, to visualize the structure of the phase space, the standard method of the Poincaré Surface of Section (PSS) is used. As the system is symmetric with respect to the rotation axis of the star, the projection of the angular momentum on this axis $L_z = r \sin\theta k_\phi$ is a constant of motion, where $k_\phi = k \cdot e_\phi$ and e_ϕ is a unit vector in the azimuthal direction. The number of degree of freedom is then reduced to $N = 2$ and the PSS is a two-dimensional surface. The chosen PSS has been constructed by computing the intersection of the phase space trajectories with the curve defined by $r_p(\theta) = r_s(\theta) - d$, situated at a small fixed radial distance d from the stellar surface $r_s(\theta)$.

The acoustic ray dynamics becomes non-integrable as soon as the rotation is not zero and undergoes a smooth transition towards chaos as the rotation increases. Dynamical systems in such a transition are said to be mixed as chaotic trajectories coexist with stable phase space structures (like island chains formed around stable periodic orbits or invariant tori). The main features of the phase space at a relatively high rotation rate are shown in Fig. 9.3 where the PSS for $L_z = 0$ trajectories is displayed together with four acoustic rays shown on the position space and on the PSS.

For such mixed systems, quantum chaos studies [29, 30] predict that the different phase space regions shown in Fig. 9.3 (the two island chains, the chaotic regions, and the whispering gallery region) are quantized independently. The frequency spectrum is then described as a superposition of independent frequency subsets associated with these phase space regions. In addition, the large number of invariant structures in the island chains regions and in the whispering gallery region enables to apply the EBK quantization method leading to regular frequency subsets. By contrast, the frequency subset associated with the chaotic region is expected to be irregular but with generic statistical properties such as described in the previous section. The island chains shown in Fig. 9.3 have been quantized in [31] to obtain:

$$\omega_{n\ell} = n\delta_n + \ell\delta_\ell + \alpha \quad \text{where} \quad \delta_n = \frac{\pi}{\int_a^b d\sigma/c_s} \tag{9.21}$$

where σ is the curvilinear coordinate along the periodic orbit and the integral is computed between the end points of the orbit (these points are shown in Fig. 9.3 for the two-period and the six-period periodic orbits and are denoted (a, b) and (a', b'), respectively). The regular spacing δ_n depends on the sound speed along the periodic orbit while δ_ℓ (whose expression is given in [31]) depends on the sound speed and on its transverse derivative along the same orbit. The integers n and ℓ are the number of nodes of the corresponding modes in the directions parallel and transverse to the orbit.

The above predictions on high-frequency p-modes have been confronted with numerically computed axisymmetric modes (using the same star model). The first prediction is that modes can be classified as chaotic modes, island modes or whispering gallery modes. This can indeed be achieved with the help of a phase-space representation of the modes. With this classification, the frequency spectrum computed in the range $[9\omega_1, 12\omega_1]$ (where ω_1 is the lowest acoustic frequency) has been split into the four subspectra shown in Fig. 9.4.

From these data, we could verified that, in accordance with the asymptotic theory, (i) the subspectra associated with the structured phase space region are regular, (ii) the theoretical expression of δ_n agrees with the empirical values within a few percent, (iii) the distribution of the consecutive frequency spacings taken from the chaotic sub-spectrum agrees reasonably well with the Wigner's distribution.

The asymptotic theory based on the acoustic ray model can thus reproduce quantitative and qualitative features of the actual high-frequency spectrum. However, there are also some limitations to the asymptotic theory that does not exist in the case of a non-rotating spherically symmetric star. Maybe the most important one is that the prediction of the chaotic subspectra concerns its statistical properties but not the individual frequencies. There exist a Fourier-like formula (called the Gutzwiller trace formula [24]) that relates all the periodic orbits of the chaotic phase space to the whole spectrum, but this formula is very delicate to use in practice. Another limitation concerns the coupling between two modes of similar frequencies associated with two dynamically independent regions of phase space. The avoided crossing effect between such modes is not taken into account by the ray dynamics and is thus expected to induce deviations from the asymptotic behavior.

Despite these limitations, the a priori information that the asymptotic theory provides on the structure of the frequency spectrum should be important to interpret the

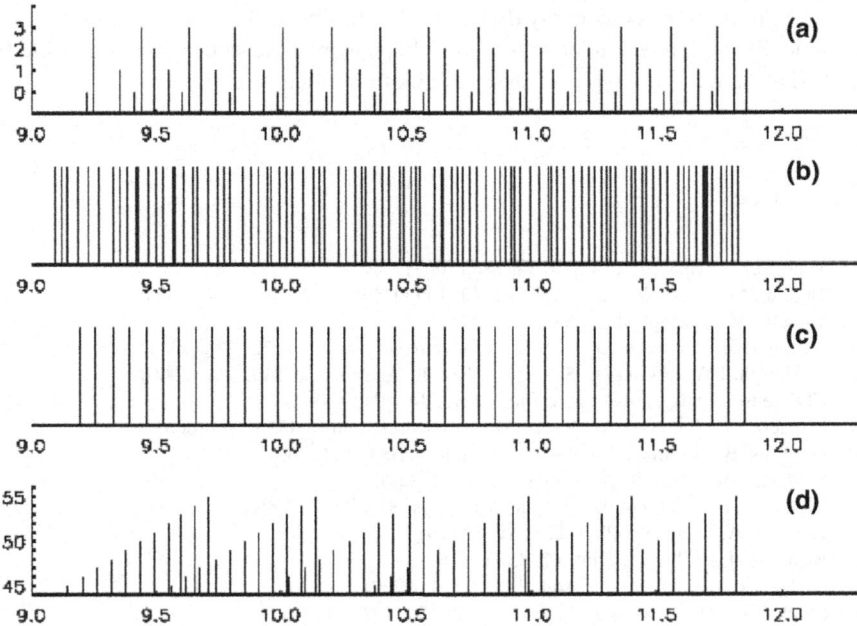

Fig. 9.4 Frequency subspectra of four classes of axisymmetric p-modes of a $\Omega = 0.59(GM/R_e^3)^{1/2}$ polytropic model of star: **a** the two-period island modes, **b** the chaotic modes antisymmetric with respect to the equator, **c** the six-period island modes, and **d** some whispering gallery modes. For the subspectra (**a**) and (**d**), the height of the vertical bar specifies one of the two quantum numbers characterizing the mode

observed frequency spectra of rapidly rotating stars. For example, synthetic spectra given by the asymptotic theory (complemented by informations on the visibility and the excitation of the modes) might be used to construct and test identification schemes. In this context, a first step would be to disentangle the regular part from the irregular part of the spectrum.

9.5 Conclusions

We have seen that ray models can be used to construct asymptotic theory of modes even when the eigenvalue problem is not separable. The methods and concepts, developed in the context of quantum physics, rely on the Hamiltonian character of the ray dynamics. For example, the structure of the frequency (or energy level) spectrum has been shown to depend on the nature of the Hamiltonian dynamics (integrable, fully chaotic, mixed). It is regular for an integrable system, irregular for a fully chaotic system, and a superposition of regular and irregular spectra for a mixed system. These methods and concept have been used to construct an asymptotic theory based on the Hamiltonian acoustic ray dynamics that has been successfully confronted to numerically computed adiabatic p-modes. In principle, the same procedure could be applied to model other types of stellar oscillation modes.

References

1. Vandakurov, Yu.V.: Astron. Zh. **44**, 786 (1967)
2. Tassoul, M.: Astrophys. J. Suppl. Ser. **43**, 469 (1980)
3. Tassoul, M.: Astrophys. J. **358**, 313 (1990)
4. Deubner, F.-L., Gough, D.O.: Annu. Rev. Astron. Astrophys. **22**, 593 (1984)
5. Roxburgh, I.W., Vorontsov, S.V.: Mon. Not. R. Astron. Soc. **317**, 141 (2000)
6. Christensen-Dalsgaard, J.: Rev. Mod. Phys. **74**, 1073 (2002)
7. Lignières, F., Rieutord, M., Reese, D.: Astron. Astrophys. **455**, 607 (2006)
8. Dintrans, B., Rieutord, M.: Astron. Astrophys. **354**, 86 (2000)
9. Saio, H.: Mon. Not. R. Astron. Soc. **360**, 1022 (2005)
10. Lignières, F., Georgeot, B.: Astron. Astrophys. **500**, 1173 (2009)
11. Einstein, A.: Deutsche Phys. Ges. **19**, 82 (1917)
12. Brillouin, L.: J. Phys. Radium **7**, 353 (1926)
13. Keller, J.B., Rubinow, S.I.: Ann. Phys. **9**, 24 (1960)
14. Stöckmann, H.-J., Stein, J.: Phys. Rev. Lett. **64**, 2215 (1990)
15. Nöckel, J.U., Stone, A.D.: Nature **385**, 45 (1997)
16. Ellegaard, C., et al.: Phys. Rev. Lett. **77**, 4918 (1996)
17. Brown, M.G., et al.: J. Acoust. Soc. Am. **113**, 2533 (2003)
18. Gough, D.O.: In: Zahn, J.-P., Zinn-Justin, J. (eds.) Les Houches Lectures Session XLVIII, pp. 399–559. North-Holland, Amsterdam (1993)
19. Lighthill, J.: Waves in Fluids. Cambridge University Press, Cambridge (1978)
20. Blümel, R., Antonsen, T.M., Georgeot, B., Ott, E., Prange, R.E.: Phys. Rev. E **53**, 3284 (1996)
21. Phillips, O.M.: Dynamics of the Upper Ocean. Cambridge University Press, Cambridge (1967)
22. Ott, E.: Chaos in Dynamical Systems. Cambridge University Press, Cambridge (1993)
23. Giannoni, M.-J., Voros, A., Zinn-Justin, J. (eds.): Les Houches Lectures Session LII. North-Holland, Amsterdam (1991)
24. Gutzwiller, M.C.: Chaos in Classical and Quantum Mechanics. Springer, Berlin (1990)
25. Weyl, H.: Math. Ann. **71**, 441 (1912)
26. Bohigas, O., Haq, R., Pandey, A.: Nuclear Data in Science and Technology, p. 809. Reidel, Dordrecht (1983)
27. Bohigas, O., Giannoni, M.-J., Schmit, C.: Phys. Rev. Lett. **52**, 1 (1984)
28. Berry, M.V., Tabor, M.: R. Soc. Lond. Proc. Ser. A **356**, 375 (1977)
29. Percival, I.C.: J. Phys. B **6**, L229 (1973)
30. Berry, M.V., Robnik, M.: J. Phys. A **17**, 2413 (1984)
31. Lignières, F., Georgeot, B.: Phys. Rev. E **78**, 016215 (2008)

Chapter 10
Angular Momentum Transport by Regular Gravito-Inertial Waves in Stellar Radiation Zones

Stéphane Mathis

Abstract In this chapter, the complete interaction between low-frequency internal gravity waves and differential rotation rotation in stably strongly stratified stellar radiation zones is examined. First, the modification of the structure of these waves due to the Coriolis acceleration is obtained. Then, their feed-back on the angular velocity profile through their induced angular momentum transport is derived. Next, the case of a weak differential rotation is studied. Finally, perspectives are discussed.

10.1 Context and Motivation

In standard models of stellar interiors, radiation zones, which are convectively stable, are postulated to be without motion other than rotation. But various observational results (e.g. surface abundances of light elements, helio- and now asteroseismology) show that these regions are the seat of transport and of mild mixing. The most likely cause of such mixing is stellar differential rotation. First, it drives a large-scale meridional circulation. Second, since in general the star rotates differentially, shear instabilities may appear (see [1] for a review of these processes). Series of models have been built that include a self-consistent evolution of the internal-rotation profile, and for massive stars they agree rather well with the observations (see [2]). The case of the Sun is somewhat different: like all other stars that have a deep surface convection zone, it has been spun down during its infancy. When only the meridional circulation and the "classical" hydrodynamic instabilities are invoked, models predict a Sun with a core rotating much faster than the surface [3–7] with a gradient which is not compatible with helioseismology [8–10]. One must conclude therefore that another, more powerful process is operating, at least in solar like stars. The most plausible

Stéphane Mathis (✉)
Laboratoire AIM, CEA/DSM-CNRS-Université Paris Diderot, IRFU/SAp Centre de Saclay,
91191 Gif-sur-Yvette , France
e-mail: stéphane.mathis@cea.fr

J.-P. Rozelot and C. Neiner (eds.), *The Pulsations of the Sun and the Stars*,
Lecture Notes in Physics 832, DOI: 10.1007/978-3-642-19928-8_10,
© Springer-Verlag Berlin Heidelberg 2011

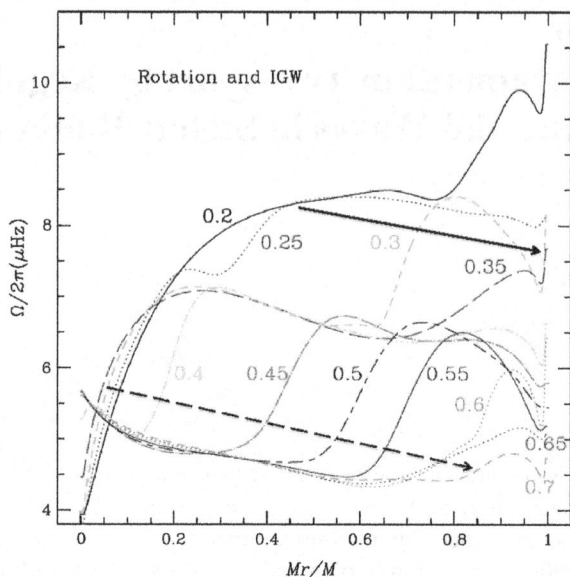

Fig. 10.1 Evolution of the (differential) rotation profile (averaged on latitudes) in a $1.2 M_\odot$ star (its metallicity is $Z = 0.02$) with an initial rotation velocity of $50\,\mathrm{km\,s^{-1}}$. This is obtained making simulations using the STAREVOL code (cf. [88] and references therein) where the transport of angular momentum by the meridional circulation, the shear vertical turbulence, and the internal gravity waves are taken into account. The *black arrows* shows the successive angular momentum extraction fronts (the first is represented by the *continuous line*, the second one by the *dashed line*), which are mainly driven by the angular momentum extraction at the surface by the wind. The curves are labeled according to the corresponding ages in Gyrs (adapted from [18], courtesy Astronomy and Astrophysics)

candidates are magnetic torquing ([11, 12] and references therein, [13, 14]) and momentum transport by Internal Gravity Waves (hereafter IGWs; [15–18]).

IGWs are thus now considered as an essential transport mechanism in (differentially) rotating stellar radiation zones, which are the seat of the mixing during star evolution (cf. [18] and Fig. 10.1).

Their treatment suffers from two major weaknesses. The first is the crude description of their generation by turbulent convection. The second is that the action of (differential) rotation on the waves is not taken into account.

This is why we have undertaken to improve the modeling of the transport by internal waves by introducing the effects of the Coriolis acceleration. Indeed, the low-frequency internal waves that are responsible for the deposition or the extraction of angular momentum (cf. [17]) can be strongly influenced by the rotation because the frequencies of the waves may be of the same order as the inertial frequency (2Ω where Ω is the star's angular velocity). Then, internal waves become gravito-inertial waves (cf. [19–21]) and the Coriolis acceleration is thus an essential restoring force for the wave dynamics as the buoyancy one associated with the stable stratification.

Moreover, IGWs are excited and propagate in regions that are differentially rotating both in the radial and in the latitudinal directions. This is the reason why the treatment of the complete interaction between the low-frequency IGWs and the differential rotation, which is chosen as generally as possible, has been undertaken.

In this lecture, we thus show how the structure of IGWs and their induced-transport are modified (Sects. 10.2 and 10.3). Then, we illustrate our purpose in the frame of the simplified case where the rotation is assumed to be only weakly differential (Sect. 10.4). Next, the question of the excited spectrum is discussed (Sect. 10.5). Finally, in Sect. 10.6, perspectives are discussed.

10.2 Structure of Low-Frequency Waves Influenced by the Coriolis Acceleration

In stellar radiation zones, the transport of angular momentum may be dominated by low-frequency waves with $\sigma \ll N$, where σ and N are respectively the wave frequency and the Brunt–Väisälä frequency, which relates to buoyancy. Moreover, in a (differentially) rotating stellar radiation zone, one must also consider the Coriolis acceleration, which is characterized by the inertial frequency (2Ω). One then has to quantify the relative importance of each restoring force in the wave dynamics, and whether the effects of the Coriolis acceleration can be treated in a perturbative way.

In the Sun, the answer is very clear for the acoustic waves which have frequencies much greater than $\Omega_\odot$. The effects of the Coriolis acceleration can be treated as a perturbation (e.g. rotationally split frequencies). However, in the case of low-frequency IGWs, which have frequencies around 1 μHz, the spin parameter ν, which measures the relative importance of rotation and stratification

$$\nu = \frac{2\Omega}{\sigma} = R_0^{-1}, \tag{10.1}$$

is of the order of unity (see Fig. 10.2, R_0 is the Rossby number). In this case, as illustrated in the diagram in Fig. 10.3 Coriolis effects cannot be treated as a perturbation. This will be also the case in rapid rotators such as young solar-type stars and massive stars.

The aim of this work is thus to examine how the improved description of wave-induced transport of angular momentum taking into account the effect of the Coriolis acceleration modifies the spatial structure of IGWs and, consequently, the angular momentum extraction/deposit by IGWs.

We first recall the main assumptions of our model and give the corresponding dynamical equations. The derivation of these equations has been presented in [22–25].

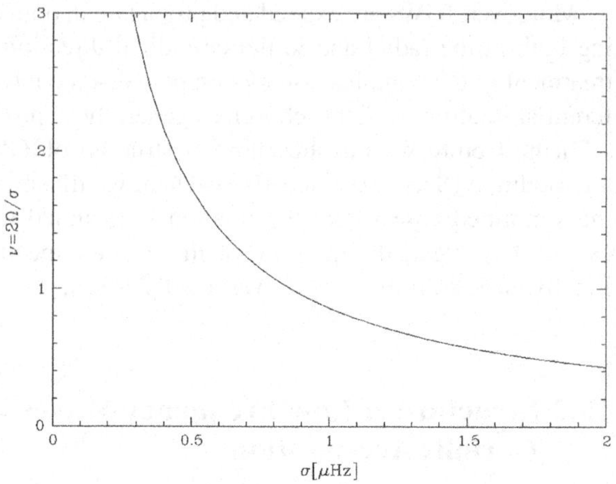

Fig. 10.2 Spin parameter $\nu = 2\Omega/\sigma$ for the Sun (we use $\frac{\Omega_\odot}{2\pi} = 430$ nHz) in the frequency range that may be relevant for the calculation of angular momentum transport (taken from [23]; courtesy Solar Physics)

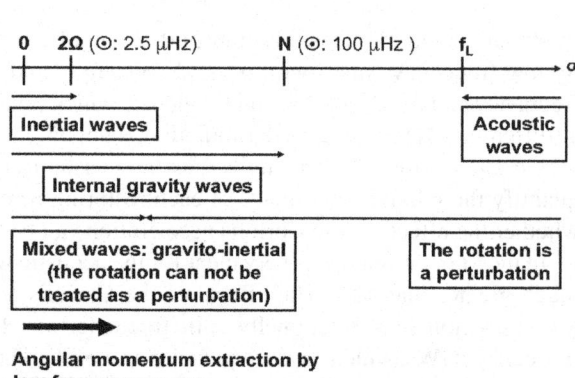

Fig. 10.3 Waves type in a differentially rotating stellar radiation zone and associated frequencies (f_L is the Lamb's frequency) (taken from [25]; courtesy Astronomy and Astrophysics)

10.2.1 Dynamical Equations

We expand the macroscopic internal velocity field in the radiative region as

$$\mathbf{V}(\mathbf{r}, t) = r \sin\theta \Omega(r, \theta)\,\hat{\mathbf{e}}_\varphi + \mathbf{u}(r, \theta, \varphi, t), \tag{10.2}$$

where the first azimuthal term is the velocity associated with the differential rotation while $\boldsymbol{u}$ is the wave velocity field. r, θ and φ are the classical spherical coordinates with their associated unit vectors $\{\hat{\mathbf{e}}_r, \hat{\mathbf{e}}_\theta, \hat{\mathbf{e}}_\varphi\}$ and t is the time. We ignore any large-scale velocity field that could be superposed (Fig. 10.4).

To treat the IGWs dynamics in a differentially rotating star, we have to solve the complete inviscid system formed by the momentum equation

$$\left(\partial_t + \Omega\partial_\varphi\right)\boldsymbol{u} + \left[2\Omega\hat{\mathbf{e}}_z \times \mathbf{u} + r\sin\theta\,(\boldsymbol{u}\cdot\nabla\Omega)\,\hat{\mathbf{e}}_\varphi\right] = -\frac{1}{\rho}\nabla P' - \nabla V' + \frac{\rho'}{\rho^2}\nabla\overline{P}, \tag{10.3}$$

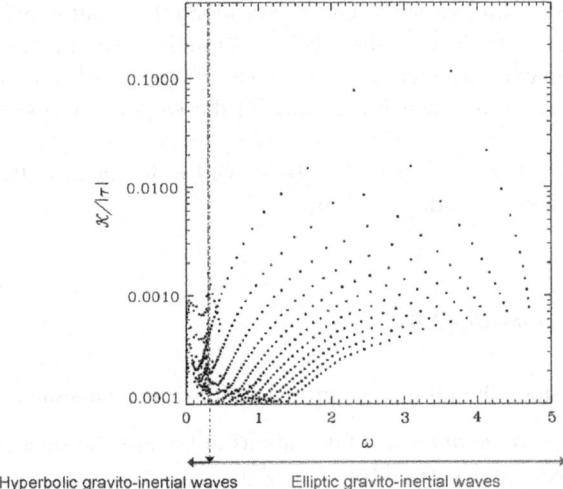

Fig. 10.4 Distribution in the complex plane $(\omega, K/|\tau|$ where K is the thermal diffusivity) of the eigenvalues of gravito-inertial modes in a $1.5\,M_\odot$ zero age main sequence star (Z=0.02), rotating with a uniform rotation ($\overline{\Omega}_s$) such that $\overline{\Omega}_s/\Omega_K = 0.3$, computed by [21]. The system of dynamical equations is solved taking into account viscous and thermal dissipation that thus leads to complex eigenfrequencies $\sigma = \omega + i\tau$, where ω corresponds to the usual mode frequency while τ is the damping. Two strongly different regimes can be isolated: in the sub-inertial regime, the eigenvalue behaviour is chaotic while it becomes regular in the super-inertial regime (adapted from [21]; courtesy Astronomy and Astrophysics)

the continuity equation

$$\left(\partial_t + \Omega\partial_\varphi\right)\rho' + \nabla\cdot\left(\overline{\rho}\boldsymbol{u}\right) = 0, \tag{10.4}$$

the energy transport equation, which we give here in the adiabatic limit

$$\left(\partial_t + \Omega\partial_\varphi\right)\left(\frac{P'}{\Gamma_1\overline{P}} - \frac{\rho'}{\overline{\rho}}\right) + \frac{N^2}{\overline{g}}u_r = 0, \tag{10.5}$$

and the Poisson's equation

$$\nabla^2 V' = 4\pi G\rho'. \tag{10.6}$$

ρ, V, P are respectively the fluid density, gravific potential ($\boldsymbol{g} = -\nabla V$ is the gravity), and pressure. Each of them has been expanded as $X(r,\theta,\varphi,t) = \overline{X}(r) + X'(r,\theta,\varphi,t)$, where $\overline{X}$ is the mean hydrostatic value of X on the isobar (which is the generalisation of the equipotential in the case of a differential rotation which disturbs the hydrostatic balance through the centrifugal force), X' being its wave's associated fluctuation. G is the universal gravity constant. $N^2 = \overline{g}\left(\frac{1}{\Gamma_1}\frac{d\ln\overline{P}}{dr} - \frac{d\ln\overline{\rho}}{dr}\right)$ is the Brunt–Väisälä frequency and $\Gamma_1 = (\partial\ln P/\partial\ln\rho)_{\overline{S}}$ (S being the macroscopic entropy) is the adiabatic exponent.

$\hat{\mathbf{e}}_z = \cos\theta\hat{\mathbf{e}}_r - \sin\theta\hat{\mathbf{e}}_\theta$ is the unit vector along the rotation axis. The terms in brackets correspond to the Coriolis acceleration in the case of a general differential rotation with the extra acceleration $r\sin\theta\,(\boldsymbol{u}\cdot\nabla\Omega)\,\hat{\mathbf{e}}_\varphi$ [16]. $D_t = \left(\partial_t + \Omega\partial_\varphi\right)$ is the Lagrangian derivative which accounts for the Doppler shift due to differential rotation.

The non-adiabaticity of waves will be treated in the next section by using the quasi-adiabatic approximation (cf. [26]).

10.2.2 Main Assumptions

To solve this system, the following approximations can be assumed:

- *the Cowling approximation*: the fluctuations of the gravitational potential associated with waves are neglected (see e.g. [27]).
- *the anelastic approximation*: since we are studied low-frequency IGWs, the anelastic approximation (i.e. $\nabla\cdot(\overline{\rho}\boldsymbol{u}) = 0$), where the acoustic waves are filtered out, is assumed.
- *the JWKB approximation*: waves which are studied here are low-frequency ones such that $\sigma \ll N$ (σ is the wave frequency in an inertial reference frame). Then, the JWKB approximation can be adopted. This also imply the *quasi—linear approximation* where the non-linear wave-wave interactions are not taken into account (see the discussions in [25, 28]).
- *the Traditional approximation*: stellar radiation zones are stably strongly stratified regions. Then, in the case where the angular velocity (Ω) is reasonably weak compared to the break-down one, $\Omega_K = \sqrt{GM/R^3}$ (M and R being respectively the star's mass and radius), we are in a situation where the centrifugal acceleration can be neglected to the first order and where $2\Omega \ll N$. This allows to adopt the Traditional approximation where the latitudinal component (along $\hat{\mathbf{e}}_\theta$) of the rotation vector $\boldsymbol{\Omega} = \Omega\hat{\mathbf{e}}_z = \Omega_V\hat{\mathbf{e}}_r + \Omega_H\hat{\mathbf{e}}_\theta$ (with $\Omega_V = \Omega\cos\theta$ and $\Omega_H = -\Omega\sin\theta$) can be neglected for all latitudes.

Let us present a brief local analysis of this approximation in the simplest case of a uniform rotation (see also [20]). The wave vector $\mathbf{k}$ and Lagrangian displacement $\boldsymbol{\xi}$ are expanded as

$$\mathbf{k} = k_V\hat{\mathbf{e}}_r + \mathbf{k}_H \quad \text{and} \quad \boldsymbol{\xi} = \xi_V\hat{\mathbf{e}}_r + \boldsymbol{\xi}_H, \tag{10.7}$$

where $\mathbf{k}_H = k_\theta\hat{\mathbf{e}}_\theta + k_\varphi\hat{\mathbf{e}}_\varphi$, $k_H = |\mathbf{k}_H|$, $\boldsymbol{\xi}_H = \xi_\theta\hat{\mathbf{e}}_\theta + \xi_\varphi\hat{\mathbf{e}}_\varphi$, $\xi_H = |\boldsymbol{\xi}_H|$ and $\boldsymbol{\xi} \propto \exp\left[i\,(\mathbf{k}\cdot\boldsymbol{r} - \sigma t)\right]$.

For low-frequency waves in radiation zones, we can write $\mathbf{k}\cdot\boldsymbol{\xi} = k_V\xi_V + \mathbf{k}_H\cdot\boldsymbol{\xi}_H \approx 0$ since $\nabla\cdot(\overline{\rho}\boldsymbol{\xi}) \approx 0$ (this is the anelastic approximation), from which we deduce that $\xi_V/\xi_H \approx -k_H/k_V$.

Next, using the results given in [29], the dispersion relation for the low-frequency gravito-inertial waves is obtained

$$\sigma^2 \approx N^2 \frac{k_H^2}{k^2} + \frac{(2\boldsymbol{\Omega} \cdot \mathbf{k})^2}{k^2},$$ (10.8)

where the two terms correspond respectively to the dispersion relations of IGWs and of inertial waves. In the case where $2\Omega \ll N$ and $\sigma \ll N$, the previous dispersion relation gives $k_H^2/k^2 \ll 1$. The vertical wave vector is then larger than the horizontal one while the displacement vector is almost horizontal: $|k_H| \ll |k_V|$, $|\xi_V| \ll |\xi_H|$. On the other hand, we get $(2\boldsymbol{\Omega} \cdot \mathbf{k})^2 \approx (2\Omega_V k_V)^2$. The latitudinal component of the rotation vector can thus be neglected in whole the sphere.

Let us now adopt a global point of view. In the general case, the operator which governs the spatial structure of waves, the Poincaré operator, is of mixed type (elliptic and hyperbolic) and not separable (for a detailed discussion we refer the reader to [21, 30, 31]). This leads to the appearance of detached shear layers associated with the underlying singularities of the adiabatic problem that could be crucial for transport and mixing processes in stellar radiation zones, since they are the seat of strong dissipation [21, 31–33].

Let us first focus on the case of a solid-body rotation ($\Omega = \overline{\Omega}_s$). In the largest part of stellar radiation zones, we are in a regime where $2\overline{\Omega}_s \ll N$. Since we are interested here in low-frequency waves ($\sigma \ll N$), the Traditional approximation, which consists in neglecting the latitudinal component of the rotation vector ($\overline{\Omega}_s \hat{\mathbf{e}}_z$), $-\overline{\Omega}_s \sin\theta \hat{\mathbf{e}}_\theta$, in the Coriolis acceleration, can be adopted in the super-inertial regime where $2\overline{\Omega}_s < \sigma \ll N$ (see e.g. [34]; for a modern description in a stellar context see [20, 24, 35]). Then, variables separation in radial and horizontal eigen-functions remains possible (c.f. [3]) that corresponds to the ergodic (regular) elliptic gravito-inertial modes family (the E_1 modes in [21, 31]; cf. Figs. 10.2 and 10.5). This approximation has however to be carefully used, as it changes the nature of the Poincaré operator, and removes the singularities and associated shear layers that appear. Then, in the sub-inertial regime, where $\sigma \leq 2\overline{\Omega}_s$, that corresponds to the equatorially trapped hyperbolic modes (the H_2 modes in [21, 31]; cf. Figs. 10.2, 10.5), the Traditional Approximation fails to reproduce the waves behaviour and the complete momentum equation has to be solved (detailed examples are given in [37, 38]).

As a first step, we thus have studied the regular elliptic waves for which the Traditional Approximation applies. Its application domain in the case of a general differential rotation will be discussed in Sect. 10.2.4

- *the quasi-adiabatic approximation*: Following [16, 26], we adopt the quasi-adiabatic approximation to treat the thermal damping of IGWs. Let us recall here that this damping is responsible for the net transport of angular momentum which is due to the bias in the wave's Doppler shift by differential rotation between retrograde ($m > 0$) and prograde waves ($m < 0$)[1] that transport respectively a negative and a positive flux of angular momentum (see (10.56) and [39]).

The wave phase is expanded as $\exp[i(m\varphi + \sigma t)]$.

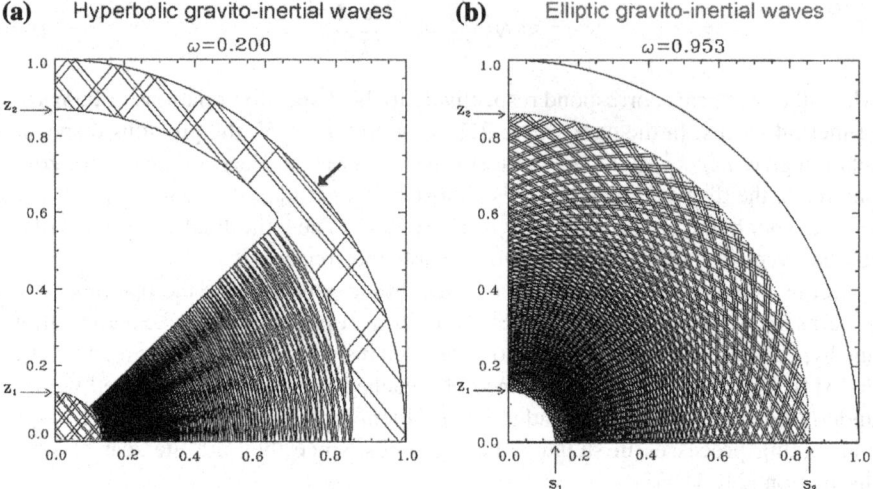

Fig. 10.5 Simulation of gravito-inertial modes in a 1.5 $M_\odot$ zero age main sequence star ($Z=0.02$), rotating with a uniform rotation ($\overline{\Omega}_s$) such that $\overline{\Omega}_s/\Omega_K = 0.3$, computed by [21] (cf. Fig10.4 4). Webs of characteristics for a H_2–hyperbolic mode (**a**) and a E_1–regular elliptic mode (**b**). The characteristics are the curve along which the energy propagates. The s_i and z_i locations correspond to $N(s_i) = \left(\omega^2 - 4\overline{\Omega}_s^2\right)^{1/2}$ and $N(z_i) = \omega$ while the arrow on the outer sphere (**a**) marks the critical latitude (θ_c) where $\theta_c = \cos^{-1}\left(\sigma/2\overline{\Omega}_s\right)$. One hundred and two hundred reflections have been respectively drawn for the considered H_2 and E_1–modes. It has to be emphasized that the H_2–mode is trapped in an equatorial belt ($\omega \leq 2\overline{\Omega}_s$) while the E_1–mode is living in the whole sphere without any critical latitude ($\omega > 2\overline{\Omega}_s$) (adapted from [21], courtesy Astronomy and Astrophysics)

10.2.3 Wave Pressure and Velocity Fields

Under those approximations, the wave's pressure and velocity field are then obtained in the case of a general differential rotation $\Omega(r, \theta)$ (the details of the derivation are given in [25]):

$$\tilde{p}(r, \theta, \varphi, t) = \sum_{\sigma, m, j} P'_{j,m}(r, \theta, \varphi, t), \qquad (10.9)$$

where

$$P'_{j,m}(r, \theta, \varphi, t) = -\overline{\rho}\, w_{j,m}\left(r, \theta; \hat{v}\right) \sin\left[\Phi_{j,m}(r, \varphi, t)\right] \exp\left[-\tau_{j,m}\left(r, \theta; \hat{v}\right)/2\right], \qquad (10.10)$$

and

$$\boldsymbol{u}(\boldsymbol{r}, t) = \sum_{k=\{r,\theta,\varphi\}} \left[\sum_{\sigma, m, j} u_{k;j,m}(r, t)\right] \hat{\mathbf{e}}_k, \qquad (10.11)$$

where

$$u_{r;j,m}(\mathbf{r}, t) = \frac{\hat{\sigma}}{N} \frac{\lambda_{j,m}^{1/2}(r; \hat{v})}{r} w_{j,m}(r, \theta; \hat{v}) \sin\left[\Phi_{j,m}(r, \varphi, t)\right]$$
$$\times \exp\left[-\tau_{j,m}(r, \theta; \hat{v})/2\right], \qquad (10.12)$$

$$u_{\theta;j,m}(\mathbf{r}, t) = -\frac{\hat{\sigma}}{r} \mathcal{G}_{j,m}^{\theta}(r, \theta; \hat{v}) \cos\left[\Phi_{j,m}(r, \varphi, t)\right] \exp\left[-\tau_{j,m}(r, \theta; \hat{v})/2\right], \qquad (10.13)$$

$$u_{\varphi;j,m}(\mathbf{r}, t) = \frac{\hat{\sigma}}{r} \mathcal{G}_{j,m}^{\varphi}(r, \theta; \hat{v}) \sin\left[\Phi_{j,m}(r, \varphi, t)\right] \exp\left[-\tau_{j,m}(r, \theta; \hat{v})/2\right]. \qquad (10.14)$$

The "local" frequency $(\hat{\sigma})$[2] which accounts for the Doppler shift by the differential rotation and the "spin parameter" (see [20]) are defined:

$$\hat{\sigma}(r, \theta) = \sigma + m\Omega(r, \theta) \quad \text{and} \quad \hat{v}(r, \theta) = \frac{2\Omega(r, \theta)}{\hat{\sigma}(r, \theta)} = \widetilde{R}_o^{-1}, \qquad (10.15)$$

where $\widetilde{R}_o$ is the local Rossby number. Unlike the particular case of uniform rotation, variables do not separate neatly anymore in the case of general differential rotations $\Omega(r)$ and $\Omega(r, \theta)$. The velocity components are thus expressed in terms of the 2D dynamical pressure $(P/\bar{\rho})$ eigenfunctions $w_{j,m}$ which are solutions of the following eigenvalue equation:

$$\mathcal{O}_{\hat{v};m}\left[w_{j,m}(r, x; \hat{v})\right] = -\lambda_{j,m}(r; \hat{v}) w_{j,m}(r, x; \hat{v}). \qquad (10.16)$$

We define the General Laplace Operator (GLO)

$$\mathcal{O}_{\hat{v};m} = \frac{1}{\hat{\sigma}} \frac{d}{dx}\left[\frac{(1-x^2)}{\hat{\sigma}\mathcal{D}(r, x; \hat{v})} \frac{d}{dx}\right] - \frac{m}{\hat{\sigma}^2 \mathcal{D}(r, x; \hat{v})}\left(1 - x^2\right) \frac{\partial_x \Omega}{\hat{\sigma}} \frac{d}{dx}$$
$$- \frac{1}{\hat{\sigma}}\left[\frac{m^2}{\hat{\sigma}\mathcal{D}(r, x; \hat{v})(1 - x^2)} + m\frac{d}{dx}\left(\frac{\hat{v}x}{\hat{\sigma}\mathcal{D}(r, x; \hat{v})}\right)\right] \qquad (10.17)$$

with

$$\mathcal{D}(r, x; \hat{v}) = 1 - \hat{v}^2 x^2 + \hat{v}\left(\partial_x \Omega/\hat{\sigma}\right) x \left(1 - x^2\right) \qquad (10.18)$$

and $x = \cos\theta$. $\mathcal{O}_{\hat{v};m}$ is the generalisation of the classical Laplace tidal operator [41], the eigenfunctions $w_{j,m}$ being thus a generalisation of the Hough functions

---

[2] Note that $\hat{\sigma}$ can vanish that corresponds to the corotation resonance. In layer(s) where this happens (which are called critical layers), a careful treatment of the complete fluid dynamics equations has to be undertaken that is beyond the scope of the present paper (see [40]).

[42, 43]. $\lambda_{j,m}\left(r; \hat{v}\right)$ are the associated eigenvalues; here, we focus on positive ones that correspond to propagative waves (cf. [43]). Furthermore, the GLO is a differential operator in x only and the $w_{j,m}$ form a complete orthogonal basis

$$\int_{-1}^{1} w_{i,m}^*\left(r, x; \hat{v}\right) w_{j,m}\left(r, x; \hat{v}\right) dx = C_{i,m}\delta_{i,j}, \tag{10.19}$$

where $C_{i,m}$ is the normalisation factor and $\delta_{i,j}$ is the usual Kronecker symbol. The boundary conditions are ruled by the regularity at the poles.

The dispersion relation is then given by

$$k_{V;j,m}^2\left(r\right) = \frac{\lambda_{j,m}\left(r; \hat{v}\right) N^2}{r^2}, \tag{10.20}$$

where $k_{V;j,m}$ is the vertical component of the wave vector ($\lambda_{j,m}$ has the dimension of $\left[t^2\right]$). That leads to the following expressions for the JWKB phase function

$$\Phi_{j,m}\left(r, \varphi, t\right) = \sigma t + \int_{r}^{r_c} k_{V;j,m} dr' + m\varphi \tag{10.21}$$

(r_c is the radius of the basis (or the top) of the adjacent convective region that excites the waves) and for the damping rate

$$\tau_{j,m} = \int_{r}^{r_c} K \frac{\lambda_{j,m}^{3/2}\left(r; \hat{v}\right) N N_T^2}{\hat{\sigma}} \frac{dr'}{r'^3}, \tag{10.22}$$

K being the thermal diffusivity. The Brunt–Väisälä frequency takes into account the effects of both the thermal and the chemical composition gradients, with the classical notations $N^2 = N_T^2 + N_\mu^2$ where $N_T^2 = \left(\overline{g}\delta/H_P\right)\left(\nabla_{ad} - \nabla\right)$ and $N_\mu^2 = \left(\overline{g}\phi/H_P\right)\nabla_\mu$, where $H_P = |dr/d\ln\overline{P}|$ $\delta = -\left(\partial\ln\overline{\rho}/\partial\ln\overline{T}\right)_{\overline{P},\overline{\mu}}$, $\phi = \left(\partial\ln\overline{\rho}/\partial\ln\overline{\mu}\right)_{\overline{P},\overline{T}}$, $\nabla_{ad} = \left(\partial\ln\overline{T}/\partial\ln\overline{P}\right)_{ad}$, $\nabla = \partial\ln\overline{T}/\partial\ln\overline{P}$ and $\nabla_\mu = \partial\ln\overline{\mu}/\partial\ln\overline{P}$, with $\overline{T}$ and $\overline{\mu}$ being respectively the mean temperature and molecular weight.

On the other hand, the latitudinal and azimuthal eigenfunctions are defined

$$\mathcal{G}_{j,m}^\theta\left(r, x; \hat{v}\right) = \frac{1}{\hat{\sigma}^2} \frac{1}{\mathcal{D}\left(r, x; \hat{v}\right)\sqrt{1-x^2}} \left[-\left(1-x^2\right)\frac{d}{dx} + m\hat{v}x\right] w_{j,m} \tag{10.23}$$

$$\mathcal{G}_{j,m}^\varphi\left(r, x; \hat{v}\right) = \frac{1}{\hat{\sigma}^2} \frac{1}{\mathcal{D}\left(r, x; \hat{v}\right)\sqrt{1-x^2}} \times \left[-\left(\hat{v}x - \left(1-x^2\right)\frac{\partial_x\Omega}{\hat{\sigma}}\right)\left(1-x^2\right)\frac{d}{dx} + m\right] w_{j,m}. \tag{10.24}$$

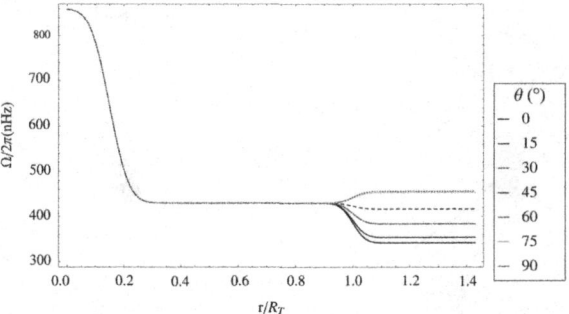

Fig. 10.6 Synthetic internal rotation profile as it may be seen in the Sun (cf. [10]): $\Omega_{syn}(r,\theta) = \Omega_{RZ} + \Omega_{RZ}A_c[1 - \text{Erf}((r - R_c)/l_c)] + 1/2[1 + \text{Erf}((r - R_T)/l_T)](A + B\cos^2\theta + C\cos^4\theta - \Omega_{RZ})$, where $\Omega_{RZ} = 430\,\text{nHz}$, $A_c = 1/2$ (such that $\Omega_{syn}(0,\theta) = 2\Omega_s$), $R_c = 0.15R_T$ $l_c = 0.075R_T$ $R_T = 0.71R_\odot$ (the position of the tachocline), $l_T = 0.05R_T$, $A = 456\,\text{nHz}$, $B = -42\,\text{nHz}$ and $C = -72\,\text{nHz}$ (we assume here a *tachocline* that is thicker than in reality) (taken from [90]; courtesy Communications in Asteroseismology)

10.2.4 The Traditional Approximation in the Case of a General Differential Rotation

As it has been emphasized by [24] and references therein, the Traditional Approximation has to be used carefully since it modifies the mathematical properties of the adiabatic wave operator. Here, in the case of a general differential rotation law, it is applicable in spherical shell(s) such that $\mathcal{D} > 0$ everywhere ($\forall r$ and $\forall\theta \in [0, \pi]$). There, the adiabatic wave operator is elliptic that corresponds to regular (elliptic) gravito-inertial waves. In the other spherical shell(s), where both $\mathcal{D} < 0$ and $\mathcal{D} > 0$, the adiabatic wave operator is hyperbolic and the Traditional Approximation cannot be applied because of the adiabatic wave's velocity field (and wave operator) singularity where $\mathcal{D} = 0$. Regularization is allowed there by thermal and viscous diffusions that lead to shear layers, the attractors, where strong dissipation occurs that may induce transport and mixing. In Fig. 10.6, we illustrate for a given chosen theoretical angular velocity profile (cf. Fig. 10.7) how those two types of spherical shells (respectively where the Traditional Approximation is allowed or forbidden) could appear.

10.3 Transport of Angular Momentum

10.3.1 Action of Angular Momentum

Since the complete wave's velocity field is derived, we focus on the induced-transport of angular momentum. The vertical and horizontal Lagrangian angular momentum

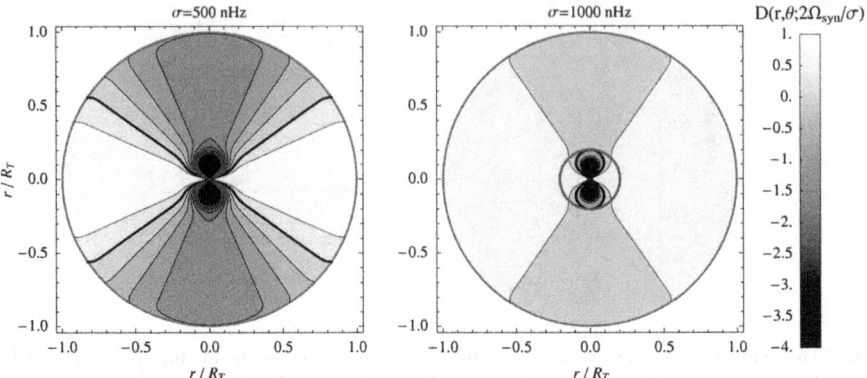

Fig. 10.7 $\mathcal{D}\left(r, \theta; 2\Omega_{\text{syn}}/\sigma\right)$ as a function of r and θ (cf. (10.18)) for $\sigma = 500$ nHz (*Left*) and $\sigma = 1000$ nHz (*Right*) for axisymmetric waves ($m = 0$). The critical surface $\mathcal{D}\left(r, \theta; 2\Omega_{\text{syn}}/\sigma\right) = 0$ is given by the *thick black line* and the iso-$\mathcal{D}$ lines such that $\mathcal{D}\left(r, \theta; 2\Omega_{\text{syn}}/\sigma\right) > 0$ and $\mathcal{D}\left(r, \theta; 2\Omega_{\text{syn}}/\sigma\right) < 0$ are respectively given by the *red* and the *blue lines*. The Traditional Approximation (T. A.) applies in spherical shell(s) such that $\mathcal{D} > 0$ everywhere ($\forall r$ and $\forall \theta \in [0, \pi]$); there, waves are regular at all latitudes. In other spherical shell(s), where both $\mathcal{D} > 0$ and $\mathcal{D} < 0$, the T. A. does not apply due to the singularity at $\mathcal{D} = 0$. Therefore, for Ω_{syn}, the T. A. does not apply for $\sigma = 500$ nHz while it applies for $\sigma = 1000$ nHz in the external spherical shell with the inner border given by the *thick red circle* (taken from [90]; courtesy Communications in Asteroseismology)

fluxes are respectively defined:

$$\mathcal{F}_V^{\text{AM}}\left(r, \theta\right) = \int_\sigma \left\langle \underbrace{\overline{\rho} r \sin \theta u_r u_\varphi}_{\text{I}} + \underbrace{\overline{\rho} r \sin \theta 2\Omega \cos \theta u_r \xi_\theta}_{\text{II}} \right\rangle_\varphi d\sigma, \tag{10.25}$$

$$\text{and} \quad \mathcal{F}_H^{\text{AM}}\left(r, \theta\right) = \overline{\rho} r \sin \theta \int_\sigma \left\langle u_\theta u_\varphi \right\rangle_\varphi d\sigma,$$

where $\langle \cdots \rangle_\varphi = (1/2\pi) \int_0^{2\pi} \cdots d\varphi$ and where we sum over the excited spectrum. The Lagrangian wave displacement is defined such that: $\boldsymbol{u} = \left(\partial_t + \Omega \partial_\varphi\right) \boldsymbol{\xi} - r \sin \theta \left(\boldsymbol{\xi} \cdot \boldsymbol{\nabla}\Omega\right) \hat{\mathbf{e}}_\varphi$. Note that in the rotating case the vertical flux ($\mathcal{F}_V^{\text{AM}}$) is the sum of the Reynolds stresses across an Eulerian surface (term I) plus a Lagrangian contribution (term II) (see [44]). Using (10.13, 10.14), we get $\mathcal{F}_H^{\text{AM}} = 0$. Then, following the methodology given in [16, 23, 24] , we get the vertical action of angular momentum (called the angular momentum luminosity in the stellar context, a term coined by [39])

$$\mathcal{L}_V^{\text{AM}}\left(r, x\right) = r^2 \mathcal{F}_V^{\text{AM}}$$

$$= -r_c^2 \int_\sigma \sum_{m,j} \left\{ \frac{\hat{m}_{j,m}\left(r_c, x; \hat{v}_c\right)}{\hat{\sigma}_{\text{CZ}}} \mathcal{F}_{V;j,m}^{\text{E}}\left(r_c, x; \hat{v}_c\right) \exp\left[-\tau_{j,m}\left(r, \theta; \hat{v}\right)\right] \right\} d\sigma. \tag{10.26}$$

which is conserved in the adiabatic limit (cf. [45, 46]). r_c is the radius of the basis (or the top) of the adjacent convective region that excites the waves while $\hat{v}_c = 2\Omega_{CZ}(r_c, \theta)/\hat{\sigma}_{CZ}$, where $\hat{\sigma}_{CZ} = \sigma + m\Omega_{CZ}(r_c, \theta)$ Ω_{CZ} being its angular velocity. On the other hand, $\mathcal{F}^E_{V;j,m}(r_c, x; \hat{v}_c)$ is the monochromatic energy flux injected by turbulent convective movements at $r = r_c$ in the studied radiation zone and

$$\hat{m}_{j,m}(r, x; \hat{v}) = \frac{\sin\theta \hat{\sigma}^2 w_{j,m}\left[\mathcal{G}^\varphi_{j,m} - \hat{v}\cos\theta\mathcal{G}^\theta_{j,m}\right]}{w^2_{j,m}} \tag{10.27}$$

is the 2D function which describes its conversion into angular momentum flux.

10.3.2 Transport of Angular Momentum by the Waves: The Shear Layer Oscillation and Secular Effects

Then, waves deposit their angular momentum inside the star as they are damped. The deposition of angular momentum is then ruled by the radial derivative of the action of angular momentum [7]

$$\overline{\rho}\frac{d}{dt}\left[r^2\langle\Omega\rangle_\theta\right] = \pm\frac{1}{r^2}\partial_r\left[\langle\mathcal{L}^{AM}_V(r, \theta)\rangle_\theta\right], \tag{10.28}$$

where $\langle\cdots\rangle_\theta = 1/2\int_0^\pi \cdots d\theta$. The "+" ("−") sign in front of the action of angular momentum corresponds to a wave traveling inward (outward).

Let us first take a look at the damping integral given in (10.22) and assume that both prograde and retrograde waves are excited with the same amplitude and have the same eigenvalue $\lambda_{j,m}$. In solid-body rotation, both waves are equally dissipated when traveling inward and there is no impact on the distribution of angular momentum. In the presence of differential rotation, the situation is different. If the interior is rotating faster than the convection zone, the local frequency of prograde waves decreases, which enhances their dissipation; the corresponding retrograde waves are then dissipated further inside. This produces an increase of the local differential rotation, and creates a double-peaked shear layer because local shears are amplified by waves, even a small perturbation triggering this (the prograde waves transport a positive flux of angular momentum while the retrograde waves transport a negative one). In the presence of shear turbulence, this layer oscillates, producing a "Shear Layer Oscillation" or SLO (cf. [47, 48]). This is the first important feature of wave-mean flow interaction.

This SLO acts as a filter, through which most low-frequency waves cannot pass. However, if the core is rotating faster than the surface, this filter is not quite symmetric, and retrograde waves will be favored. As a result, a net *negative* flux of angular momentum will result, and produce a spin down of the core [17]. This is the filtered angular momentum action, which contributes to the secular evolution of angular momentum (for details, see [18]). It plays a key role in flatening the rotation profile as observed in the present Sun [49].

Then, following [50], averaging over latitudes Ω and $\mathcal{L}_V^{AM}$ in spherical shell(s) where the Traditional Approximation applies, and expanding the vertical action of angular momentum as $\mathcal{L}_V^{AM} = \sum_l \mathcal{L}_{V;l}^{AM}(r) \sin^2\theta \, P_l(\cos\theta)$, we get for the averaged rotation rate on an isobar $(\langle\Omega\rangle_\theta)$

$$\bar{\rho}\frac{d}{dt}\left(r^2\langle\Omega\rangle_\theta\right) - \frac{1}{5r^2}\partial_r\left(\bar{\rho}r^4\langle\Omega\rangle_\theta U_2\right)$$

$$= \frac{1}{r^2}\partial_r\left(\bar{\rho}\nu_V r^4\partial_r\langle\Omega\rangle_\theta\right) - \frac{1}{r^2}\partial_r\left[\langle\mathcal{L}_V^{AM;fil}\rangle_\theta(r)\right] + \langle\Gamma_{\mathcal{F}_L}\rangle_\theta(r), \qquad (10.29)$$

and for the first mode of the latitudinal differential rotation

$$\bar{\rho}\frac{d}{dt}\left(r^2\Omega_2\right) - 2\bar{\rho}\langle\Omega\rangle_\theta\left[2V_2 - \frac{1}{2}\frac{d\ln\left(r^2\langle\Omega\rangle_\theta\right)}{d\ln r}U_2\right]$$

$$= \frac{1}{r^2}\partial_r\left(\bar{\rho}\nu_V r^4\partial_r\Omega_2\right) - 10\bar{\rho}\nu_H\Omega_2 - \frac{1}{r^2}\partial_r\left[\mathcal{L}_{V;2}^{AM}(r)\right] + \Gamma_{\mathcal{F}_L};2(r), \quad (10.30)$$

where $\Omega = \langle\Omega\rangle_\theta + \tilde{\Omega}_2(r,\theta)$ and $\tilde{\Omega}_2 = \Omega_2(r)[P_2(\cos\theta) + 1/5]$. d/dt is the Lagrangian derivative that accounts for the contractions and the dilatations of the star during its evolution. The meridional circulation is expanded in Legendre polynomials as

$$\mathcal{U}_M(r,\theta) = \sum_{l>0}\left\{U_l(r)\,P_l(\cos\theta)\,\hat{\mathbf{e}}_r + V_l(r)\,\partial_\theta P_l(\cos\theta)\,\hat{\mathbf{e}}_\theta\right\}, \qquad (10.31)$$

where $V_l = 1/[l(l+1)\bar{\rho}r]\,d\left(\bar{\rho}r^2U_l\right)/dr$ is obtained assuming the anelastic approximation; U_2 is the only mode which leads to a net transport of angular momentum on an isobar as demonstrated in [51]. (ν_V, ν_H) are respectively the vertical and the horizontal turbulent viscosities. Finally, $\Gamma_{\mathcal{F}_L}$ and $\Gamma_{\mathcal{F}_L};2$ are the horizontal average of the torque of the Lorentz force and its first latitudinal mode.

These equations give the evolution of the differential rotation, both in the radial and in the latitudinal directions, in the spherical shell(s) where the Traditional Approximation can be applied. The evolution equations for the differential rotation (both in r and θ) capturing gravito-inertial waves feedback is thus derived, taking into account the modification of IGWs through the Coriolis acceleration and their feedback on the angular velocity profile through the net induced transport of angular momentum due to the differential damping of retrograde and prograde waves.

Then, the associated boundary conditions at $r = r_b$ and $r = r_t$, where r_b and r_t are respectively the radius of the base and of the top of the considered radiative region, are given by:

$$\frac{d}{dt}\left[\int_0^{r_b} r^4\bar{\rho}\langle\Omega\rangle_\theta dr\right] = \frac{1}{5}r^4\bar{\rho}\langle\Omega\rangle_\theta U_2 - \mathcal{F}_B(r_b) - \langle\mathcal{L}_V^{AM;fil.}\rangle_\theta(r_b) \qquad (10.32)$$

and

$$\frac{d}{dt}\left[\int_{r_t}^{R} r^4\overline{\rho}\langle\Omega\rangle_\theta dr\right] = -\frac{1}{5}r^4\overline{\rho}\langle\Omega\rangle_\theta U_2 - \mathcal{F}_\Omega + \mathcal{F}_B(r_t) + \langle\mathcal{L}_V^{AM;fil.}\rangle_\theta(r_t), \quad (10.33)$$

$\mathcal{F}_\Omega$ and $\mathcal{F}_B$ being respectively the flux of angular momentum loss at the surface and the magnetic angular momentum flux through the interfaces. In the solar case, $r_b = 0$ and $r_t = r_{SLO}$.

We are thus now in a position where we are able to get a coherent picture of solar and stellar radiation zone dynamics, taking into account the highly non-linear interaction between the differential rotation, the associated meridional circulation, the vertical and horizontal shear-induced turbulence, a potential fossil magnetic field, and the low-frequency waves where the action of the rotation on the waves through the Coriolis acceleration and their feed-back on the angular velocity distribution are treated in a coherent way. In the following diagram, those interactions are summarized with the action of each process:

- the *meridional circulation*, which is due to structural adjustments and to the extraction of angular momentum at the surface by the wind [52, 53], advects angular momentum, chemical elements and the magnetic field;
- the *shear-induced turbulence* acts to suppress its cause, namely the vertical and the horizontal gradients of angular velocity;
- the *fossil magnetic field*, which is sheared by the differential rotation, advected by the meridional circulation, and diffused by ohmic effects, transports angular momentum through the large-scale Lorentz torque and the Maxwell stresses associated with MHD instabilities;
- the *low-frequency internal waves* generated by the convective movements transport angular momentum as well as magnetic field, that modifies the angular velocity distribution and the associated mixing;
- the *braking due to the wind* in the early phases, forces an extraction of angular momentum at the surface, which drives the behaviour of the meridional circulation and the potential wave-induced fronts of angular momentum extraction. Fig. (10.8).

10.4 The Weak Differential Rotation Case

10.4.1 Definitions

Let us begin by looking at the differential rotation law. First, from now on, we restrict ourselves to a "shelllular" rotation $\overline{\Omega}[r(P)]$ (in other words, the angular velocity is constant on an isobar) due to the anisotropic turbulence in a strongly stratified

Fig. 10.8 Dynamical transport processes in stellar radiation zones: those regions are the seat of highly non-linear interactions between the differential rotation, the meridional circulation, the turbulence, the magnetic field, and the IGWs (taken from [89]; courtesy Solar Physics)

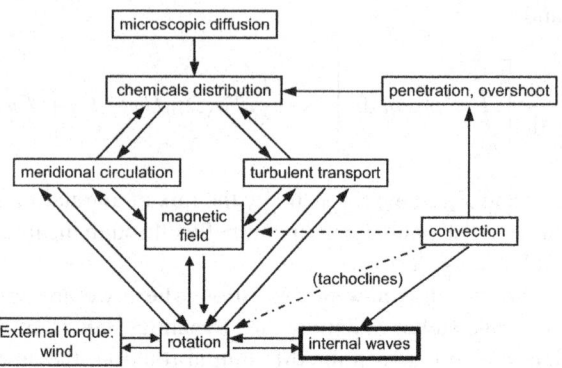

star [52]; $r\left(P\right)$ is the radius of the isobar, which is the generalization of the equipotential in the case of differential rotation. Next, we split this "shellular" rotation law into a solid body rotation, $\overline{\Omega}_s$, and a (small) differential rotation fluctuation, $\delta\overline{\Omega}\left(r\right)$. This hypothesis will allow us to separate neatly the variables in the treatment of the dynamical equations and the formalism presented hereafter remains valid only in the case of "reasonable" values of the fluctuation of the angular velocity $\left(\delta\overline{\Omega}\right)$ around its mean value $\left(\overline{\Omega}_s\right)$ and of the radial gradient of Ω. Thus, we write:

$$\Omega\left(r,\theta\right) \approx \overline{\Omega}\left(r\right) = \overline{\Omega}_s + \delta\overline{\Omega}\left(r\right) \quad \text{where } \delta\overline{\Omega}\left(r\right) \ll \overline{\Omega}_s. \qquad (10.34)$$

$\overline{\Omega}_s$ will be taken into account for the calculation of the structure of the low-frequency adiabatic waves, while $\delta\overline{\Omega}$ will be accounted for only in the treatment of the damping due to dissipative processes. This is the *"weak differential rotation case"*.

We now define the frequencies we shall use in this case:

$$\begin{cases} \sigma_s = \sigma + m\overline{\Omega}_s \\ \hat{\sigma}\left(r\right) = \tilde{\sigma}\left(r\right) = \sigma + m\overline{\Omega}\left(r\right) = \sigma_s + m\delta\overline{\Omega}\left(r\right) \end{cases}, \qquad (10.35)$$

σ is the wave's frequency in the inertial frame, while σ_s is given in the corotating frame (with the uniform rotation angular velocity $\overline{\Omega}_s$). $\tilde{\sigma}\left(r\right)$ is the local frequency Doppler-shifted by the differential rotation. Finally, the spin parameter becomes

$$\hat{\nu}_s = \nu_s = \frac{2\overline{\Omega}_s}{\sigma_s}. \qquad (10.36)$$

Remember that we chose the sign of m such that the prograde waves have $m < 0$ while the retrograde waves have $m > 0$.

10.4.2 Wave Pressure and Velocity Fields

In the "weak differential rotation case", a neat variable separation in r and θ is obtained (cf. [20, 22, 24]) and we get:

$$w_{j,m}(r,\theta;\hat{v}) = \frac{\tilde{p}_{j,m}(r)}{\overline{\rho}}\Theta_{j,m}(\cos\theta;v_s),\tag{10.37}$$

and

$$\mathcal{G}^{\theta}_{j,m}(r,\theta;\hat{v}) = \frac{1}{\sigma_s^2}\frac{\tilde{p}_{j,m}(r)}{\overline{\rho}}\mathcal{H}^{\theta}_{j,m}(\cos\theta;v_s),\tag{10.38}$$

$$\mathcal{G}^{\varphi}_{j,m}(r,\theta;\hat{v}) = \frac{1}{\sigma_s^2}\frac{\tilde{p}_{j,m}(r)}{\overline{\rho}}\mathcal{H}^{\varphi}_{j,m}(\cos\theta;v_s).\tag{10.39}$$

Moreover, the pressure fluctuation becomes:

$$P'(r,\theta,\varphi,t) = \sum_{m,j}P'_{j,m}(r,\theta,\varphi,t)\tag{10.40}$$

with

$$P'_{j,m} = \tilde{p}_{j,m}(r)\Theta_{j,m}(\cos\theta;v_s)\exp[i(m\varphi+\sigma t)].\tag{10.41}$$

The Hough functions ($\Theta_{j,m}$, $\mathcal{H}^{\theta}_{j,m}$ and $\mathcal{H}^{\varphi}_{j,m}$) will be discussed later in this section. Using the approximations described in Sect. 10.2.2 (the anelastic, the quasi-adiabatic, the Traditional and the JWKB approximations), the wave velocity field is then expanded as:

$$\mathbf{u} = \begin{cases} \sum_{m,j}u_{r;j,m}(r,\theta,\varphi,t)\\ \sum_{m,j}u_{\theta;j,m}(r,\theta,\varphi,t)\\ \sum_{m,j}u_{\varphi;j,m}(r,\theta,\varphi,t) \end{cases}\tag{10.42}$$

where the monochromatic radial, latitudinal, and azimuthal components are given by:

$$u_{r;j,m}(r,\theta,\varphi,t) = \mathcal{E}_{j,m}(r)\sin[\Phi_{j,m}(r,\varphi,t)]\Theta_{j,m}(\cos\theta;v_s)$$
$$\times \exp[-\tau_{j,m}(r;v_s)/2],\tag{10.43}$$

$$u_{\theta;j,m}(r,\theta,\varphi,t) = -\frac{rk_{V;j,m}}{\Lambda_{j,m}(v_s)}\mathcal{E}_{j,m}(r)\cos[\Phi_{j,m}(r,\varphi,t)]\mathcal{H}^{\theta}_{j,m}(\cos\theta;v_s)$$
$$\times \exp[-\tau_{j,m}(r;v_s)/2],$$
$$\tag{10.44}$$

$$u_{\varphi;j,m}(r,\theta,\varphi,t) = \frac{rk_{V;j,m}}{\Lambda_{j,m}(v_s)}\mathcal{E}_{j,m}(r)\sin[\Phi_{j,m}(r,\varphi,t)]\mathcal{H}^{\varphi}_{j,m}(\cos\theta;v_s)$$
$$\times \exp[-\tau_{j,m}(r;v_s)/2],$$
$$\tag{10.45}$$

where the attenuation $\tau_{j,m}$ will be made explicit in (10.56). In the JWKB regime, the phase function ($\Phi_{j,m}$) is given by:

$$\Phi_{j,m}(r,\varphi,t) = \sigma t + \int_r^{r_c} k_{V:j,m} dr' + m\varphi,\tag{10.46}$$

where the vertical wave vector

$$k_{V:j,m}^2 = \left(\frac{N^2}{\tilde{\sigma}^2(r)}\right)\frac{\Lambda_{j,m}(\nu_s)}{r^2}\quad\text{with}\quad\Lambda_{j,m}(\nu_s) = \sigma_s^2\lambda_{j,m}(\hat{\nu}_s),\tag{10.47}$$

has been drawn from the equation for the radial component of the Lagrangian displacement ξ, where $u = d\xi/dt$:

$$\frac{d^2}{dr^2}\left(\overline{\rho}^{1/2}r^2\xi_{r:j,m}\right) + \left[\left(\frac{N^2}{\tilde{\sigma}^2(r)}\right)\frac{\Lambda_{j,m}(\nu_s)}{r^2}\right]\left(\overline{\rho}^{1/2}r^2\xi_{r:j,m}\right) = 0.\tag{10.48}$$

This equation, obtained after variable separation in r and θ in (10.3–10.5), has been derived using the anelastic approximation. The JWKB amplitude function ($\mathcal{E}_{j,m}(r)$) is given by

$$\mathcal{E}_{j,m}(r) = \mathcal{A}_{j,m}r^{-\frac{3}{2}}\overline{\rho}^{-\frac{1}{2}}\left(\frac{N^2}{\tilde{\sigma}^2}\right)^{-\frac{1}{4}},\tag{10.49}$$

where the amplitude of the wave ($\mathcal{A}_{j,m}$) must be determined from boundary conditions.Fig. 10.9.

This separation of variables is allowed by the Traditional Approximation and by the "weak differential rotation approximation"; it leads to an equation that depends only on θ for the angular function of $P'_{j,m}$ and of $u_{r:j,m}$ $\Theta_{j,m}$:

$$\sigma_s^2\mathcal{O}_{\hat{\nu}_s;m}\left[\Theta_{j,m}(x;\nu_s)\right] = \mathcal{L}_{\nu_s;m}\left[\Theta_{j,m}(x;\nu_s)\right] = -\Lambda_{j,m}(\nu_s)\Theta_{j,m}(x;\nu_s),\tag{10.50}$$

the GLO (cf. (10.17)) reducing to

$$\mathcal{L}_{\nu_s;m} = \frac{d}{dx}\left(\frac{1-x^2}{1-\nu_s^2x^2}\frac{d}{dx}\right) - \frac{1}{1-\nu_s^2x^2}\left(\frac{m^2}{1-x^2} + m\nu_s\frac{1+\nu_s^2x^2}{1-\nu_s^2x^2}\right),\tag{10.51}$$

where $x = \cos\theta$. The differential rotation fluctuation $(\delta\overline{\Omega})$ is only taken into account for the Doppler shift of the waves and their dissipation (cf. (10.56)), but not in the derivation of their horizontal spatial structure. In fact, even in the case of a general shelllular rotation law $\overline{\Omega}(r)$, $\hat{\nu} = 2\overline{\Omega}/\hat{\sigma}$ depends on r as well as the eigenvalues ($\lambda_{j,m}$) and the variables no longer separate. Equation (10.50) is the so-called Laplace equation (cf. [41]) while the $\Theta_{j,m}$ are the Hough functions (cf. [42, 54]). In the non-rotating case, the Laplace operator ($\mathcal{L}_{\nu_s;m}$) is equivalent to the classical horizontal spherical laplacian, and the Hough functions reduce to the associated Legendre polynomials.

Fig. 10.9 Hough functions:
(continuous lines)
$v = 0$, $m = 2$;
$v = 0.86$, $\sigma = 1\,\mu\text{Hz}$
(*dotted lines*) $m = 2$ (*dashed lines*) $m = -2$ (taken from
[89]; courtesy Solar Physics)

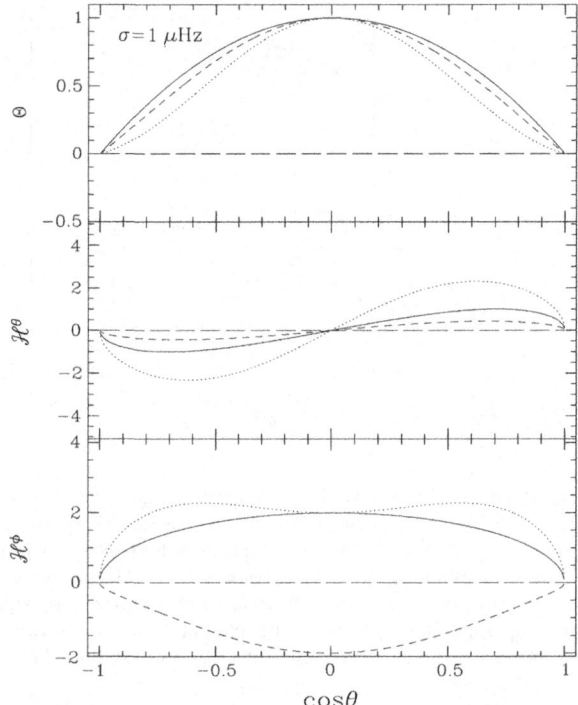

Let us briefly describe the main features of these Hough functions. First, since $\mathcal{L}_{v_s;m}$ depends explicitly on m, we have $\Lambda_{j,-m} \neq \Lambda_{j,m}$ and $\Theta_{j,-m} \neq \Theta_{j,m}$. In other words, for a given j, prograde and retrograde waves have a different horizontal spatial structure (see Fig. 10.10). This is crucial for the transport of angular momentum by those waves that depends on the subtle balance between prograde and retrograde waves (cf. [18] and references therein). Moreover, the transmission of the kinetic energy flux of the turbulent motions, which are at the origin of the generation of the waves at the interface between convection and radiation, is modified (cf. (10.27)).

Now, by using the latitudinal and the azimuthal components of the momentum equation (10.3), and eliminating the pressure fluctuation, we obtain the respective angular functions for $u_{\theta;j,m}$ and $u_{\varphi;j,m}$, $\mathcal{H}^{\theta}_{j,m}(x; v_s)$ and $\mathcal{H}^{\varphi}_{j,m}(x; v_s)$:

$$\mathcal{H}^{\theta}_{j,m}(x; v_s) = \sigma_s^2 \mathcal{O}^{\theta}_{\hat{v}_s;m}\left[\Theta_{j,m}(x; v_s)\right] = \mathcal{L}^{\theta}_{v_s;m}\left[\Theta_{j,m}(x; v_s)\right], \tag{10.52}$$

where (cf. (10.23))

$$\mathcal{L}^{\theta}_{v_s;m} = \frac{1}{\left(1 - x^2 v_s^2\right)\sqrt{1 - x^2}} \times \left[-(1 - x^2)\frac{\mathrm{d}}{\mathrm{d}x} + m v_s x\right], \tag{10.53}$$

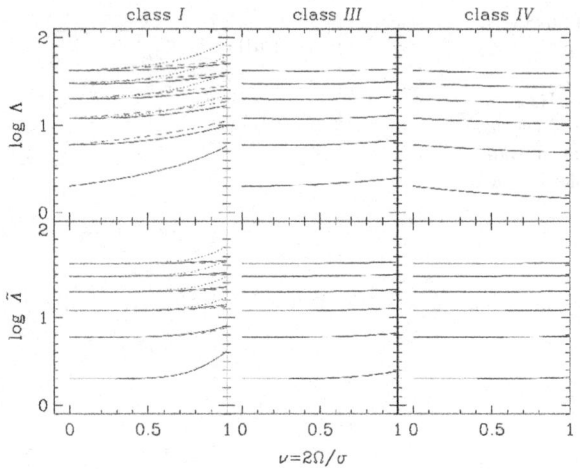

Fig. 10.10 (*top*) Eigenvalues Λ of Laplace's tidal equation in the presence of rotation, in the range that is relevant for a solar model and the Traditional Approximation. In the absence of rotation, one has $\Lambda = \ell(\ell+1)$. (*bottom*) Equivalent horizontal eigenvalue $\tilde{\Lambda}$ (see text for details). (*left*) Class I waves with $1 \le \ell \le 6$, and $m = -\ell+2$ (*black continuous line*), $m = 0$ (*blue dotted line*), $m = \ell$ (*red dashed line*). (*middle*) Class III waves ($s = 0$) with negatives values of m (are shown $m = 0, \ldots, -5$) are present in the relevant range of ν. (*right*) Class IV waves ($s = -1$) have indices $m = -1, \ldots, -6$ (taken from [89]; courtesy Solar Physics)

and

$$\mathcal{H}^{\varphi}_{j,m}(x; \nu_s) = \sigma_s^2 \mathcal{O}^{\varphi}_{\hat{\nu}_s;m}\left[\Theta_{j,m}(x; \nu_s)\right] = \mathcal{L}^{\varphi}_{\nu_s;m}\left[\Theta_{j,m}(x; \nu_s)\right], \qquad (10.54)$$

where (cf. (10.24))

$$\mathcal{L}^{\varphi}_{\nu_s;m} = \frac{1}{\left(1 - x^2\nu_s^2\right)\sqrt{1 - x^2}} \times \left[-\nu_s x\left(1 - x^2\right)\frac{\mathrm{d}}{\mathrm{d}x} + m\right]. \qquad (10.55)$$

As we have for the Hough functions, we get $\mathcal{H}^{\theta}_{j,-m} \neq \mathcal{H}^{\theta}_{j,m}$ and $\mathcal{H}^{\varphi}_{j,-m} \neq \mathcal{H}^{\varphi}_{j,m}$ since $\mathcal{L}^{\theta}_{\nu_s;-m} \neq \mathcal{L}^{\theta}_{\nu_s;m}$, $\mathcal{L}^{\varphi}_{\nu_s;-m} \neq \mathcal{L}^{\varphi}_{\nu_s;m}$ and $\Theta_{j,-m} \neq \Theta_{j,m}$.

These properties are illustrated in Figs. 10.9 and 10.10, where the eigenfunctions $\Theta_{j,m}, \mathcal{H}^{\theta}_{j,m}, \mathcal{H}^{\varphi}_{j,m}$ and their associated eigenvalues $\Lambda_{j,m}$ are given for prograde and retrograde waves in the case where $\nu_s = 0.86$ ($\sigma = 1\mu$Hz in the solar case).

Finally, the radiative damping is given by:

$$\tau_{j,m}(r; \nu_s) = \Lambda^{3/2}_{j,m}(\nu_s) \int_r^{r_c} K \frac{N N_T^2}{\tilde{\sigma}^4} \frac{\mathrm{d}r'}{r'^3}. \qquad (10.56)$$

As has been previously emphasized, $\Lambda_{j,-m} \neq \Lambda_{j,m}$; the respective radiative damping associated to the Doppler-shift due to differential rotation is thus modified as well as the deposition and extraction of angular momentum.

Keeping the radial component of the momentum equation (10.3) and the energy equation (10.5) and using once again the anelastic approximation, the monochromatic Eulerian pressure fluctuation ($P'_{j,m}$) is derived:

$$P'_{j,m}(r,\theta,\varphi,t) = -\bar{\rho}\frac{\tilde{\sigma}}{k_{V;j,m}}\left(\frac{N^2}{\tilde{\sigma}^2}\right)\mathcal{E}_{j,m}(r)\sin[\phi_{j,m}(r,\varphi,t)]$$

$$\times \Theta_{j,m}(\cos\theta; v_s)\exp[-\tau_{j,m}(r; v_s)/2]. \tag{10.57}$$

Finally, following [26], we define a horizontal wave number given by:

$$k_{H;j,m} = \frac{\tilde{\Lambda}_{j,m}^{1/2}(v_s)}{r} \quad \text{where } \tilde{\Lambda}_{j,m}^2(v_s) = \frac{\langle|r^2\nabla_H^2\Theta_{j,m}(\cos\theta; v_s)|^2\rangle_\theta}{\langle|\Theta_{j,m}(\cos\theta, v_s)|^2\rangle_\theta}, \tag{10.58}$$

where ∇_H^2 is the horizontal spherical laplacian. In the absence of rotation, we recover $\tilde{\Lambda}_{j,m}(v_s) = \Lambda_{j,m}(v_s) = l(l+1)$.

When the Coriolis acceleration is taken into account in a rotating stably stratified radiative region, low-frequency IGWs are thus modified and become gravito-inertial waves. Under the Traditional and the "weak differential rotation" approximations, four types of gravito-inertial waves may be identified [55–57]:

- *Class I waves*: they are internal gravity waves, which exist in the non-rotating case, that are modified by the Coriolis acceleration; rotation increases their eigenvalues ($\Lambda_{j,m}$), and hence their radial wave number and their damping (cf. Fig. 10.10). These waves are thus deposited closer to their excitation region than when the Coriolis acceleration is ignored. They can be treated using the Traditional Approximation as long as their frequencies are super-inertial ($\sigma > 2\bar{\Omega}_s, v_s < 1$).
- *Class II waves*: they are purely retrograde waves ($m > 0$), which exist only in the case of rapid rotation. Their dynamics is driven by the conservation of specific vorticity combined with the effects of curvature. However, due to their sub-inertial frequency range ($\sigma \leq 2\bar{\Omega}_s, v_s \geq 1$), they can not be treated using the Traditional Approximation. They are sometimes called "quasi-inertial" waves that corresponds to the geophysical Rossby waves (cf. [58]).
- *Class III waves*: they are mixed class *I* and class *II* waves. $m \leq 0$ waves exist in the absence of rotation. $m > 0$ appear when $v_s = m + 1$ with small eigenvalues while their horizontal eigenfunctions are $\Theta_{j,m}(v_s = m + 1; x) = P_{m+1}^m(x)$. When they appear and have small eigenvalues, they behave mostly like class *II* waves; $m \leq 0$ and $m > 0$ waves with large eigenvalues behave rather like class *I* waves. Their eigenvalues are much smaller than those of class *I* waves. Thus, they will be damped farther from the excitation region and over a more larger portion of the stellar radiative region. As for class *I* waves, they can be treated using the Traditional Approximation as long as their frequencies are super-inertial. They may be identified with the geophysical Yanai waves [59].
- *Class IV waves*: they are purely prograde waves ($m < 0$) whose characteristics change little with rotation, their displacement in the θ direction being very small. Like class *II* waves, their dynamics is driven by the conservation of specific

vorticity, but here it is combined with the stratification effects; their eigenvalues are smaller than those of both class I and class III waves. Hence, they are damped somewhat deeper in the stellar radiation zone where they deposit their positive angular momentum. As for class I and class III waves, they can be treated using the Traditional Approximation as long as their frequencies are super-inertial. They may be indentified with the geophysical Kelvin waves.

If we define a new index s by

$$s = \ell - m + 1 \text{ for } m > 0 \text{ or } s = \ell + m - 1 \text{ for } m \leq 0. \tag{10.59}$$

Class IV waves have $s = -1$, class III waves, $s = 0$, and class I waves have $s = 1, 2, 3, \ldots$

In the presence of the Coriolis acceleration, the spatial structure and damping of the low-frequency IGWs are thus modified. We shall now study the mean energy and angular momentum transported by these waves in the "weak differential rotation" case.

10.4.3 Angular Momentum Transport in the "Weak Differential Rotation" Case

First, we consider the mean flux of kinetic energy associated with a monochromatic wave. Following [26], it can be expressed as the product of the mean volumic density of kinetic energy of the wave on an isobar and of its vertical group velocity:

$$\langle \mathcal{F}^{K}_{V;j,m} \rangle_{\theta}(r) = \frac{1}{2} \langle \mathcal{F}^{E}_{V;j,m} \rangle_{\theta}(r) = \frac{1}{2} \overline{\rho} < \mathbf{u}^2_{j,m} >_{\theta,\varphi} V^{V}_{g;j,m}. \tag{10.60}$$

The group velocity is derived from (10.47):

$$V^{V}_{g;j,m} = \frac{d\tilde{\sigma}}{dk_{V;j,m}} = -\frac{\tilde{\sigma}}{k_{V;j,m}} = -V^{V}_{p;j,m}, \tag{10.61}$$

where $\langle u^2_{j,m} \rangle_{\theta,\varphi}$ is the mean value of the squared velocity on an isobar

$$\langle \mathbf{u}^2_{j,m} \rangle_{\theta,\varphi} = \frac{1}{4\pi} \int_{\Omega=4\pi} \left(u^2_{r;j,m} + u^2_{\theta;j,m} + u^2_{\varphi;j,m} \right) d\Omega, \tag{10.62}$$

$\int_{\Omega=4\pi} \cdots d\Omega = \int_0^{2\pi} \int_0^{\pi} \cdots \sin\theta d\theta d\varphi$ being the horizontal average over colatitudes (θ) and longitudes (φ), and $V^{V}_{p;j,m}$ is the phase velocity. Using the final expression of the wave velocity field given in (10.42, 10.45), we finally find the following expressions as in [22, 23]:

$$\langle \mathcal{F}^K_{V;j,m} \rangle_\theta = -\frac{1}{2}\bar{\rho}\frac{\mathcal{E}_{j,m^2}(r)}{2}\left(\langle \Theta^2_{j,m}(\cos\theta; \nu_s) \rangle_\theta + \frac{r^2 k^2_{V;j,m}}{\Lambda^2_{j,m}(\nu_s)} \mathcal{J}_{H;j,m}(\nu_s) \right)$$

$$\times \frac{\tilde{\sigma}^2}{N}\frac{1}{k_{H;j,m}}\exp\left[-\tau_{j,m}(r; \nu_s)\right],$$

$$(10.63)$$

where

$$\mathcal{J}_{H;j,m}(\nu_s) = \left\langle \left[\mathcal{H}^\theta_{j,m}(\cos\theta; \nu_s)\right]^2 \right\rangle_\theta + \left\langle \left[\mathcal{H}^\varphi_{j,m}(\cos\theta; \nu_s)\right]^2 \right\rangle_\theta. \qquad (10.64)$$

Next, using (10.25), we obtain for the mean vertical angular momentum flux:

$$\left\langle \mathcal{F}^{AM}_{V;j,m} \right\rangle_\theta (r) = \frac{1}{2}\bar{\rho}r\frac{rk_{V;j,m}}{\Lambda_{j,m}(\nu_s)}\mathcal{E}^2_{j,m}(r)\left[\mathcal{J}_{I;j,m}(\nu_s) - \nu_s\mathcal{J}_{II;j,m}(\nu_s) \right]$$

$$\times \exp\left[-\tau_{j,m}(r; \nu_s)\right], \qquad (10.65)$$

where the angular integrals $\mathcal{J}_{I;j,m}(\nu_s)$ and $\mathcal{J}_{II;j,m}(\nu_s)$ are given by:

$$\begin{cases} \mathcal{J}_{I;j,m}(\nu_s) = \left\langle \Theta_{j,m}(\cos\theta; \nu_s)\mathcal{H}^\varphi_{j,m}(\cos\theta; \nu_s)\sin\theta \right\rangle_\theta \\ \mathcal{J}_{II\,j,m}(\nu_s) = \left\langle \Theta_{j,m}(\cos\theta; \nu_s)\mathcal{H}^\theta_{j,m}(\cos\theta; \nu_s)\cos\theta\,\sin\theta \right\rangle_\theta \end{cases} \qquad (10.66)$$

Then, using the approximation $r^2 k_{V;j,m} \gg \Lambda_{j,m}$, we derive the relation between $\mathcal{F}^{AM}_{V;j,m}$ and $\mathcal{F}^K_{V;j,m}$:

$$\left\langle \mathcal{F}^{AM}_{V;j,m} \right\rangle_\theta (r) = -2\frac{m'(\nu_s)}{\tilde{\sigma}}\langle \mathcal{F}^K_{V;j,m} \rangle_\theta (r), \qquad (10.67)$$

where m' is given by:

$$m'(\nu_s) = \Lambda_{j,m}(\nu_s)\frac{\mathcal{J}_{I;j,m}(\nu_s) - \nu_s\mathcal{J}_{II;j,m}(\nu_s)}{\mathcal{J}_{H;j,m}(\nu_s)}. \qquad (10.68)$$

This relation links the mean flux of angular momentum carried by a monochromatic wave on an isobar to that of the kinetic energy as in (10.26). In the non-rotating case, we retrieve $m'(\nu_s = 0) = m$.

In the case of class I gravito-inertial waves, $m'(\nu_s)$ is decreased compared to the non-rotating case. Heuristically, this is a consequence of the lesser horizontal extent of the eigenfunctions. Hence, the ability of those waves to transport angular momentum is decreased compared to the non-rotating case. However, we can see that if the prograde and retrograde waves are equally excited, we should expect that the damping of these waves could produce a Shear Layer Oscillation (SLO) similar to the one obtained in the case where the Coriolis acceleration is not taken into account (cf. [18]) . In the case of class III waves, we obtain the same behaviour, but with a slower convergence rate. Finally, in the case of class IV waves, $m'(\nu_s)$ varies only

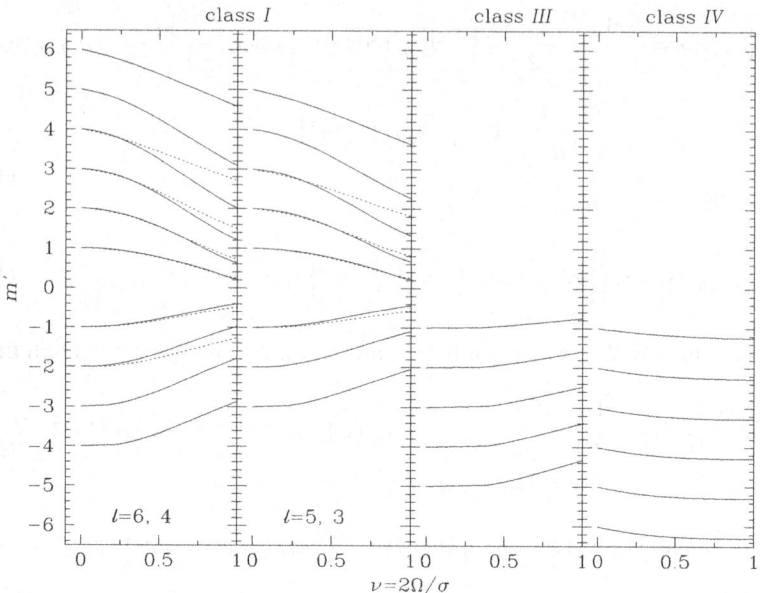

Fig. 10.11 Ratio $m' = -\frac{\sigma}{2}\frac{\mathcal{F}_V^{AM}}{\mathcal{F}_V^K}$ for various modes. Class I waves corresponding to $\ell = 5, 6$ (*continuous lines*) $\ell = 3, 4$ (*dotted lines*). Class III waves of order $m = -5, \ldots, -1$. Class IV waves of order $m \geq -6$ (taken from [89]; courtesy Solar Physics)

slightly with rotation and remains close to m, their angular momentum flux being always positive so that they could induce a deposition of angular momentum where they are damped.

The relation between the kinetic energy and the angular momentum mean vertical fluxes being now established, the vertical action of angular momentum can be derived. Applying (10.26) to the "weak differential rotation case" we finally get

$$
\langle \mathcal{L}_V^{AM}\rangle_\theta (r) = r^2 \langle \mathcal{F}_V^{AM}\rangle_\theta
$$
$$
= \int_\sigma \sum_{m,j} \left\{ \mathcal{L}_{V;j,m}^{AM}(r_c) \exp\left[-\tau_{j,m}\left(r, \delta\overline{\Omega}(r); v_s\right)\right] \right\} d\sigma, \tag{10.69}
$$

where we identify

$$
\langle \mathcal{L}_{V;j,m}^{AM}\rangle_\theta (r_c) = -r_c^2 \frac{2}{\overline{\sigma}} m'(v_s) \langle \mathcal{F}_{V;j,m}^K\rangle_\theta (r_c), \tag{10.70}
$$

$\overline{\Omega}_s$ thus corresponding to the mean rotation rate at r_c: $\overline{\Omega}_s = \overline{\Omega}(r_c)$. (Fig. 10.11).

10.5 On the Track of the Excited Spectrum

The search of prescription for the excitation is one of the nowadays major unsolved and debated question of the wave-induced transport theory. To achieve this aim, different approaches have been adopted.

The first analytical one consists in deriving, using phenomenological prescritions, the energy flux transmission between the turbulent convective movements and the IGWs using the matching of the wave pressure fluctuation with that of the turbulent convection. The Kolmogorov turbulent energy spectrum is assumed. This procedure is described in details in [16, 26, 60] in the non-rotating case and by [23] in the one where the Coriolis acceleration is taken into account.

The second semi-analytical approach consists in deriving, in the most consistent way as possible, the wave amplitude with describing their stochastic volumetric excitation by the convective Reynolds stresses and the turbulent entropy advection. This method takes into account both the spatial and the temporal correlations between turbulent eddies and waves. Formalisms follow the first work by [61] which was devoted to solar p-modes and adapted to IGWs by [48] . Those first contributions assumed the Kolmogorov energy spectrum. Those works were then generalized by [62, 63] in order to take into account a general turbulent energy spectrum, which can be extracted from realistic 3D numerical simulations of turbulent convection, and by [64, 65] who derived a rigorous treatment of the excitation with accounting for the non-radial character of the modes crucial in the case of IGWs for which the displacement is mostly horizontal. Finally, the Coriolis acceleration is now taken into account [24, 66] and the generalized formalism has now to be applied to gravito-inertial waves.

Penetrative convection is also an efficient process to generate IGWs. This was first investigated by [67, 68] in the case of atmospheric flows. Then, in the stellar context, [69–71] , following [68] , used several models for wave excitation by plumes in order to study the problem of light elements mixing induced by IGWs (see also the work by [72, 73] for the case of convective cores). However, they considered that waves are generated solely by turbulence inside plumes and they did not investigate the generation of waves caused by the impact of plumes on the stably stratified region that is now undertaken (cf. [66]).

The major approach to obtain prescription for the wave energy spectrum in this case consists in computing numerical simulations of turbulent penetrative convection at the interface between convective and radiative regions. Such simulations have shown IGWs excitation (see for example [74–80]) but specific work has to be undertaken to provide a quantitative estimate of the amplitude and of the spectrum of waves.

First work dedicated to such study has been completed in 2D Cartesian geometry by [81] . In this work, the assumed stratification is polytropic and authors add a viscous boundary layer at the bottom of the stable zone in order to avoid the reflexion of excited waves and thus the appearance of normal modes in the simulation box. Their main results are that phenomenological semi-analytical models (the Garcià-

Lopez and Spruit's one, hereafter GLS91, or the plume model by [82] , hereafter RZ95) significantly underestimate the flux of IGWs by a factor 100 (GLS91) and 10 (RZ95) compared to 2D direct numerical simulations. On the other hand, in the domain (σ, k_H), the numerically obtained wave spectrum is much broader than those predicted using GLS91 which results in lack of high frequency waves and RZ95 where low frequencies are missing. However, authors emphasized that 2-D simulations probably produce stronger downflows compared to more realistic 3-D simulations. This is the reason why [83] revisited their own work comparing their previous results with those obtained in 3-D Cartesian box using the same stratification where downdrafts are significantly less vigorous. On one hand, the excited IGWs have lower amplitude. On the other hand, the wave energy flux increases with the depth of the convective layer.

In the same way, [84] proposed a quantitative investigation of the spectrum, the amplitude and the life-time of IGWs excited by penetrative convection in solar-like stars using 2-D numerical simulations of compressible convection assuming that the gas is monoatomic and perfect. The wave generation is studied from the linear response of the radiative zone to the plumes penetration using projections onto the g-modes linear eigenfunctions. Authors show that up to 40% of the total kinetic energy is transmitted to IGWs during times of significant excitation.

Finally, work is now undertaken to take into account realistic stratification, the global geometry, and the (differential) rotation. In this way, [85, 86] computed integrated models of the Sun interior (both the convective envelope and the radiative core) in 2-D polar geometry that represents the equatorial plane of the Sun using a realistic stratification given by a solar model. As in the work by [81] , the found frequency spectrum is broader than those determined using semi-analytical models with a more uniform distribution between low and high frequencies. On the other hand, it is shown that non-linear effects have to be taken into account. These effects broaden the frequency ridges in the dispersion relation.

Furthermore, just under the convection zone, the energy is increased by two orders of magnitude over what the linear dispersion relation would predict for energy in waves. Progresses on such type of numerical simulations is now under progress in 3-D spherical geometry with using the Anelastic Spherical Harmonics code (see [78, 87] for the code description and A.-S. Brun, 2009, private communication).

Therefore, all those possible sources of prescription for the wave excited spectrum have to be carefully examined given its uncertainty; this will be studied in the application of our formalism.

10.6 Conclusion

In this lecture, a complete formalism to treat the dynamics of regular (elliptic) low-frequency gravito-inertial waves in stably strongly stratified differentially rotating stellar radiation zones, from tachocline(s) where they are excited to their bulk, has been presented. Then, the feedback on the angular velocity profile through the induced-angular momentum transport has been established.

The Traditional Approximation has been used to treat the impact of the Coriolis acceleration on IGWs in its domain of validity. Two main effects are observed: first, the horizontal structure of waves is modified, their dynamics being now driven simultaneously by the stratification, as in the non-rotating case, and by the Coriolis acceleration. Hence, the amount of angular momentum carried by a wave and its damping are modified. In the simplified case of the "weak differential rotation case" a classification of regular gravito-inertial waves has been established. In the case of class I gravito-inertial waves, the main effect of the Coriolis acceleration is to modify the horizontal functions, the prograde and the retrograde waves being now different. Furthermore, a reduction of the effectiveness of these waves in transporting angular momentum is obtained. Furthermore, rotation increases their damping, these waves being thus deposited closer to their excitation region than gravity waves in the non-rotating case. The class III waves have the same behaviour for m' but with a slower convergence rate. Their eigenvalues are however quite smaller and hence they are damped farther from the convection zone border. Finally, class IV waves have a different behaviour. For these waves the main restoring force is the conservation of vorticity combined with stratification. They have $m' = m$ and their eigenvalues are smaller than those of both class I and class III waves. Hence, they are damped deeper in the core in the case of solar-like stars and nearer the stellar surface in massive stars where they deposit their positive angular momentum.

Complete numerical simulations remain to be performed to verify and understand the net effects of all those gravito-inertial waves on the angular momentum transport in stellar interiors. Furthermore, the hyperbolic regime (sub-inertial in the case of a uniform rotation), which has to be treated taking into account the complete Coriolis acceleration, has to be examined to get a global picture of the gravito-inertial wave transport. In parallel, works must be devoted to its implementation in existing dynamical stellar evolution codes and to its application to different type of stars and evolution stages. This effort will lead to the building of more and more realistic stellar models, which will benefit from new constraints provided by the development of asteroseismology both on the ground and in space.

Acknowledgments S. Mathis thanks warmly the organisers of the school, C. Neiner and J.-P. Rozelot for their invitation to give this lecture. This work was supported in part by the Programme National de Physique Stellaire and the Centre National d'Etudes Spatiales (GOLF instrument).

References

1. Talon, S.: EAS Publications Series **32**, 81 (2008)
2. Maeder, A., Meynet, G.: Ann. Rev. Astronom. Astrophys. **38**, 143 (2000)
3. Pinsonneault, M.H., Kawaler, S.D., Sofia, S., Demarque, P.: Astrophys J. **338**, 424 (1989)
4. Chaboyer, B., Demarque, P., Pinsonneault, M.H.: Astrophys. J. **441**, 865 (1995)
5. Talon, S.: Ph. D. Thesis, Université Paris VII, (1997)

6. Matias, J., Zahn, J-P.: IAU Symposium 18. In: Provost, J., Schmider, F.-X. (eds.) Observatoire de Nice, Nice poster volume, p. 103 (1997)
7. Talon, S., Zahn, J-P.: Astron. Astrophys. **329**, 315 (1998)
8. Brown, T.M., Christensen-Dalsgaard, J., Dziembowski, W.A., Goode, P., Gough, D.O., Morrow, C.A.: Astrophys. J. **343**, 526 (1989)
9. Turck-Chièze, S., Couvidat, S., Piau, L., Ferguson, J., Lambert, P., Ballot, J., García, R.A., Nghiem, P.: Phys. Rev. Lett. **93**, id. 211102, (2004)
10. García, R.A., Turck-Chièze, S., Jiménez-Reyes, S.J., Ballot, J., Pallé, P.L., Eff-Darwich, A., Mathur, S., Provost, J.: Science **316**, 1591 (2007)
11. Gough, D.O., McIntyre, M.E.: Nature **394**, 567 (1998)
12. Garaud, P.: Mounthly Notices R. Astron. Soc. **329**, 1 (2002)
13. Brun, A-S., Zahn, J-P.: Astron. Astrophys. **457**, 665 (2006)
14. Garaud, P., Garaud, J-D.: Mounthly Notices R. Astron. Soc. **391**, 1239 (2008)
15. Schatzman, E.: Astron. Astrophys. **279**, 431 (1993)
16. Zahn, J.-P., Talon, S., Matias, J.: Astron. Astrophys. **322**, 320 (1997)
17. Talon, S., Kumar, P., Zahn, J-P.: Astrophys. J. **574**, L175 (2002)
18. Talon, S., Charbonnel, C.: Astron. Astrophys. **440**, 981 (2005)
19. Berthomieu, G., Gonczi, G., Graff, P., Provost, J., Rocca, A.: Astron. Astrophys. **70**, 597 (1978)
20. Lee, U., Saio, H.: Astrophys J. **491**, 839 (1997)
21. Dintrans, B., Rieutord, M.: Astron. Astrophys. **354**, 86 (2000)
22. Mathis, S.: Ph. D. Thesis, Université Paris XI (2005)
23. Pantillon, F.P., Talon, S., Charbonnel, C.: Astron. Astrophys. **474**, 155 (2007)
24. Mathis, S., Belkacem, K., Goupil, M-J., Samadi, R.: Commun. Asteroseismol. **157**, 144 (2008)
25. Mathis, S.: Astron. Astrophys. **506**, 811 (2009)
26. Press, W.H.: Astrophys J. **245**, 286 (1981)
27. Cowling, T.G.: Mounthly Notices R. Astron. Soc. **101**, 367 (1941)
28. Rogers, T.M., MacGregor, K.B., Glatzmaier, G.A.: Mounthly Notices R. Astron. Soc. **387**, 616 (2008)
29. Unno, W., Osaki, Y., Ando, H., Saio, H., Shibahashi, H.: Non radial oscillation of stars, 2nd edn. University of Tokyo Press (1989)
30. Friedlander, S., Siegmann, W.L.: Geophys. Astrophys. Fluid Dyn. **19**, 267 (1982)
31. Dintrans, B., Rieutord, M., Valdettaro, L.: J. Fluid Mech. **398**, 271 (1999)
32. Stewartson, K., Richard, J.: J. Fluid Mech. **35**, 759 (1969)
33. Stewartson, K., Walton, I.C.: Proc. Roy. Soc. Lond. **349**, 141 (1976)
34. Eckart, C., Hydrodynamics of Oceans and Atmospheres. Pergamon Press, Oxford (1960)
35. Bildsten, L., Ushomirsky, G., Cutler, C.: Astrophys. J. **460**, 827 (1996)
36. Friedlander, S.: Geophys. J. Roy. Astr. Soc. **89**, 637 (1987)
37. Gerkema, T., Shrira, V.I.: J. Fluid Mech. **529**, 195 (2005)
38. Gerkema, T., Zimmerman, J.T.F., Mass, L.R.M., Van Haren, H.: Rev. Geophys.(2009). doi:10.1029/2006RG000220
39. Goldreich, P., Nicholson, P.D.: Astrophys. J. **342**, 1079 (1989)
40. Booker, J., Bretherton, F.: J. Fluid Mech. **27**, 513 (1967)
41. Laplace, P.-S.: Mécanique Céleste: Bureau des Longitudes, Paris (1799)
42. Hough, S.S.: Phil. Trans. Roy. Soc. **191**, 139 (1898)
43. Ogilvie, G.I., Lin, D.N.C.: Astrophys. J. **610**, 477 (2004)
44. Bretherton, F.P.: Quat. J. Roy. Met. Soc. **95**, 213 (1969)
45. Hayes, W.D.: Proc. Roy. Soc. Lond. **320**, 187 (1970)
46. Goldreich, P., Nicholson, P.D.: Astrophys. J. **342**, 1075 (1989)
47. Ringot, O.: Astron. Astrophys. **335**, L89 (1998)
48. Kumar, P., Talon, S., Zahn, J-P.: Astrophys J. **520**, 859 (1999)
49. Charbonnel, C., Talon, S.: Science **309**, 2189 (2005)
50. Mathis, S., Zahn, J-P.: Astron. Astrophys. **440**, 653 (2005)
51. Mathis, S., Zahn, J.-.P.: Astron. Astrophys. **425**, 229 (2004)

52. Zahn, J-P.: Astron. Astrophys. **265**, 115 (1992)
53. Decressin, T., Mathis, S., Palacios, A., Siess, L., Talon, S., Charbonnel, C., Zahn, J.P.: Astron. Astrophys. **495**, 271 (2009)
54. Longuet-Higgins, F.R.S.: Phil. Trans. Roy. Soc. **262**, 511 (1968)
55. Miles, J.W.: Proc. Roy. Soc. Lond. **353**, 377 (1977)
56. Pedlosky, J., Geophysical Fluid Dynamics, 2nd edn. Springer, New York (1987)
57. Townsend, R.H.D.: Mounthly Notices R. Astron. Soc. **340**, 1020 (2003)
58. Provost, J., Berthomieu, G., Rocca, A.: Astron. Astrophys. **94**, 126 (1981)
59. Yanai, M., Maruyama, T.: J. Meteorol. Soc. Jpn **44**, 291 (1966)
60. García-López, R.J., Spruit, H.C.: Astrophys. J. **377**, 268 (1991)
61. Goldreich, P., Murray, N., Kumar, P.: Astrophys. J. **424**, 466 (1994)
62. Samadi, R., Goupil, M-J.: Astron. Astrophys. **370**, 136 (2001)
63. Samadi, R., Goupil, M.-J., Lebreton, Y.: Astron. Astrophys. **370**, 147 (2001)
64. Belkacem, K., Samadi, R., Goupil, M.-J., Dupret, M.-.A.: Astron. Astrophys. **478**, 163 (2008)
65. Belkacem, K., Samadi, R., Goupil, M.-J., Dupret, M.A., Brun, A.S., Baudin, F.: Astron. Astrophys. **494**, 191 (2009)
66. Belkacem, K.: Ph. D. Thesis, Observatoire de Paris (2008)
67. Townsend, A.A.: J. Fluid Mech. **22**, 241 (1965)
68. Townsend, A.A.: J. Fluid Mech. **24**, 307 (1966)
69. Montalbán, J.: Astron. Astrophys. **281**, 421 (1994)
70. Montalbán, J., Schatzman, E.: Astron. Astrophys. **305**, 513 (1996)
71. Montalbán, J., Schatzman, E.: Astron. Astrophys. **354**, 943 (2000)
72. Lo, Y.-C., Schatzman, E.: Astron. Astrophys. **322**, 545 (1997)
73. Lo, Y.-C.: Ph. D. Thesis, Université Louis Pasteur - Observatoire de Strasbourg (1997)
74. Hurlburt, N.E., Toomre, J., Massaguer, J.M.: Astrophys J. **311**, 563 (1986)
75. Hurlburt, N.E., Toomre, J., Massaguer, J.M., Zahn, J.-P.: Astrophys J. **421**, 245 (1994)
76. Andersen, B.N.: Sol. Phys. **152**, 241 (1994)
77. Brummell, N.H., Clune, T.L., Toomre, J.: Astrophys. J.**570**, 825 (2002)
78. Browning, M.K., Brun, A.S., Toomre, J.: Astrophys. J. **601**, 512 (2004)
79. Rogers, T.M., Glatzmaier, G.A.: Astrophys. J. **620**, 432 (2005)
80. Rogers, T.M., Glatzmaier, G.A., Jones, C.A.: Astrophys. J. **653**, 765 (2006)
81. Kiraga, M., Jahn, K., Stepien, K., Zahn, J.-P.: Acta Astronomica **53**, 321 (2003)
82. Rieutord, M., Zahn, J.-P.: Astron. Astrophys. **296**, 127 (1995)
83. Kiraga, M., Stepien, K., Jahn, K.: Acta Astronomica **55**, 205 (2005)
84. Dintrans, B., Brandenburg, A., Nordlund, A., Stein R.F.: Astron. Astrophys. **438**, 365 (2005)
85. Rogers, T.M., Glatzmaier, G.A.: Mounthly Notices R. Astron. Soc. **364**, 1135 (2005)
86. Rogers, T.M., Glatzmaier, G.A.: Astrophys. J. **653**, 756 (2006)
87. Clune, T.L., et al.: Parallel Comput. **25**, 361 (1999)
88. Palacios, A., Talon, S., Charbonnel, C., Forestini, M.: Astron. Astrophys. **399**, 603 (2003)
89. Mathis, S., Talon, S., Pantillon, F.P., Zahn, J.-P.: Sol. Phys. **251**, 101 (2008)
90. Mathis, S.: Commun. Asteroseismol. **157**, 209 (2008)

Chapter 11
Stochastic Excitation of Acoustic Modes in Stars

R. Samadi

Abstract For more than ten years, solar-like oscillations have been detected and frequencies measured for a growing number of stars with various characteristics (e.g. different evolutionary stages, effective temperatures, gravities, metal abundances...). Excitation of such oscillations is attributed to turbulent convection and takes place in the uppermost part of the convective envelope. Since the pioneering work of Goldreich and Keeley (APJ, 211:934, 1977; 212:243, 1977) more sophisticated theoretical models of stochastic excitation were developed, which differ from each other both by the way turbulent convection is modeled and by the assumed sources of excitation. We review here these different models and their underlying approximations and assumptions. We emphasize how the computed mode excitation rates crucially depend on the way turbulent convection is described but also on the stratification and the metal abundance of the upper layers of the star. In turn we will show how the seismic measurements collected so far allow us to infer properties of turbulent convection in stars.

11.1 Introduction

Solar p-modes are known to have finite lifetimes (a few days) and very low amplitudes (a few cm/s in velocity and a few ppm in intensity). Their finite lifetimes result from several complex damping processes that are so far not clearly understood. Their excitation is attributed to turbulent convection and takes place in the upper-most part of the Sun, which is the place of vigorous and turbulent motions. Since the pioneering work of Lighthill [1], we know that a turbulent medium generates incoherent acoustic pressure fluctuations (also called acoustic "noise"). A very small fraction of the associated kinetic energy goes into to the normal modes of the solar cavity. This

R. Samadi (✉)
Observatoire de Paris, LESIA, CNRS UMR 8109 92195 Meudon, France
e-mail: reza.samadi@obspm.fr

.-P. Rozelot and C. Neiner (eds.), *The Pulsations of the Sun and the Stars*,
Lecture Notes in Physics 832, DOI: 10.1007/978-3-642-19928-8_11,
© Springer-Verlag Berlin Heidelberg 2011

small amount of energy then is responsible for the small observed amplitudes of the solar acoustic modes (p modes).

In the last decade, solar-like oscillations have been detected in numerous stars, in different evolutionary stages and with different metallicity (see recent review by Bedding and Kjeldsen [2]). As in the Sun, these oscillations have rather small amplitudes and have finite lifetimes. The excitation of such solar-like oscillations is attributed to turbulent convection and takes place in the outer layers of stars having a convective envelope.

Measuring mode amplitudes and the mode lifetimes permits us to infer $\mathcal{P}$, the energy supplied per unit time into the acoustic modes. Deriving $\mathcal{P}$ puts constraints on the theoretical models of mode excitation by turbulent convection [3]. However, as pointed-out by Baudin et al. [4], even for the Sun, inferring $\mathcal{P}$ from the seismic data is not a trivial task. For stellar seismic data, this is even more difficult [5]. We discuss here the problems we face in deriving reliable seismic constraints on $\mathcal{P}$.

A first attempt to explain the observed solar five minute oscillations was carried out by Unno and Kato [6]. They have considered monopole[1] and dipole[2] source terms that arise from an isothermal stratified atmosphere. Stein [7] has generalised Lighthill [1]'s approach to a stratified atmosphere. He found that monopole source terms have a negligible contribution to the noise generation compared to the quadrupole source term.[3] Among the quadrupole source terms, the Reynolds stress was expected to be the major source of acoustic *wave* generation. It was only at the beginning of the 1970s that solar five minutes oscillations have been clearly identified as global resonant modes [8–10]. A few years later, Goldreich and Keeley [11] have proposed the first theoretical model of stochastic excitation of acoustic *modes* by the Reynolds stress. Since this pioneering work, different improved models have been developed [12–19]. These approaches differ from each other either in the way turbulent convection is described or by the excitation process.

In the present paper, we briefly review the different main formulations and discuss the main assumptions and approximations on which these models are based. As shown by Samadi et al. [17], the energy supplied per time unit to the modes by turbulent convection crucially depends on the way eddies are temporally correlated. A realistic modeling of the eddy time-correlation at various scale lengths then is an important issue, which is discussed in detail here. We will also highlight how the mean structure and the chemical composition of the upper convective envelope influence the mode driving. Finally, we will summarize how the seismic measurements obtained so far from the ground allow us to distinguish between different dynamical descriptions of turbulent convection.

---

[1]   A monopole term is associated with a fluctuation of density.

[2]   A dipole term is associated with a fluctuation of a force.

[3]   A quadrupole term is associated with a shear.

11.2 Mode Energy

We will show below how the energy of a solar-like oscillation is related to the driving and damping process. The mode total energy (potential plus kinetic) is by definition the quantity:

$$E_{osc}(t) = \int d^3x \rho_0 v_{osc}^2(\mathbf{r}, t) \tag{11.1}$$

where v_{osc} is the mode velocity at the position $\mathbf{x}$, and ρ_0 the mean density.

Mode damping occurs over a time-scale much longer than that associated with the driving. Accordingly, damping and driving can be completely decoupled in time. Furthermore, we assume a constant and linear damping such that

$$\frac{d v_{osc}(t)}{dt} = -\eta v_{osc}(t) \tag{11.2}$$

where η is the (constant) damping rate. The time derivative in (11.2) is performed over a time scale much larger than the characteristic time over which the driving occurs.

Let $\mathcal{P}$ be the amount of energy injected per unit time into a mode by an arbitrary source of driving (which acts over a time scale much shorter than $1/\eta$). According to (11.1) and (11.2), the variation of E_{osc} with time is given by:

$$\frac{d E_{osc}}{dt}(t) = \mathcal{P} - 2\eta E_{osc}(t). \tag{11.3}$$

Solar-like oscillations are known to be stable modes. As a consequence, their energy cannot growth on a time scale much longer than the time scales associated with the damping and driving process. Accordingly, averaging (11.3) over a long time scale gives:

$$\overline{\frac{d E_{osc}}{dt}}(t) = 0, \tag{11.4}$$

where $\overline{()}$ refer to a time average. From (11.3) and (11.4), we immediately derive:

$$\overline{E}_{osc} = \frac{\overline{\mathcal{P}}}{2\eta.} \tag{11.5}$$

We then clearly see with (11.5) that a stable mode has its energy (and thus its amplitude) controlled by the balance between the driving ($\mathcal{P}$) and the damping (η). Then, the major difficulties are to model the processes that are at the origin of the driving and the damping. For ease of notation, we will drop from now on the symbol $\overline{()}$ from E_{osc} and $\mathcal{P}$.

11.3 Seismic Constraints

As we shall see later, the mode displacement, δr_{osc}, can be written in terms of the adiabatic eigen-displacement $\boldsymbol{\xi}$, and an instantaneous amplitude $A(t)$:

$$\delta r_{\text{osc}} \equiv \frac{1}{2}\left(A(t)\boldsymbol{\xi}(r)e^{-i\omega_{\text{osc}}t} + cc\right) \tag{11.6}$$

where cc means complex conjugate, ω_{osc} is the mode eigenfrequency, and $A(t)$ is the instantaneous amplitude resulting from both the driving and the damping. Note that, since the normalization of $\boldsymbol{\xi}$ is arbitrary, the actual *intrinsic* mode amplitude is fixed by the term $A(t)$, which remains to be determined. The mode velocity, $\boldsymbol{v}_{\text{osc}}$, is then given by:

$$\boldsymbol{v}_{\text{osc}}(r, t) = \frac{d\delta r_{\text{osc}}}{dt} = \frac{1}{2}(-i\omega_{\text{osc}}A(t)\boldsymbol{\xi}(r)e^{-i\omega_{\text{osc}}t} + cc) \tag{11.7}$$

where cc means complex conjugate. Note that we have neglected in (11.7) the time derivative of A. This is justified since the mode period $(2\pi/\omega_{\text{osc}})$ is in general much shorter than the mode lifetime $(\sim 1/\eta)$

From (11.7) and (11.1), we derive the expression for the mean mode energy:

$$E_{\text{osc}} = \int d^3x\rho_0\overline{v_{\text{osc}}^2} = \frac{1}{2}\overline{|A|^2}I\omega_{\text{osc}}^2, \tag{11.8}$$

where

$$I \equiv \int_0^M d^3x\rho_0\boldsymbol{\xi}^* \cdot \boldsymbol{\xi} \tag{11.9}$$

is the mode inertia. For the sake of simplicity, we will from now on only consider radial modes. According to (11.7), the mean-square surface velocity associated with a *radial* mode measured at the radius r_h, is then given by the relation

$$v_s^2(r_h) = \frac{1}{2}\overline{|A|^2}\omega_{\text{osc}}^2 \mid \xi_r(r_h) \mid^2 \tag{11.10}$$

where ξ_r is the radial component of the mode eigenfunction. It is convenient and common to define the mode mass as the quantity:

$$\mathcal{M}(r_h) \equiv \frac{I}{\mid \xi_r(r_h) \mid^2} \tag{11.11}$$

where r_h is the radius in the atmosphere where the mode is measured in velocity. According to (11.8),(11.10), and (11.11), we derive the following relation:

$$E_{\text{osc}} = \mathcal{M}v_s^2 \tag{11.12}$$

It should be noticed, that although $\mathcal{M}$ and v_s depend on the choice for the radius r_h, E_{osc} is by definition intrinsic to the mode (see 11.1) and hence is independent of r_h.

Using (11.5), and (11.12), we finally derive:

$$v_s^2(r_h, \omega_{\mathrm{osc}}) = \frac{\mathcal{P}}{2\pi \mathcal{M} \Gamma} \tag{11.13}$$

where $\Gamma = \eta/\pi$ is the mode linewidth, and η the mode damping rate. From (11.13), one again sees that the mode surface velocity is the result of the balance between excitation ($\mathcal{P}$) and the damping ($\eta = \Gamma\pi$). However, it also depends on the mode mass ($\mathcal{M}$): For a given driving ($\mathcal{P}$) and damping (Γ), the larger the mode mass (or the mode inertia), the smaller the mode velocity.

When the frequency resolution and the signal-to-noise are high enough, it is possible to resolve the mode profile and then to measure *both* Γ and the mode height H in the power spectral density (generally given in m^2/Hz). In that case v_s is given by the relation (see e.g. [4]):

$$v_s^2(r_h, \omega_{\mathrm{osc}}) = \pi C_{\mathrm{obs}} H \Gamma \tag{11.14}$$

where the constant C_{obs} takes the observational technique and geometrical effects into account (see [4]). From (11.13) and (11.14), one can then infer from the observations the mode excitation rates ($\mathcal{P}$) as:

$$\mathcal{P}(\omega) = 2\pi \mathcal{M} \Gamma v_s^2 = 2\pi^2 \mathcal{M} C_{\mathrm{obs}} H \Gamma^2 \tag{11.15}$$

Provided that we can measure Γ and H, it is then possible to constraint $\mathcal{P}$. However, we point out that the derivation of $\mathcal{P}$ from the observations is also based on models since $\mathcal{M}$ is required. Furthermore, there is a strong anti-correlation between H and Γ (see e.g. [20], [21]) that can introduce important bias. This anti-correlation vanishes when considering the squared mode amplitude, v_s^2, since $v_s^2 \propto H\Gamma$ (see 11.14). However, $\mathcal{P}$ still depends on Γ, which is strongly anti-correlated with H.

As an alternative to comparing theoretical results and observational data, Chaplin et al. [16] proposed to derive H from the theoretical excitation rates, $\mathcal{P}$, and the observed mode line width, Γ, according to the relation:

$$H = \frac{\mathcal{P}}{2\pi^2 \mathcal{M} C_{\mathrm{obs}} \Gamma^2} \tag{11.16}$$

However, as pointed-out by Belkacem et al. [18], H strongly depends on the observation technique. The quantity $C_{\mathrm{obs}} H$, is less dependent on the observational data but still depends on the instrument since different instruments probe different layers of the atmosphere (see below). Therefore, one has difficulty to compare values of $H C_{\mathrm{obs}}$ coming from different instruments.

11.3.1 Solar Seismic Constraints

Baudin et al. [4] have inferred the solar p-mode excitation rates from different instruments, namely GOLF on-board SOHO, the BiSON and GONG networks. As pointed out by Baudin et al. [4], the layer (r_h) where the mode mass is evaluated must be properly estimated to derive correct values of the excitation rates from (11.15). Indeed solar seismic observations in Doppler velocity are usually measured from a given spectral line. The layer where the oscillations are measured then depends on the height (r_h) in the atmosphere where the line is formed. Different instruments use different solar lines and then probe different regions of the atmosphere. For instance, the BiSON instruments use the KI line whose height of formation is estimated at the optical depth $\tau \approx 0.013$. The optical depth associated with the different spectral lines used in helioseismology are given in Houdek [22] with associated references.

Solar p-mode excitation rates, $\mathcal{P}$, derived by Baudin et al. [4] are shown in Fig. 11.1 (left panel). For $\nu \lesssim 3.2$ mHz $\mathcal{P}^{\mathrm{GONG}}$ and $\mathcal{P}^{\mathrm{BiSON}}$ are consistent with each other, whereas $\mathcal{P}^{\mathrm{GOLF}}$ is systematically smaller than $\mathcal{P}^{\mathrm{GONG}}$ and $\mathcal{P}^{\mathrm{BiSON}}$, although the discrepancy remains within 1-σ. At high frequency, differences between the different data sets are more important. This can be partially attributed to the choice of the layers r_h where $\mathcal{M}$ are evaluated. Indeed, the sensitivity of $\mathcal{M}$ to r_h is the larger at high frequency. On the other hand, low-frequency mode masses are much less sensitive to the choice of r_h. Accordingly, the discrepancy seen at low frequency between GOLF and the other data sets suggests that the absolute calibration of the GOLF data may not be correct (see [4]). In Fig. 11.1, we then present $\mathcal{P}$ derived from GOLF data after multiplying them by a factor in order that they match at low frequency $\mathcal{P}^{\mathrm{GONG}}$ and $\mathcal{P}^{\mathrm{BiSON}}$. We find a rather good agreement between $\mathcal{P}^{\mathrm{GOLF}}$ and $\mathcal{P}^{\mathrm{BiSON}}$ whereas, at high frequency, $\mathcal{P}^{\mathrm{GONG}}$ are systematically lower than $\mathcal{P}^{\mathrm{GOLF}}$ ore $\mathcal{P}^{\mathrm{BiSON}}$. The residual high-frequency discrepancy is likely due to an incorrect determination of the layer r_h where the different seismic measurements originate (see a detailled discussion in Baudin et al. [4]).

11.3.2 Stellar Seismic Constraints

Seismic observations in Doppler velocity of solar-like pulsators are performed using spectrographs dedicated to stellar seismic measurements (e.g. UCLES, UVES, HARPS). Such spectrographs use a large number of spectral lines in order to reach a high enough signal-to-noise ratio. In the case of stellar seismic measurements, it is then more difficult than for helioseismic observations to estimate the effective height r_h. As discussed in detail in Samadi et al. [5], the computed mode surface velocities v_s, depend significantly on the choice of the height, h, in the atmosphere where the mode masses are evaluated. This is illustrated in Fig. 11.2 for the case of the star α Cen A.

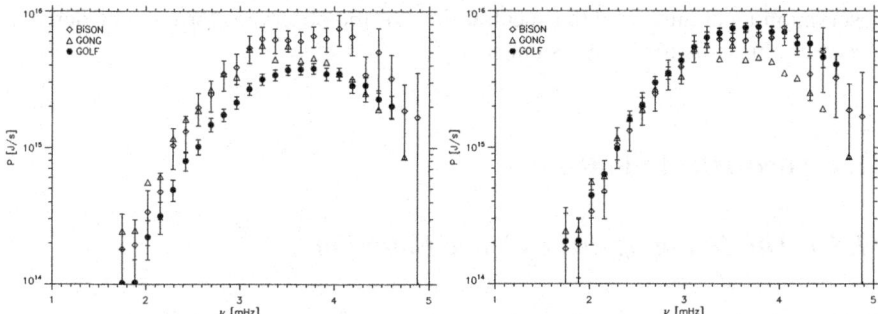

Fig. 11.1 *Left:* Solar *p*-mode excitation rates, $\mathcal{P}$, as a function frequency and derived from different instruments. The *filled circles* correspond to seismic data from SOHO /GOLF, the *diamonds* to seismic data from the BiSon network, and the *triangles* to seismic data from the GONG network. *Right:* Same as left panel. $\mathcal{P}$ derived from GOLF data multiplied by a factor in order that they match at low frequency the $\mathcal{P}$ derived from GONG or BiSON

Fig. 11.2 Mode mass evaluated for the case of α Cen A at different heights *h* above the photosphere. The *upper curve* corresponds to the photosphere ($h = 0$) and the *lower curve* to the top of the atmosphere ($h = 1000$ km). The step in *h* is 200 km

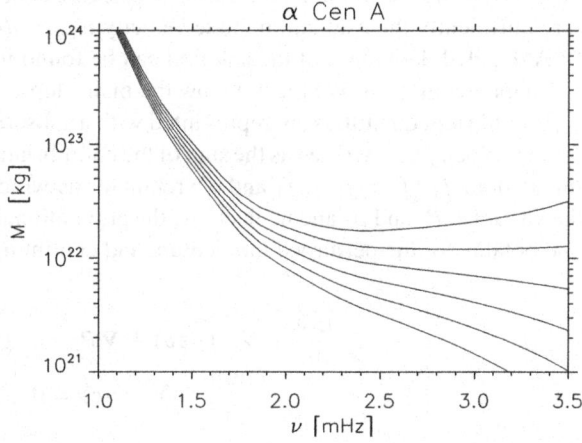

A recent work by Kjeldsen et al. [23] allows us to estimate the value for an effective r_h. Indeed, the authors have found that solar modes measured with the UCLES spectrograph have amplitudes slightly smaller than those measured by the BiSON network. The instruments of the BiSON network use the potassium (K) resonance line, which is formed at an optical depth $\tau_{500\,\mathrm{nm}} \simeq 0.013$. Kjeldsen et al. [23]'s results then suggest that acoustic modes measured by UCLES or an equivalent spectrograph (e.g. HARPS) are measured at an effective height (r_h) slightly below the formation depth of the K line, i.e. at optical depth slightly above $\tau_{500\,\mathrm{nm}} \simeq 0.013$. Accordingly, in the case of stellar seismic observations we will evaluated the mode masses at that optical depth. A more rigorous approach would be to compute an effective mode mass by weighting appropriately the different mode masses associated with the different spectral lines that contribute to the seismic measure. In order to infer accurate mode excitation rates from the stellar seismic data, the mode masses representative of the observation technique *and* the spectral lines of the observed star must be derived. However, this calls for further studies.

11.4 Theoretical Models

11.4.1 The Inhomogeneous Wave Equation

Most of the theoretical models of stochastic excitation adopt GK's approach. It consists to solve first, with appropriate boundary conditions, the equation that governs the *adiabatic* wave propagation (also called the homogeneous wave equation). This provides the well-known adiabatic displacement eigenvectors ($\xi(r, t)$). Then, we include in the wave equation of propagation turbulent sources of driving as well as a linear damping. The complete equation (so-called inhomogeneous wave equation) is then solved and the solution corresponds to the forced mode displacement, $\delta r_{osc}(r, t)$ (or equivalently the oscillation mode velocity $v_{osc} = d\delta r_{osc}/dt$).

A detailed derivation of the solution can be found in Samadi and Goupil [15] or in Chaplin et al. [16]. We recall below the main steps.

Equilibrium quantities are represented with a subscript 0. Each variable f, except for the velocity v, is written as the sum of the equilibrium quantity, f_0 and an Eulerian fluctuation, f_1, $f = f_0 + f_1$ and we retain terms which are linear and quadratic in the variables P_1 and ρ_1 and neglect, g_1, the gravitational perturbation.[4] Accordingly, one obtains for the perturbed momentum and continuity equations:

$$\frac{\partial \rho v}{\partial t} + \nabla : (\rho v v) + \nabla P_1 - \rho_1 g_0 = 0 \qquad (11.17)$$

$$\frac{\partial \rho_1}{\partial t} + \nabla \cdot (\rho v) = 0 \qquad (11.18)$$

where ω_{osc} is the mode frequency, P, ρ, v and g denote respectively the gas pressure, density, velocity and gravity.

The perturbed equation of state to second order in a Eulerian description is given by:

$$P_1 = c_s^2 \rho_1 + \alpha_s s_1 + \alpha_{\rho\rho} \rho_1^2 + \alpha_{ss} s_1^2 + \alpha_{\rho s} \rho_1 s_1 \qquad (11.19)$$

where s is the entropy, $\alpha_s = (\partial P/\partial s)_\rho$, $c_s = \Gamma_1 P_0/\rho_0$ denotes the average sound speed, $\Gamma_1 = (\partial \ln P/\partial \ln \rho)_s$ is the adiabatic exponent and $\alpha_{\rho\rho}$, α_{ss} and $\alpha_{\rho s}$ are the second partial derivatives of P versus s and ρ. Note that (11.19) assumes a constant chemical composition (this is indeed the case in the outer convective layers) but also constant ionisation rates.

---

[4] Neglecting the perturbation of the gravity corresponds to Cowling [24]'s approximation. This approximation remains valid for modes with a high n radial order.

The velocity field v is split into a component due to the pulsational displacement δr_{osc} and a turbulent component u as

$$v = v_{osc} + u \tag{11.20}$$

Linearisation of (11.17–11.19) yields for the velocity field, in the absence of turbulence ($u = 0$), the homogeneous wave equation

$$\left(\frac{\partial^2}{\partial t^2} - L \right) v_{osc} = 0 \tag{11.21}$$

where L is the linear wave operator (see its expression in SG). With appropriate boundary conditions [25] one recovers the usual eigenvalue problem:

$$L(\xi(r, t)) = -\omega_{osc}^2 \xi(r, t) \tag{11.22}$$

where ω_{osc} is the mode eigenfrequency and $\xi(r, t) \equiv e^{-i\omega_{osc} t} \xi(r)$ is the adiabatic displacement eigenvector.

In the presence of turbulence, the pulsational displacement (δr_{osc}) is written in terms of the above adiabatic solution $\xi(r, t)$ and an instantaneous amplitude $A(t)$ according to (11.6). Under the assumption of a slowly varying intrinsic amplitude $A(t)$, the velocity (v_{osc}) is related to $A(t)$ and δr_{osc} according to (11.7).

Differentiating (11.17) with respect to t, subtracting the time averaged equation of motion, neglecting non-linear terms in v_{osc}, assuming an incompressible turbulence ($\nabla \cdot u = 0$) and using (11.18) and (11.19) yields the inhomogeneous wave equation

$$\rho_0 \left(\frac{\partial^2}{\partial t^2} - L \right) [v_{osc}] + D [v_{osc}] = \frac{\partial}{\partial t} S - C \tag{11.23}$$

with

$$S \equiv S_R + S_S \tag{11.24}$$

$$S_R = \nabla : (\rho_0 \, uu) - \nabla : (\langle \rho_0 uu \rangle) \tag{11.25}$$

$$S_S = -\nabla (\bar{\alpha}_s s_t) \tag{11.26}$$

where s_t is the *Eulerian* turbulent entropy fluctuations and $\bar{\alpha}_s = \overline{(\partial P / \partial \rho)_s}$. The terms S_R (11.25) and S_S (11.26) are two driving sources, namely the Reynolds stress tensor and a term that involves the Eulerian entropy fluctuations. The last term C in the RHS of (11.23) gathers terms that involve ρ_1 as well as the second order terms of (11.19). C can in principle contribute to the driving. However, one can show that its contribution is negligible compared to S_R and S_S (see SG, GK).

The operator D in the LHS of (11.23) involves both the turbulent velocity field (u) and the pulsational velocity. This term contributes to the dynamical linear damping.

As we will see later, it is more convenient to decompose the Eulerian entropy fluctuations in terms of the Lagrangian ones, that is as:

$$\frac{\partial s_t}{\partial t} = \frac{d\delta s_t}{dt} - \boldsymbol{u} \cdot \nabla(s_0 + s_t) \tag{11.27}$$

where s_0 is the mean entropy. Accordingly, S_S is such that:

$$\frac{\partial S_S}{\partial t} = -\nabla \left(\frac{d}{dt} (\bar{\alpha}_s \delta s_t) - \bar{\alpha}_s \boldsymbol{u} \cdot \nabla s_t \right) \tag{11.28}$$

where we have dropped the term $\boldsymbol{u} \cdot \nabla s_0$ since it does not contribute to the driving (GK, see also SG). Integration of (11.28) with respect to time then gives S_S.

11.4.2 General Solution

Substituting (11.7) into (11.23), yields, with the help of (11.21), a differential equation for $A(t)$. This latter equation is straightforwardly solved and one obtains the solution for A:

$$A(t) = \frac{ie^{-\eta t}}{2\omega_{\mathrm{osc}} I} \int_{-\infty}^{t} dt' \int_{\mathcal{V}} d^3 x e^{(\eta + i\omega_{\mathrm{osc}})t'} \boldsymbol{\xi}^*(\boldsymbol{x}) \cdot \boldsymbol{S}(\boldsymbol{x}, t') \tag{11.29}$$

where I is the mode inertia (which expression is given in (11.9)) and the spatial integration is performed over the stellar volume, $\mathcal{V}$. As the sources are random, A can only be calculated in square average, $\langle |A|^2 \rangle$. This statistical average is performed over a large set of realizations. From (11.29) and with the help of some simplifications as detailed in SG, one finds:

$$\left\langle |A|^2 \right\rangle = \frac{C^2}{8\eta(\omega_{\mathrm{osc}} I)^2} \tag{11.30}$$

with

$$C^2 \equiv \int_{\mathcal{V}} d^3 x_0 \int_{-\infty}^{+\infty} d^3 r d\tau e^{-i\omega_{\mathrm{osc}}\tau} \left\langle \boldsymbol{\xi}^* \cdot \boldsymbol{S}_1 \boldsymbol{\xi} \cdot \boldsymbol{S}_2 \right\rangle \tag{11.31}$$

where η is the mode damping rate (which can be derived from seismic data) I the mode inertia (11.9), $\boldsymbol{x}_0$ the position in the star where the stochastic excitation is integrated, $\mathcal{V}$ is the volume of the convective region, $\boldsymbol{S}$ represents the different driving terms, $\boldsymbol{r}$, and τ are the spatial correlation and temporal correlation lengths associated with the local turbulence, subscripts 1 and 2 refer to quantities that are evaluated at the spatial and temporal positions $[x_0 - \frac{r}{2}, -\frac{\tau}{2}]$ and $[x_0 + \frac{r}{2}, \frac{\tau}{2}]$ respectively, and finally $\langle \cdot \rangle$ refers to a statistical average.

According to (11.5), (11.8) and (11.30), the theoretical mode excitation rate, $\mathcal{P}$, is then given by the expression:

$$\mathcal{P} = \frac{C^2}{8I} \qquad (11.32)$$

11.4.3 Driving Sources

The Reynolds stress tensor (11.25) was identified early on by Lighthill [1] as a source of acoustic noise and then as a source of *mode* excitation (GK). This term represents a mechanical source of driving and is considered by most of the theoretical formulations as the dominant contribution to the mode excitation [11–13, 16, 17, 26]. However, as pointed out by Osaki [27], the first calculations by GK's significantly under-estimate the power going to the solar modes compared to the observations.

In order to explain the mode excitation rates derived from the observations, Goldreich et al. [14] identified the *Lagrangian* entropy fluctuations, i.e. the term δs_t in (11.28), as an additional driving source. These authors claimed that this term is the dominant source of driving. However, GMK assumed that entropy fluctuations (s_t) behave as a passive scalar. A passive scalar f is a quantity that obeys an equation of diffusion (see e.g. [28]):

$$\frac{df}{dt} = \frac{\partial f}{\partial t} + u \cdot \nabla f = \chi \nabla^2 f \qquad (11.33)$$

where χ is a diffusion coefficient. As shown by SG, assuming as GMK that δs_t is a passive scalar leads to a vanishing contribution. On the other hand, SG have shown that the term $\bar{\alpha}_s \, u \cdot \nabla s_t$ in the RHS of (11.28) contributes effectively to the mode driving. In SG formulation, the so-called entropy source term is then:

$$\frac{\partial}{\partial t} S_S = \nabla \left(\bar{\alpha}_s u \cdot \nabla s_t \right) \qquad (11.34)$$

The term $u \cdot \nabla s_t$ in the RHS of (11.34) is an advective term. Since it involves the entropy fluctuations it can be considered as a thermal source of driving. The source term of (11.34) was also identified by GK, but was considered as negligible. It must also be pointed out that the theoretical formalisms by Balmforth [13] and Chaplin et al. [16] did not consider this source term. According to Samadi et al. [17], this term is not negligible (about ~15% of the total power) but nevertheless small compared to the Reynolds stress source term (S_R) in the case of the Sun.

Finally, as seen in (11.31), S_R and S_S lead to cross terms. However, assuming as GMK that s_t behaves as a passive scalar and an *incompressible* turbulence (i.e. $\nabla \cdot u = 0$), SG have shown that the crossing term between S_R and S_S vanishes. Hence, in the framework of those assumptions, there is *no canceling* between the two contributions (but see Sect. 11.8).

11.4.4 Length Scale Separation

As seen in the RHS of (11.31), the eigen-displacement $\boldsymbol{\xi}(\boldsymbol{r})$ is coupled spatially
with the source function, S. In order to derive a theoretical formulation that can be
evaluated, it is necessary to spatially decouple $\boldsymbol{\xi}(\boldsymbol{r})$ from S. This is the reason why all
theoretical formulations explicitly or implicitly assume that eddies that effectively
contribute to the driving have a characteristic length scale smaller than the mode
wavelength. Indeed, provided this is the case, $\boldsymbol{\xi}(\boldsymbol{r})$ can be removed from the integral
over r and τ that appears in the RHS of (11.30) (see SG). This assumption is justified
for low turbulent Mach numbers M_t ($M_t \propto u/c_s$ where c_s is the sound speed).
However, at the top of the solar convective zone, that is in the super-adiabatic region,
M_t is no longer small ($M_t \sim 0.3$). Furthermore, for G and F stars lying on the
main sequence, M_t is expected to increase with the effective temperature and to
reach a maximum for $M \sim 1.6\,M_{\odot}$ (see Houdek et al. [29]). Hence, for F type
stars, significantly hotter than the Sun, the length scale separation becomes a more
questionable approximation (see the discussion in Sect. 11.11).

11.4.5 Closure Models

The second integral in RHS of (11.30) involves the term $\langle S_1 S_2 \rangle$, which is a two-
point spatial *and* temporal correlation products of the source terms. Hence, the
Reynolds stress source term (11.25) leads to the two-point correlation product of
the form $\langle (\boldsymbol{uu})_1 (\boldsymbol{uu})_2 \rangle$. In the same way, the entropy source term (11.34) leads to
the two-point correlation product of the form $\langle (\boldsymbol{u}s_t)_1 (\boldsymbol{u}s_t)_2 \rangle$. In both case, we deal
with fourth-order two-point correlation product involving turbulent quantities (that
is $\boldsymbol{u}$ and s_t). Fourth-order moments are solutions of equations involving fifth-order
moments. In turn, fifth-order moments are expressed in term of six-order moments. . .
and so on. This is the well known closure problem. A simple closure model is the
quasi-normal approximation (QNA hereafter) that permits one to express fourth order
moments in term of second order ones (see details in e.g. [28]), that is:

$$\langle (u_i u_j)_1 (u_k u_l)_2 \rangle(\boldsymbol{r}, \tau) = \langle (u_i u_j)_1 \rangle \langle (u_k u_l)_2 \rangle + \langle (u_i)_1 (u_l)_2 \rangle \langle (u_j)_1 (u_k)_2 \rangle$$
$$+ \langle (u_i)_1 (u_k)_2 \rangle \langle (u_j)_1 (u_l)_2 \rangle \tag{11.35}$$

The decomposition of (11.35) is strictly valid when the velocity is normally dis-
tributed. The first term in the RHS of (11.35) cancels the term $\langle \boldsymbol{uu} \rangle$ in (11.25) (see
details in Chaplin et al. [16]). An expression similar to (11.35) is derived for the
correlation product $\langle (\boldsymbol{u}s_t)_1 (\boldsymbol{u}s_t)_2 \rangle$ (see SG).

11.4.6 Adopted Model of Turbulence

It is usually more convenient to express (11.35) in the frequency (ω) and wavenumber (k) domains. We then define $\phi_{i,j}$ as the temporal and spatial Fourier transform of $\langle (u_i)_1 (u_j)_2 \rangle$. For an inhomogeneous, incompressible, isotropic and stationary turbulence, there is a relation between $\phi_{i,j}$ and the kinetic energy spectrum E, which is [30]:

$$\phi_{ij}(\boldsymbol{k}, \omega) = \frac{E(k, \omega)}{4\pi k^2} \left(\delta_{ij} - \frac{k_i k_j}{k^2} \right) \tag{11.36}$$

where k and ω are the wavenumber and frequency respectively associated with the turbulent elements, and $\delta_{i,j}$ is the Kronecker symbol. Following Stein [7], it is possible to split for each layers $E(k, \omega)$ as:

$$E(k, \omega) = E(k) \chi_k(\omega) \tag{11.37}$$

where $E(k)$ is the time averaged kinetic energy spectrum and $\chi_k(\omega)$ is the frequency component of $E(k, \omega)$. In other words, $\chi_k(\omega)$ measures—in the frequency and k wavenumber domains—the temporal correlation between eddies. As discussed in Sect. 11.2), the way the eddy time-correlation is modeled has an important consequence on the efficiency of the mode driving. A decomposition similar to that of (11.37) is performed for the spectrum associated with the entropy fluctuations ($E_s(k, \omega)$).

Note that $\chi_k(\omega)$ and $E(k)$ satisfy by definition the following normalisation conditions:

$$\int_{-\infty}^{+\infty} d\omega \chi_k(\omega) = 1 \tag{11.38}$$

$$\int_{0}^{\infty} dk\, E(k) = \frac{1}{2} \langle u^2 \rangle = \frac{\Phi}{2} \langle u_z^2 \rangle \equiv \frac{3}{2} u_0^2, \tag{11.39}$$

where u_z is the vertical component of the velocity, $\Phi \equiv \langle u^2 \rangle / \langle u_z^2 \rangle$ is the anisotropy factor introduced by Gough [31], and u_0 is a characteristic velocity introduced for convenience. A normalisation condition similar to (11.39) is introduced for $E_s(k)$ (see details in SG).

11.4.7 Complete Formulation

On the basis of the different assumptions mentioned above, SG then derive for *radial* modes the following theoretical expression for $\mathcal{P}$:

$$P = \frac{1}{8I} \left(C_R^2 + C_S^2 \right) \tag{11.40}$$

where C_R^2 and C_S^2 are the turbulent Reynolds stress and entropy contributions respectively. There expressions are (see SG):

$$C_R^2 = 4\pi^3 \mathcal{G} \int_0^M dm \rho_0 \left| \frac{d\xi_r}{dr} \right|^2 S_R(m, \omega_{\mathrm{osc}}) \tag{11.41}$$

$$C_S^2 = \frac{4\pi^3 \mathcal{H}}{\omega_{\mathrm{osc}}^2} \int_0^M dm \frac{\bar{\alpha}_s^2}{\rho_0} g_r(\xi_r, m) S_S(m, \omega_{\mathrm{osc}}) \tag{11.42}$$

with S_R and S_S are the source terms associated with the Reynolds stress and entropy fluctuations respectively:

$$S_R = \int_0^\infty dk \frac{E^2(k, m)}{k^2} \int_{-\infty}^{+\infty} d\omega \chi_k(\omega_{\mathrm{osc}} + \omega, m) \chi_k(\omega, m) \tag{11.43}$$

$$S_S = \int_0^\infty dk \frac{E_s(k, m) E(k, m)}{k^2} \int_{-\infty}^{+\infty} d\omega \chi_k(\omega_{\mathrm{osc}} + \omega, m) \chi_k(\omega, m) \tag{11.44}$$

In (11.41) and (11.42), ρ_0 is the mean density, $\mathcal{G}$ and $\mathcal{H}$ are two anisotropic factors (see their expressions in SG), and finally $g_r(\xi_r, m)$ is a function that involves the first and the second derivatives of ξ_r, its expression is:

$$g_r(\xi_r, m) = \left(\frac{1}{\alpha_s} \frac{d\alpha_s}{dr} \frac{d\xi_r}{dr} - \frac{d^2 \xi_r}{dr^2} \right)^2 \tag{11.45}$$

It is in general more convenient to rewrite (11.41) and (11.42) in the following forms:

$$C_R^2 = 4\pi^3 \mathcal{G} \int_0^M dm \frac{\rho_0 u_0^4}{k_0^3 \omega_0} \left| \frac{d\xi_r}{dr} \right|^2 \tilde{S}_R(m, \omega_{\mathrm{osc}}) \tag{11.46}$$

$$C_S^2 = \frac{4\pi^3 \mathcal{H}}{\omega_{\mathrm{osc}}^2} \int_0^M dm \frac{(\bar{\alpha}_s \tilde{s} u_0)^2}{\rho_0 k_0^3 \omega_0} g_r(\xi_r, m) \tilde{S}_s(m, \omega_{\mathrm{osc}}) \tag{11.47}$$

where we have defined the dimensionless source functions $\tilde{S}_R \equiv (k_0^3 \omega_0 / u_0^4) S_R$ and $\tilde{S}_s \equiv (k_0^3 \omega_0 / (u_0^2 \tilde{s})) S_R$, $\tilde{s}$ and where $\tilde{s}$ is the rms of the entropy fluctuations. We have introduced for convenience the characteristic frequency ω_0 and the characteristic wavenumber k_0; they are defined as:

$$\omega_0 \equiv k_0 u_0 \tag{11.48}$$

$$k_0 \equiv \frac{2\pi}{\Lambda} \tag{11.49}$$

where Λ is a characteristic size derived from $E(k)$ and u_0 is the characteristic velocity given by (11.39). For future use, it is also convenient to define a characteristic time τ_0 as:

$$\tau_0 = \frac{2\pi}{k_0 u_0} = \frac{\Lambda}{u_0} \tag{11.50}$$

From (11.46) we can show that the driving by the Reynolds stress is locally proportional to the kinetic energy flux. Indeed, the flux of kinetic energy in the vertical direction is by definition:

$$F_{\text{kin}} \equiv w E_{\text{kin}} = w \left(\frac{1}{2} \rho_0 \, u^2 \right) = \frac{3}{2} \sqrt{\frac{3}{\Phi}} \rho_0 \, u_0^3, \tag{11.51}$$

where $E_{\text{kin}} \equiv (1/2) \rho_0 u^2$ is the kinetic energy per unit volume. Substituting (11.51) into (11.46) yields the relation:

$$C_R^2 \propto \int_0^M dm \, F_{\text{kin}} \Lambda^4 \left| \frac{d\xi_{\text{r}}}{dr} \right|^2 \tilde{S}_R(m, \omega_{\text{osc}}). \tag{11.52}$$

Concerning the driving by the entropy fluctuations, we can show that locally this driving does not only depend on F_{kin} but also on the convective flux (F_c). Indeed, lets define as GMK the quantity:

$$\mathcal{R} \equiv \frac{\alpha_s \tilde{s}}{\rho_0 u_0^2} \tag{11.53}$$

Substituting (11.53) into (11.47) yields the relation:

$$C_S^2 \propto \int_0^M dm \, F_{\text{kin}} \Lambda^4 \mathcal{R}^2 \mathcal{F}^2 \left(\frac{\omega_0}{\omega_{\text{osc}}} \right)^2 \tilde{S}_S(m, \omega_{\text{osc}}) \tag{11.54}$$

where we have defined as in SG the quantity $\mathcal{F}^2 \equiv \Lambda^2 g_{\text{r}}$. Finally, since $\mathcal{R} \propto F_c/F_{\text{kin}}$ (see Samadi et al. [32]), we can conclude that locally the driving by entropy source term is proportional to F_{kin} and to the square of the ratio $\mathcal{R} \propto F_c/F_{\text{kin}}$.

11.5 Turbulent Spectrum

As seen in Sect. 11.4.6, the model of stochastic excitation developed by SG involves $E(k, \omega)$, the turbulent kinetic spectrum as well as $E_s(k, \omega)$, the spectrum associated with the turbulent entropy fluctuations. Both spectra are split in terms of a time averaged spectrum ($E(k)$ for the velocity and $E_s(k)$ for the entropy fluctuations), and a frequency component $\chi_k(\omega)$ (see Sect. 11.4.6). Different prescriptions were investigated for both components. The results of these investigations are summarized in Sect. 11.5.1 for $E(k)$ and in Sect. 11.5.2 for $\chi_k(\omega)$.

11.5.1 Time Averaged Spectrum, $E(k)$

Two approaches are commonly adopted for prescribing $E(k)$. The classic one is to assume an analytical function derived either from theoretical considerations or empirical ones. The more commonly used analytical spectrum is the so-called Kolmogorov spectrum [33], which derives originally from Oboukhov [34]'s postulate that energy is transferred from the large scales to the small scales at a constant rate. Other theoretical spectra, such as the so-called Spiegel's spectrum [35], or purely empirical spectra, such as those proposed by Musielak et al. [36], were also considered. All of these analytical functions differ from each other by the way $E(k)$ varies with k. But for all of them, it is required to set a priori the characteristic wavenumber, k_0, at which energy is injected into the turbulent cascade. The second approach consists to obtain $E(k)$ directly from hydrodynamical 3D simulations. This method has two advantages: it provides both the k dependence of $E(k)$ and the characteristic wavenumber k_0. On the other hand, the inconvenient is that such method depends on the quality of the 3D hydrodynamical simulation.

These two approaches have been compared in Samadi et al. [37]. Among the different analytical functions tested, the best agreement with a solar 3D simulation was found with the so-called "Extended Kolmogorov Spectrum" defined by Musielak et al. [36]. This spectrum increases at low scales as k^{+1} and decreases at low scales according to the Kolmogorov spectum, i.e. as $k^{-5/3}$. However, due to the limited spatial resolution of the solar simulation used, the Kolmogorov scaling is validated over a limited range only. Nevertheless, the major part of the excitation arises from the most-energetic eddies, also refered to the energy bearing eddies. Accordingly, the contribution of the small scales, that are not resolved by the present 3D simulations, are expected to be relatively small. However, to confirm this, a quantitative estimate must be undertaken.

More important is the choice for the characteristic wavenumber k_0. Indeed, the integrands of (11.46) and (11.47) are both proportional to k_0^{-4}. Accordingly, the computed $\mathcal{P}$ are very sensitive to the choice for k_0. This characteristic wavenumber can be obtained from 3D simulations. However, by default, one usually relates k_0 to the mixing-length Λ_{MLT} according to:

Fig. 11.3 Eddy time-correlation function, χ_k, as a function of frequency ν for the layer where the radial component of the velocity is maximum. The *filled dots* represent χ_k obtained from a solar 3D simulation with an horizontal resolution of $\simeq 25\,\mathrm{km}$ [17]. χ_k is shown here for the wavenumber k at which $E(k)$ peaks. The *solid line* represents a Lorentzian function and the *dashed line* a Gaussian function

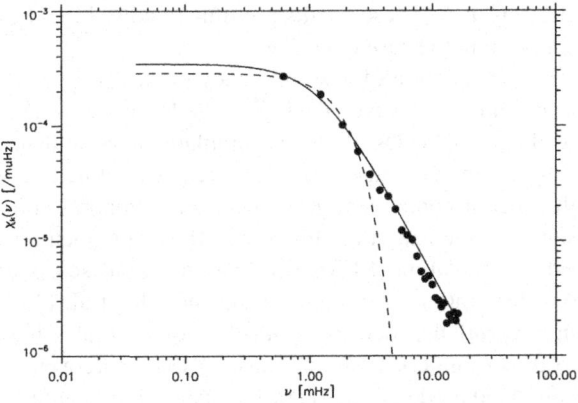

$$k_0 = k_0^{\mathrm{MLT}} \equiv \frac{2\pi}{\beta \Lambda_{\mathrm{MLT}}} \tag{11.55}$$

where $\Lambda_{\mathrm{MLT}} = \alpha H_p$ is the mixing-length, α the mixing-length parameter, H_p the pressure scale height, and β a free parameter, which is usually set to a value of the order of one. The solar 3D simulation used by Samadi et al. [37] indicates that in the Sun $k_0 \simeq 3.6\,\mathrm{Mm}^{-1}$ at the top of the excitation region. This characteristic wavenumber corresponds to an horizontal size of the granules of $\Lambda_g = 2\pi/k_0 \sim 2\,\mathrm{Mm}$. This horizontal size is reached at the top of the excitation region with a value of β that depends on the adopted value for α and the solar 1D model used. For other stars, 3D simulations are rarely available. In that case, one usually assumes for β the same value that the one adopted for the Sun. Hence, an open and important question is whether or not the parameter β can be kept the same for other stars as for the Sun.

11.5.2 Eddy Time-Correlation, $\chi_k(\omega)$

Most of the theoretical formulations explicitly or implicitly assume a Gaussian function for $\chi_k(\omega)$ [11–14], 16, 38]. However, 3D hydrodynamical simulations of the outer layers of the Sun show that, at the length associated with the energy bearing eddies, χ_k is rather Lorentzian [17]. This is well illustrated in Fig. 11.3. As pointed-out by Chaplin et al. [16], a Lorentzian χ_k is also a result predicted for the largest, most-energetic eddies by the time-dependent mixing -length formulation derived by Gough [31]. Therefore, there is some numerical and theoretical evidences that χ_k is rather Lorentzian at the length scale of the energy bearing eddies.

As shown by Samadi et al. [17], calculation of the mode excitation rates based on a Gaussian χ_k results for the Sun in a significant under-estimation of the maximum of $\mathcal{P}$ whereas a better agreement with the observations is found when a Lorentzian χ_k is used. A similar conclusion is reached by Samadi et al. [5] in the case of the star α Cen A. These results are illustrated in Fig. 11.5 in the case of the Sun and in Fig. 11.6 in the case of α Cen A.

The excitation of low-frequency modes ($\nu \lesssim 3$ mHz) is mainly due to the large scale eddies. However, the higher the frequency the more important the contribution of the small scales. 3D solar simulations show that, at small scales, χ_k is neither Lorentzian nor Gaussian [39]. Hence, according to [39], it is impossible to separate the spatial component $E(k)$ from the temporal component at all scales with the same simple analytical functions. However, such results are obtained using Large Eddy Simulation (LES). The way the small scales are treated in LES can affects our description of turbulence. Indeed, He et al. [40] have shown that LES results in a $\chi_k(\omega)$ that decreases at all resolved scales too rapidly with ω with respect to direct numerical simulations (DNS). Moreover, Jacoutot et al. [41] found that computed mode excitation rates depend significantly on the adopted sub-grid model. Furthermore, Samadi et al. [42] have shown that, at a given length scale, χ_k tends toward a Gaussian when the spatial resolution is decreased. This is illustrated in Fig. 11.4 by comparison with Fig. 11.3. In summary, the numerical resolution or the sub-grid model can substantially affect our description of the small scales. Improving the modeling of the excitation of the high frequency modes then requires more realistic and more resolved hydrodynamical 3D simulations.

Up to now, only analytical functions were assumed for $\chi_k(\omega)$. We have here implemented, for the calculation of $\mathcal{P}$, the eddy time-correlation function derived *directly* from long time series of 3D simulation realizations with an intermediate horizontal resolution $\simeq 50$ km As shown in Figs. 11.5 and 11.6, the mode excitation rates, $\mathcal{P}$, obtained from χ_k^{3D}, are found to be comparable to that obtained assuming a Lorentzian χ_k, except at high frequency in the case of the Sun. This is obviously the direct consequence of the fact that a Lorentzian χ_k reproduces rather well χ_k^{3D} (see Fig. 11.3), except at high frequency where χ_k^{3D} decreases more rapidly than the Lorentzian function (see Fig. 11.4, left). At high frequency, calculations based on a Lorentzian χ_k result in larger $\mathcal{P}$ and reproduce better the helioseismic constraints than those based on χ_k^{3D} (see Fig. 11.5). This indicates perhaps that χ_k^{3D} decreases more rapidly with frequency than it should. This is consistent with He et al. [40]'s results who found that LES predict a too rapidly decrease with ν compared to the DNS (see above).

Chaplin et al. [16] also found that the use of a Gaussian χ_k severely under-estimates the observed solar mode excitation rates. However, in contrast with Samadi et al. [17], they mention that a Lorentzian χ_k results in a severe over-estimation for the low-frequency modes. In order to illustrate the results by Chaplin et al. [16], we have computed the solar mode excitation rates using their formalism and a solar envelope equilibrium model similar to the one considered by these authors (see [37]). The result is shown in Fig. 11.7. We clearly see that the mode excitation rates computed using a Gaussian χ_k overestimate by ~ 20 the seismic constraints. This result is consistent with this found by Samadi et al. [17]. On the other hand, in contrast with Samadi et al. [17], the modes with frequency below $\nu \sim 2$ mHz are severely over-estimated when a Lorentzian χ_k is assumed. It should be pointed out that the excitation of

Fig. 11.4 *Top:* Same as Fig. 11.3 for a solar 3D simulation with an horizontal resolution of $\simeq 50\,\mathrm{km}$ [41]. *Bottom:* Same as top for a solar 3D simulation with an horizontal resolution of $\simeq 120\,\mathrm{km}$

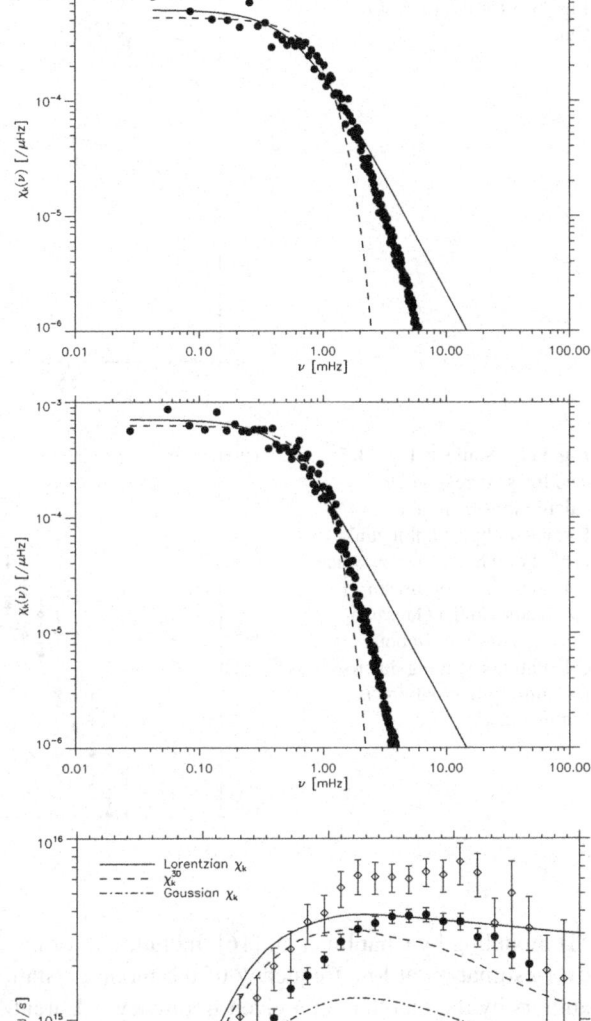

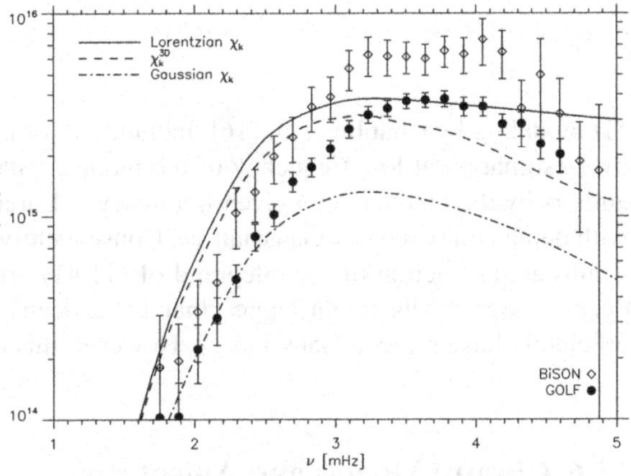

Fig. 11.5 Solar p-mode excitation rates as a function of ν. Filled circles and diamonds correspond as in Fig. 11.1 to seismic data from SOHO /GOLF and BiSON network respectively. The lines correspond to semi-theoretical calculations based on different choices for χ_k: Lorentzian χ_k (*solid line*), χ_k3D i.e. χ_k derived directly from the solar 3D simulation (*dashed line*), and a Gaussian χ_k (*dot-dashed line*)

modes with frequency $\nu \lesssim 2\,\mathrm{mHz}$ occurs in a region more extended than covered by the solar 3D simulation used by Samadi et al. [37]. On the other hand, the pure

Fig. 11.6 Same as in
Fig. 11.5 for the case of α
Cen A

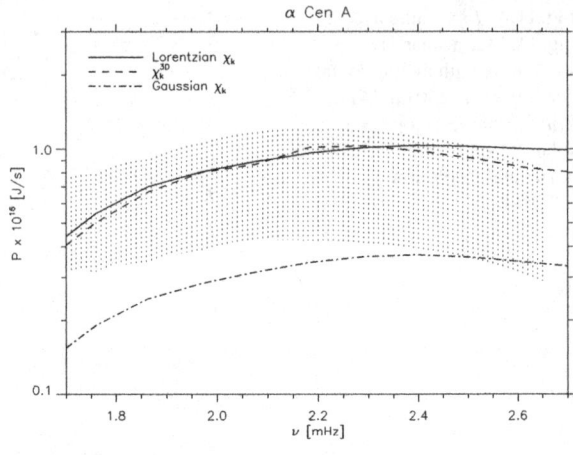

Fig. 11.7 Same as Fig. 11.5.
The lines correspond to
calculations using the
formalism by Chaplin et al.
[16]. Two choices for χ_k was
considered: a Lorentzian χ_k
(*solid line*) and a Gaussian
χ_k (*dashed line*). In both
calculations, driving due to
the entropy fluctuations is
not included

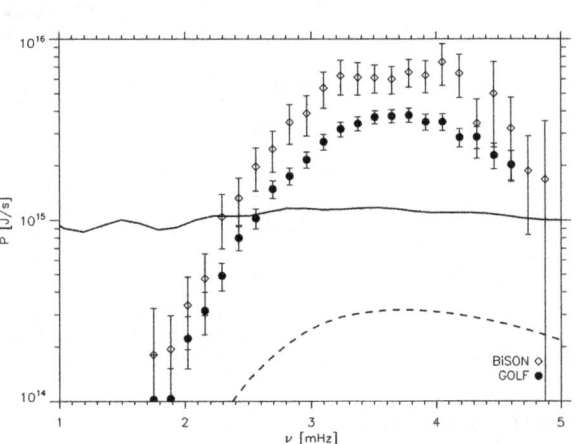

1D modeling by Chaplin et al. [16], includes all of the convective zone. The severe over-estimation at low frequency of the mode excitation rates, is explained by the authors by the fact that, at a given frequency, a Lorentzian χ_k decreases too slowly with depth compared to a Gaussian χ_k. Consequently, for the low-frequency modes, a substantial fraction of the integrand of (11.41) arises from large eddies situated deep in the Sun. This might suggest that, in the deep layers, the eddies that contribute efficiently have rather a Gaussian χ_k. However, this remains an open issue.

11.6 Closure Models and Anisotropy

The decomposition of (11.35) assumes the quasi-normal approximation (QNA). However, it is well known that the departure from the QNA is important in a strongly turbulent medium. In addition, a closure model based on the QNA does not ensure the positiveness of the energy (see details in e.g. [28]). Furthermore, the QNA is strictly valid only for normally distributed fluctuating quantities with zero mean. However, the upper-most part of the convection zone is a turbulent convective medium composed of essentially two flows that are asymmetric with respect to each other. Hence, in such a medium, the probability distribution function of the fluctuations of the vertical velocity and temperature do not follow a Gaussian law. As verified by Belkacem et al. [43] and Kupka and Robinson [44], departure from the QNA is important in the upper part of the solar convective zone. Indeed, this approximation under estimates, in the quasi-adiabatic region, by $\approx 50\%$ the fourth-order moment of the vertical velocity derived from a solar 3D simulation.

The term in the LHS of (11.35) corresponds to a two-point correlation product involving the velocity, i.e. $\langle (u_i u_j)_1 (u_k u_l)_2 \rangle (r, \tau)$ where r and τ are the spatial correlation and temporal correlation lengths respectively. For $r \rightarrow 0$ and $\tau \rightarrow 0$, this term reduces to a *one-point* correlation product, $\langle u_i u_j u_k u_l \rangle$, also referred as a fourth-order moment (FOM hereafter). As we shall see below, it is possible to derive an improved closure model for this term that does not rely on the QNA. However, we still require a prescription for the *two point* correlation products involving the velocity $(\langle (u_i u_j)_1 (u_k u_l)_2 \rangle (r, \tau))$ and the entropy fluctuations $(\langle (u s_t)_1 (u s_t)_2 \rangle (r, \tau))$. For radial modes or low ℓ order modes, only the radial component of the velocity (w) matters. Hence, for these modes we require a prescription for $\langle w_1^2 w_2^2 \rangle (r, \tau)$ and $\langle (w s_t)_1 (w s_t)_2 \rangle (r, \tau)$. By default, Belkacem et al. [18] have proposed that $\langle w_1^2 w_2^2 \rangle (r, \tau)$ varies with r and τ in the same way than in the QNA (11.35), that is:

$$\langle w_1^2 w_2^2 \rangle = \frac{\mathcal{K}_w}{3} \langle w_1^2, w_2^2 \rangle_{\text{QNA}} \tag{11.56}$$

where $\mathcal{K}_w$ is a constant and $\langle w_1^2 w_2^2 \rangle_{\text{QNA}}$ is the two-point correlation product given for w according to the QNA (11.35). Accordingly, the contribution of the Reynolds stress $(C_R^2, (11.41)$ is modified as:

$$C_R^2 = 4\pi^3 \mathcal{G} \int_0^M dm \rho_0 \left(\frac{d\xi_r}{dr} \right)^2 \frac{\mathcal{K}_w}{3} S_R(m, \omega_{\text{osc}}) \tag{11.57}$$

Note that the contribution of the entropy fluctuations $(C_S^2, (11.42)$ still assumes the QNA. This inconsistency has a small impact on computed mode excitation rates since C_S^2 is significantly smaller than C_R^2, at least for stars that are not too hot (but see Sect. 11.8).

The constant $\mathcal{K}_w$ is determined in the limit case where $r \rightarrow 0$ and $\tau \rightarrow 0$. Indeed, when $r \rightarrow 0$ and $\tau \rightarrow 0$, we have:

$$\langle w^4 \rangle = \frac{\mathcal{K}_w}{3} \langle w^4 \rangle_{\text{QNA}}, \tag{11.58}$$

where $\langle w^4 \rangle$ is by definition the fourth-order moment (FOM hereafter) associated with w and $\langle w^4 \rangle_{QNA}$ is the one given by the QNA. In the same way, (11.35) gives:

$$\langle w^4 \rangle_{QNA} = 3\langle w^2 \rangle^2. \tag{11.59}$$

Using (11.58) and (11.59), we then derive the constant $\mathcal{K}_w$:

$$\mathcal{K}_w = 3\frac{\langle w^4 \rangle}{\langle w^4 \rangle_{QNA}} = \frac{\langle w^4 \rangle}{\langle w^2 \rangle^2}, \tag{11.60}$$

which is by definition the Kurtosis. This quantity measures the oblateness of the probability density function (see e.g. [43]). For normally distributed w we have $\mathcal{K}_w = 3$. The Kurtosis then measures the departure of the FOM from the QNA.

closure models more sophisticated than the QNA can be used. Among those, the two-scale mass flux model [45, 46] takes the asymmetries in the medium into account but is only applicable for quasi-laminar flows. For $\mathcal{K}_w$, Gryanik and Hartmann [44] obtained the following expression:

$$\mathcal{K}_w = (1 + S_w^2) \tag{11.61}$$

with the skewness, S_w, given by:

$$S_w \equiv \frac{\langle w^3 \rangle}{\langle w \rangle^{3/2}} = \frac{1 - 2a}{\sqrt{a(1-a)}} \tag{11.62}$$

where a is the mean fractional area occupied by the updrafts in the horizontal plane. In the QNA limit, i.e. when the random quantities are distributed according to a Normal distribution with zero mean, we necessarily have $S_w = 0$. Hence, in the QNA limit, (11.61) does not match the expected value i.e. $\mathcal{K}_w = 3$. Then, Gryanik and Hartmann [46] proposed to modify (11.61) as follows:

$$\mathcal{K}_w = 3(1 + \frac{1}{3}S_w^2). \tag{11.63}$$

Figure 11.8 shows that the FOM based on (11.63) with S_w given by (11.62), results in a negligible improvement with respect to the QNA. However, when S_w is derived directly from the 3D simulation and plugged into (11.63), (11.63) is then a very good evaluation of the FOM derived from a 3D simulation of the outer layer of the Sun as verified by Belkacem et al. [43] and Kupka and Robinson [44].

Belkacem et al. [18] have generalized Gryanik and Hartmann [46]'s approach by taking the skewness introduced by the presence of up- and down-drafts *and* the turbulent properties of each flow into account. Accordingly, they have derived a more accurate expression for S_w (see the expression in [43]). As shown in Fig. 11.8, calculations of the FOM based on (11.63) and their expression for S_w reproduce rather well—in the quasi-adiabatic region—the FOM derived from the solar 3D simulation.

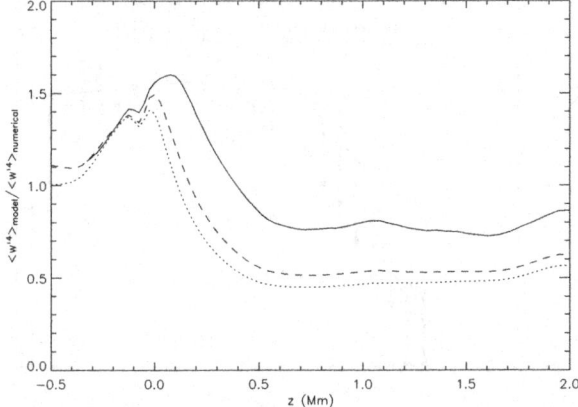

Fig. 11.8 Fourth-order moment (FOM) of the velocity, $\langle w4 \rangle = \mathcal{K}_w \langle w2 \rangle^2$, as a function of depth z, normalized to the FOM derived from the 3D simulation. In all cases the Kurtosis $\mathcal{K}_w$, (11.60) is calculated according to (11.63) but with different skewness, S_w. The *solid line* S_w is computed according to Belkacem et al. [43] closure model, the *dashed line* assumes Gryanik & Hartmann [46] expression for S_w (11.62) and finally the dotted line assumes the QNA, that is $S_w = 0$ and $\mathcal{K}_w = 3$

Belkacem et al. [18] have computed mode excitation rates, $\mathcal{P}$ according to (11.57) with the Kurtosis $\mathcal{K}_w$ given by (11.63) and with the skewness S_w computed according to Belkacem et al. [43] closure model. The maximum in $\mathcal{P}$ is found about 30% larger than in calculations based on the QNA and fits better the maximum in $\mathcal{P}$ derived from the helioseismic data. This increase is significantly larger than the entropy contribution (the term $\mathcal{S}_S$ in (11.42), which is of the order of $\sim$15%, see Sect. 11.8). We stress that, however, 30% is of the same order as the difference between seismic constraints of different origins (SOHO /GOLF, GONG, BiSON). These results are illustrated in Fig. 11.9.

11.7 Importance of the Stellar Stratification and Chemical Composition

11.7.1 Role of the Turbulent Pressure

Rosenthal et al. [47] have shown that taking the turbulent pressure into account in a realistic way in the 1D global solar models results in a much better agreement between observed and theoretical mode frequencies of the Sun. Following Rosenthal et al. [47], Samadi et al. [5] have studied the importance for the calculation of the mode excitation rates of taking the turbulent pressure into account in the averaged 1D model. For this purpose, they have built two 1D models representative of the star α

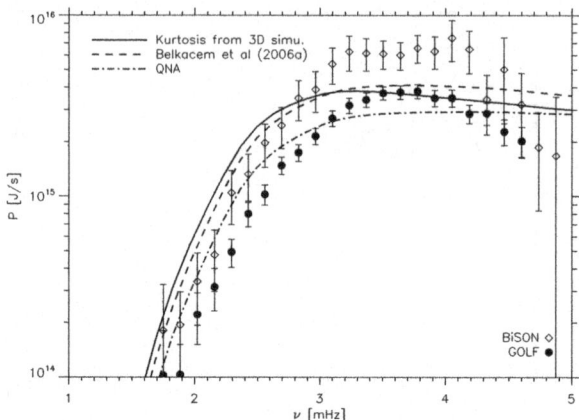

Fig. 11.9 Same as in Fig. 11.5. The thick lines correspond to calculations where the Reynolds stress contribution is computed according to (11.57). The Kurtosis (K_w) is computed here in a different manner: for the *solid line* K_w is obtained directly from a 3D solar simulation, for the *dashed line* the Kurtosis is calculated according to (11.63) where the skewness (S_w) is obtained from Belkacem et al. [43] closure model, and finally for the *dot-dashed line* we have assumed the QNA, that is $S_w = 0$ and $K_w = 3$

Cen A. One model (here refered as the "patched" model), has its surface layers taken directly from a fully compressible 3D hydrodynamical numerical model. A second model (here refered as "standard" model), has its surface layers computed using standard physics, in particular convection is described according to Böhm-Vitense [48]'s mixing -length local theory of convection (MLT) and turbulent pressure is ignored.

Samadi et al. [5] found that the calculations of $\mathcal{P}$ involving eigenfunctions computed on the basis of the "patched" global 1D model reproduce much better the seismic data derived for α Cen A than calculations based on the eigenfunctions computed with the "standard" stellar model, i.e. built with the MLT and ignoring turbulent pressure. This is because a model that includes turbulent pressure results in *lower* mode masses $\mathcal{M}$ than a model that ignores turbulent pressure. This can be understood as follows: Within the super-adiabatic region, a model that includes turbulent pressure provides an additional support against gravity, hence has a lower gas pressure and density than a model that does not include turbulent pressure (see also [47, 49]). As a consequence, mode inertia (11.9) or equivalently mode masses (11.11) are then *lower* in a model that includes turbulent pressure.

11.7.2 Role of the Surface Metal Abundance

Samadi et al. [50] have recently studied the role of the surface metal abundance on the efficiency of the stochastic driving. For this purpose, they have computed two 3D

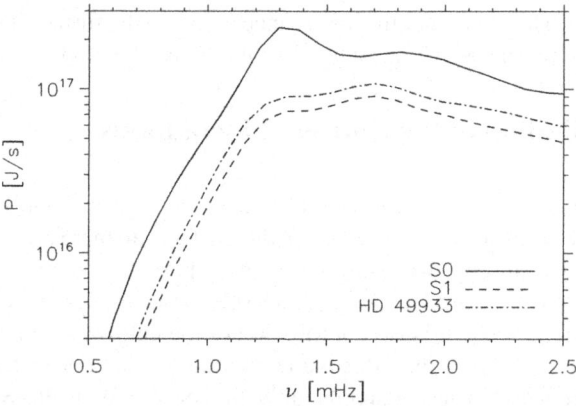

Fig. 11.10 Mode excitation rates, $\mathcal{P}$, as a function of the mode frequency (ν) obtained for two 3D models with the effective temperature and the surface gravity of HD 49933 but with two different surface metal abundances (see Sect. 11.7.2) and Samadi et al. [50]. The *solid line* corresponds to the 3D model with the metal abundance (S0) and the *dashed line* to metal poor 3D model (S1). The *dot-dashed line* corresponds to the mode excitation rates derived for the specific case of HD 49933 as explained in Samadi et al. [50]

hydrodynamical simulations representative—in effective temperature and gravity—of the surface layers of HD 49933, a star which is rather metal poor compared to the Sun since its surface iron-to-hydrogen abundance is [Fe/H] = −0.37. One 3D simulation (hereafter labeled as S0) has a solar metal abundance and the other (hereafter labeled as S1) has [Fe/H] ten times smaller. For each 3D simulation they have build a "patched" model in the manner of Samadi et al. [5] and computed the acoustic modes associated with the "patched" model.

As seen in Fig. 11.10, the mode excitation rates $\mathcal{P}$ associated with S1 are found to be about *three times smaller* than those associated with S0. This difference is related to the fact that a lower surface metallicity results in a lower opacity, and accordingly in an higher surface density. In turn, the higher the density, the smaller are the convective velocities to transport by convection the same amount of energy. Finally, smaller convective velocities result in a less efficient driving (for details see [50]). This conclusion is qualitatively consistent with that by Houdek et al. [29] who—on the basis of a mixing-length approach—also found that the mode amplitudes decrease with decreasing metal abundance.

Using the seismic determinations of the mode linewidths measured by CoRoT for HD 49933 [51] and the theoretical mode excitation rates computed for the specific case of HD 49933, Samadi et al. [52] have derived the theoretical mode amplitudes of the acoustic modes of HD 49933. Except at rather high frequency ($\nu \gtrsim 1.9\,\mathrm{mHz}$), their amplitude calculations are within approximately 1-σ in agreement with the mode amplitudes derived from the CoRoT data Samadi et al. [52]. They also show that assuming a solar metal abundance rather than the observed metal abundance of the star would result in larger mode amplitudes and hence in a larger discrepancy with the seismic data. This illustrates the importance of taking the surface metal abundance of the solar-like pulsators into account when modeling the mode excitation.

11.8 Contribution of the Entropy Fluctuations

Using the method summarised in Sect. 11.9.2, Stein and Nordlund [25] have computed *directly* from a 3D simulation of the surface of the Sun the contribution of the incoherent entropy fluctuations (11.26). They also found that the entropy fluctuation is small compared to the Reynolds stress contribution. However, as shown by Samadi et al. [42], the relative contribution of the entropy to the total excitation rate increases rapidly with the effective temperature, T_{eff}. For instance, the solar-like pulsator HD 49333 has a significantly higher T_{eff} than the Sun. Samadi et al. [50] found that for this star the entropy fluctuations contributes up to $\sim 30\%$ while it is only about 15% in the case of the Sun (see [42]) and in the case of α Cen A (see [5]).

As pointed-out by Houdek [22], the solar and stellar 3D simulations performed by Stein et al. [53] show some partial canceling between the Reynolds stress contribution (S_R, (11.25) and contribution due to the entropy (S_S, (11.26)). This cancelation increases with increasing T_{eff} (see [53]). In the theoretical model of stochastic excitation, the cross terms between the entropy fluctuations and the Reynolds stresses vanish (see Sect. 11.2). As originally suggested by Houdek [22] and discussed in Samadi et al. [52], the existence of a partial canceling can decrease the mode amplitude and improve the agreement with the seismic observations. However, there is currently no theoretical modeling of the interference between these two terms (see the discussion in Sect. 11.11) and in [52]).

11.9 Alternative Approaches

11.9.1 Energy Equipartition

Under certain conditions that we will emphasize below, GK have shown that there is an equipartition of kinetic energy between an acoustic mode and the resonant eddy. To derive this principle, GK assume that the acoustic modes are damped by turbulent viscosity *and* excited by the Reynolds stresses. We reproduce here their demonstration. For the sake of simplicity, we will consider modes with $\omega_{osc} \tau_0 \lesssim 1$ where τ_0 is the characteristic time of the energy bearing eddies typically located in the upper part of the convective zone, that is the region where the driving is the most vigorous. Furthermore, we neglect as did GK the driving by the entropy fluctuations (11.42). According to (11.40) and (11.52), we have roughly for acoustic modes with $\omega_{osc} \tau_0 \lesssim 1$:

$$\mathcal{P} \propto \frac{1}{I} \int dm \left| \frac{d\xi_r}{dr} \right|^2 E_{\text{eddy}} \Lambda u_0, \tag{11.64}$$

where Λ is the characteristic size of the energy bearing eddies, u_0 their characteristic velocity (11.39), $\tau_0 = \Lambda/u_0$ their characteristic lifetime (11.50), and $E_{\text{eddy}} = (3/2)\rho_0 u_0^2 \Lambda^3$ their total kinetic energy. Let k_{osc} be the vertical oscillation wave number. We have then $d\xi_r/dr = ik_{\text{osc}}\xi_r$. We further assume that—in the driving region—the acoustic waves are purely propagating. This assumption then implies $\omega_{\text{osc}} = k_{\text{osc}} c_s$ where c_s is the sound speed. Accordingly, we can simplified (11.64) as:

$$\mathcal{P} \propto \frac{\omega_{\text{osc}}^2}{I} \int dm \left(\frac{\xi_r}{c_s} \right)^2 E_{\text{eddy}} \Lambda u_0. \tag{11.65}$$

In the region where the mode are excited, E_{eddy}, u_0, and c_s vary quite rapidly. However, again for the sake of simplicity we will assume that these quantities are constant and evaluate them at the layer where the excitation is the most efficient, i.e. at the peak of the super-adiabatic temperature gradient. The integration of (11.64) can be approximated as

$$\mathcal{P} \propto \frac{1}{I} \left(\frac{\omega_{\text{osc}}}{c_s} \right)^2 E_{\text{eddy}} \Lambda u_0 \int dm \xi_r^2. \tag{11.66}$$

Using the expression of the mode inertia (11.9), we can finally simplify (11.66) as:

$$\mathcal{P} \propto \left(\frac{\omega_{\text{osc}}}{c_s} \right)^2 E_{\text{eddy}} \Lambda u_0. \tag{11.67}$$

Modes damped by turbulent viscosity have their damping rates η given by [54, 55],

$$\eta \propto \frac{1}{3I} \int dm \nu_t \left| r \frac{d}{dr} \left(\frac{\xi_r}{r} \right) \right|^2, \tag{11.68}$$

where ν_t is the turbulent viscosity. The simplest prescription for ν_t is the concept of eddy-viscosity. This consists in assuming $\nu_t = u_0 \lambda = \tau_0 u_0^2$. Obviously the turbulent medium is characterized by eddies with a large spectrum of size. However, only the eddies for which $\omega_{\text{osc}} \tau_\lambda \approx 1$ are expected to efficiently damp the mode with frequency ω_{osc}. Since we are looking at the modes such that $\omega_{\text{osc}} \tau_\lambda \lesssim 1$, only the largest eddies efficiently damp the mode, that is the eddies with size Λ. Accordingly, we adopt $\nu_t = u_0 \Lambda$. With the same simplifications and assumptions as those used for deriving (11.67), we can simplify (11.68) as:

$$\eta \propto \left(\frac{\omega_{\text{osc}}}{c_s} \right)^2 \Lambda u_0. \tag{11.69}$$

From (11.5), (11.67) and (11.69), we then derive the mode kinetic energy:

$$E_{osc} \propto E_{eddy}. \tag{11.70}$$

Equation (11.70) highlights an equipartition of kinetic energy between an acoustic mode and the resonant eddies. Christensen-Dalsgaard and Frandsen [56] used this "equipartition principle" to derive the first quantitative estimate of solar-like oscillations in stars. The relation of (11.70) was derived by assuming that modes are damped by turbulent viscosity. However, as pointed-out by Osaki [26], theoretical mode line-widths, $\Gamma = \eta/\pi$, computed in the manner of Goldreich and Keeley [55], i.e. assuming a viscous damping, are underestimated compared to the observations. Gough [57] proposed a different prescription for ν_t. Nevertheless, assuming Gough [57]'s prescription also results in similar Γ (see [58]). On the other hand, Xiong et al. [59] report that the turbulent viscosity is the dominant source of damping of the radial p modes. As discussed recently by Houdek [60], there is currently no consensus about the physical processes that contribute dominantly to the damping of p modes. If the damping due to turbulent viscosity turns out to be negligible, then there is no reason that the balance between the mode kinetic energy and the kinetic energy of resonant eddies holds in general.

11.9.2 "Direct" Calculation

The model of stochastic excitation presented in Sect. 11.4 is based on several simplifications and assumptions concerning the turbulence and the source terms. There is an alternative approach proposed by Nordlund and Stein [61] that does not rely on such simplifications and assumptions. In such approach, the rate at which energy is stochastically injected into the acoustic modes is obtained *directly* from 3D simulations of the outer layers of a star by computing the (incoherent) work performed on the acoustic mode by turbulent convection. In their approach, the energy input per unit time into a given acoustic mode is calculated numerically according to (11.74) of Nordlund and Stein [61] multiplied by S, the area of the simulation box, to get the excitation rate (in Js^{-1})

$$\mathcal{P}_{3D}(\omega_{osc}) = \frac{\omega_{osc}^2 S}{8 \Delta \nu \mathcal{E}_{\omega_{osc}}} \left| \int_r dr \, \Delta \hat{P}_{nad}(r, \omega_{osc}) \frac{\partial \xi_r}{\partial r} \right|^2 \tag{11.71}$$

where $\Delta \hat{P}_{nad}(r, \omega)$ is the discrete Fourier component of the non-adiabatic pressure fluctuations, $\Delta P_{nad}(r, t)$, estimated at the mode eigenfrequency $\omega_{osc} = 2\pi \nu_0$, ξ_r is the radial component of the mode displacement eigenfunction, $\Delta \nu = 1/T_s$ the frequency resolution corresponding to the total simulation time T_s and $\mathcal{E}_{\omega_{osc}}$ is the normalised mode energy per unit surface area defined in Nordlund and Stein ([61], their (11.63) as:

$$\mathcal{E}_{\omega_{osc}} = \frac{1}{2}\omega_{osc}^2 \int_r dr \xi_r^2 \rho \left(\frac{r}{R}\right)^2. \tag{11.72}$$

Equation (11.71) corresponds to the calculation of the $P\,dV$ work associated with the non-adiabatic gas and turbulent pressure (Reynolds stress) fluctuations. In contrast to the pure theoretical models (see Sect. 11.4), the derivation of (11.71) does not rely on a simplified model of turbulence. For instance, the relation of (11.36) is no longer required. Furthermore, they do not assume that entropy fluctuations behave as a passive scalar (11.33). However, as for the theoretical models, it is assumed that ξ_r varies on a scale-length larger than the eddies that contributes effectively to the driving (this is the so-called "length-scale separation", see Sect. 11.4). In addition, (11.71) implicitly assumes the quasi-Normal approximation (11.35).

The expression of (11.71) has been applied to the case of the Sun by Stein and Nordlund [61]. These authors obtain a rather good agreement between $\mathcal{P}_{3D}$ (11.71) and the solar mode excitation rates derived from the GOLF instrument by Roca Cortés et al. [62]. However, solar mode excitation rates derived by Stein and Nordlund [61] from the seismic analysis by Roca Cortés et al. [62] are—for a reason that remains to be understood—systematically lower than those derived from the seismic analysis by Baudin et al. [4]. Stein et al. [53] have computed $\mathcal{P}_{3D}$ (11.71) for a set of stars located near the main sequence from K to F and a subgiant K IV star. The comparison between these calculations and those based on SG's formalism has been undertaken by Samadi et al. [42]. The maximum in $\mathcal{P}_{3D}$ was found systematically lower than those from calculations based on SG's formalism (11.40–11.44). These systematic differences were attributed by Samadi et al. [42] to the low spatial resolution of the hydrodynamical 3D simulations computed by Stein et al. [53].

11.10 Stochastic Excitation Across the HR Diagram

11.10.1 Mode Excitation Rates

Using several 3D simulations of the surface of main sequence stars, Samadi et al. [40] have shown that the maximum of the mode excitation rates, $\mathcal{P}_{max}$, varies with the ratio L/M as $(L/M)^\alpha$ where L and M are the luminosity and the mass of the star respectively and α is the slope of this scaling law. Furthermore, they found that the slope α is rather sensitive to the adopted function for χ_k : $\alpha = 3.1$ for a Gaussian χ_k and $\alpha = 2.6$ for a Lorentzian one.

The increase of $\mathcal{P}_{max}$ with L/M is not surprising: it should first be noticed that, even though the ratio L/M is the ratio of two global stellar quantities, it nevertheless essentially characterizes the properties of the stellar surface layers where the mode excitation is located since $L/M \propto T_{eff}^4/g$. Indeed, by definition of the effective temperature, T_{eff}, and the stellar radius R, the total luminosity of the star, L, is given by the Steffan's law: $L = 4\pi\sigma T_{eff}^4 R^2$ where σ is Steffan's constant. Furthermore, the surface gravity is $g = GM/R^2$ where G is the gravitational constant. Accordingly, $L/M \propto T_{\text{eff}}^4/g$.

Second, as we will show now, it is possible to roughly explain the dependence of $\mathcal{P}_{\text{max}}$ with g and T_{eff}. (11.67) can be rewritten as:

$$\mathcal{P} \propto \left(\frac{\omega_{\text{osc}}}{c_s}\right)^2 F_{\text{kin}} \Lambda^4 \tag{11.73}$$

where

$$F_{\text{kin}} = \frac{3}{2}\rho_0 u_0^3 \tag{11.74}$$

is by definition the flux of kinetic energy per unit volume[5] and u_0 is the characteristic velocity given by (11.39).

The characteristic size Λ is approximately proportional to the pressure scale height H_p (see e.g. [5]). From hydrostatic equilibrium, we have $P = \rho g H_p$. Assuming now the equation of state of a perfect gas, we then derive $H_p \propto T/g$. The sound speed is given by the relation $c_s^2 = \Gamma_1 P/\rho$. Accordingly, using again the perfect gas equation, we then have $c_s^2 \propto T$. From these simplifications, we can simplify (11.73) as:

$$\mathcal{P} \propto \omega_{\text{osc}}^2 F_{\text{kin}} T^3 g^{-4} \tag{11.75}$$

In the framework of the mixing -length approach, it can be shown that F_{kin} is roughly proportional to the convective flux F_c. Indeed, in this framework, the eddies are accelerated by the buoyancy force over a distance equal to the mixing-length $\Lambda = \alpha H_p$ where α is the mixing-length parameter. Accordingly, the kinetic energy of the eddies, E_{eddy}, is given by (see the lecture notes by Bohm-Vitense [63])

$$E_{\text{eddy}} \equiv \frac{3}{2}\rho u_0^2 \Lambda^3 = g(\Delta\rho\Lambda^3)\Lambda \tag{11.76}$$

where $\Delta\rho$ is the difference between the density of the eddy and its surroundings. In the Boussinesq approximation, the perturbation of the equation of state gives:

$$\frac{\Delta\rho}{\rho} \propto \frac{\Delta T}{T} \tag{11.77}$$

where ΔT is the difference between the temperature of the eddy and its surrounding. Now, the convective flux (also referred to as the enthalpy flux) is by definition the quantity:

$$F_c \equiv u_0\left(\rho C_p \Delta T\right) \tag{11.78}$$

---

[5] for the sake of simplicity we assume here an isotropic medium, accordingly the flux of kinetic energy is the same in any direction.

where $c_p = (\partial s/\partial \ln T)_p$. Finally, from the definition of (11.74) and the set of (11.76–11.78), one derives $F_{\text{kin}} \propto g\Lambda/T F_c$ and, since $\Lambda \propto T/g$, we show finally that $F_{\text{kin}} \propto F_c$.

In the region where the driving is the most efficient, the total energy flux, F_{tot}, is no longer totally transported by convection (that is $F_c \langle F_{\text{tot}}\rangle$). However, in order to derive an expression that depends only on the surface parameters of the star, we will assume that all of the energy is transported by convection ; that is $F_c \approx F_{\text{tot}} = \sigma T_{\text{eff}}^4 \propto g(L/M)$ where σ is the Steffan's constant. Accordingly, (11.75) can be further simplified as:

$$\mathcal{P} \propto \omega_{\text{osc}}^2 T_{\text{eff}}^4 T^3 g^{-4} \approx \omega_{\text{osc}}^2 T_{\text{eff}}^7 g^{-4}, \qquad (11.79)$$

where we have assumed $T = T_{\text{eff}}$.

Let now defines $\nu_{\text{max}} = \omega_{\text{osc}}^{\text{max}}/2\pi$ the peak frequency associated with $\mathcal{P}$. This characteristic frequency can be estimated according to:

$$\nu_{\text{max}} \approx u_0/\Lambda \qquad (11.80)$$

where the quantity u_0/Λ is estimated in the layer where u_0 is maximum. Using similar simplifications as used previously for $\mathcal{P}$, we can show that

$$\nu_{\text{max}} \propto g(T_{\text{eff}}/\bar{\rho})^{1/3}, \qquad (11.81)$$

where $\bar{\rho}$ is the mean density at the photosphere. We assume that $\bar{\rho}$ is equal to the star mean density, that is $\bar{\rho} \approx M/R^3 \propto g/R$. Accordingly, we then derive from (11.79) and (11.81):

$$\mathcal{P}_{\text{max}} \propto \left(T_{\text{eff}}^4\right)^{23/12} g^{-3} M^{1/3}, \qquad (11.82)$$

where M is the stellar mass. For main sequence stars lying in the domain where solar-like oscillations are expected, $M^{1/3}$ varies very slowly such that it can be ignored in (11.82). Then, (11.82) can finally be simplified as:

$$\mathcal{P}_{\text{max}} \propto \left(T_{\text{eff}}^4\right)^2 g^{-3}. \qquad (11.83)$$

We now clearly see from (11.83) that $\mathcal{P}_{\text{max}}$ as expected increases with increasing $F_{\text{tot}} = \sigma T_{\text{eff}}^4$ and decreases with increasing g.

11.10.2 Mode Surface Velocity

Prior to the CoRoT mission, only crude and indirect derivations of the averaged mode linewidth had been proposed for a few stars (see [23, 64, 65]). However, for the majority of solar-like pulsators observed so far from the ground in Doppler velocity,

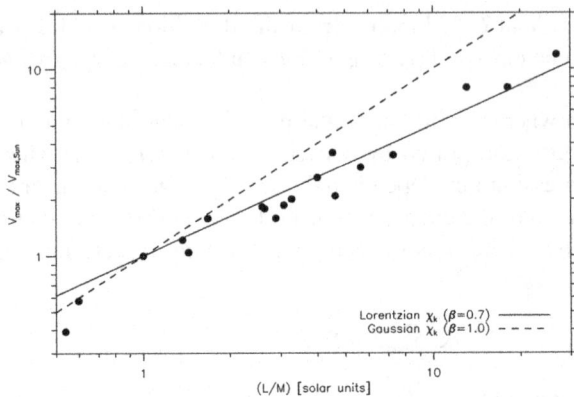

Fig. 11.11 Ratio between V_{max} the maximum of the mode velocity relative to the observed solar value ($V_{max}^{\odot} = 25.2$ cm/s for $\ell = 1$ modes, see [23]. *Filled dots* correspond to the stars for which solar-like oscillations have been detected in Doppler velocity (see a detailed list of references in [72].The *lines*—except the *dot-dashed line*—correspond to the power laws obtained from the predicted scaling laws for $\mathcal{P}_{max}$ and estimated values of the damping rates η_{max} (see text for details). Results for two different eddy time-correlation functions, χ_k, are presented: Lorentzian χ_k (*solid line*) and Gaussian χ_k (*dashed line*)

such measurements are not available, only the maximum of the mode surface velocity (V_{max} hereafter) is in general accessible. For the numerous solar-like pulsators observed from the ground, we must compute the mode surface velocity according to (11.13), which requires the knowledge of not only $\mathcal{P}$ but also of the mode damping rates ($\eta = \pi \Gamma$).

Houdek et al. [29] have computed η for a large set of main sequence models. Using Balmforth [13]'s formulation of stochastic excitation, they have also computed the mode excitation rates ($\mathcal{P}$). From their theoretical computations of $\mathcal{P}$ and $\Gamma = \eta/\pi$, they have derived v_s according to (11.13). Their theoretical calculations for V_{max} result in a scaling law of the form $(L/M)^{\beta}$ with a exponent $\beta = 1.5$ (see [29]).

We have plotted in Fig. 11.11 the quantity V_{max} associated with the solar-like pulsators observed so far in Doppler velocity. Clearly, V_{max} increases as $(L/M)^{\beta}$ where the exponent $\beta \simeq 0.7$. A similar scaling law with the exponent $\beta = 1$ was earlier derived by Kjeldsen and Bedding [66] from the theoretical calculations by Christensen-Dalsgaard and Frandsen [56]. Houdek et al. [29]'s scaling law significantly over-estimates the mode amplitudes in F-type stars. For instance for Procyon ($T_{eff} \simeq 6480$ K, $L \simeq 6.9 L_{\odot}$ and $L/M \simeq 4.6$), this scaling law over-estimates V_{max} by a factor ~ 4.

Samadi et al. [42] have derived V_{max} using mode damping rates computed by Houdek et al. [29] and the different scaling laws found for $\mathcal{P}_{max} \propto (L/M)^{\alpha}$. They also found that V_{max} scales as $(L/M)^{\beta}$. This is not surprising since $\mathcal{P}_{max}$ varies as $(L/M)^{\alpha}$. Furthermore, the exponent β is found to depend significantly on the choice of χ_k : $\beta = 0.7$ for a Lorentzian χ_k and $\beta = 1$ for a Gaussian χ_k. As shown in Fig. 11.11, the best agreement with the observations is found when a Lorentzian

χ_k is assumed. On the other hand, assuming a Gaussian χ_k results in a larger exponent β. When theoretical mode amplitudes are calibrated with respect to the solar mode amplitudes, calculations based on a Gaussian χ_k over-estimate the amplitudes of solar-like pulsators significantly more luminous than the Sun.

Theoretical calculations by Houdek et al. [29] assume a Gaussian χ_k. Then according to Samadi et al. [42]'s results, the too large value found for β by Houdek et al. [29] can partially be explained by the use of a Gaussian χ_k. However, according to Houdek [22], their too high value of β might be explained essentially by the mode damping rates that could be under-estimated by a factor $\sim$1.8.

11.11 Discussion and Perspectives

The way mode excitation by turbulent convection is modeled is still very simplified. As discussed below, several approximations must be improved, some assumptions or hypothesis must be removed.

As seen in Sect. 11.2, the driving efficiency crucially depends on the eddy time-correlation (χ_k). Current models assume that χ_k varies with ω in the same way at any length scale. At the length scale of the energy bearing eddy, there are some strong indications that χ_k is Lorentzian rather than Gaussian. However, at smaller scale, it is not yet clear what is the correct description for χ_k.

Use of more realistic 3D simulations would be very helpful to represent the correct dynamic behavior of the small-scales.

Current theoretical models that include the entropy fluctuations in the driving assume that the entropy fluctuations behave as a passive scalar (see Sect. 11.3). As a consequence, cross terms between S_R and S_S vanish. This is a *strong hypothesis* that is unlikely to be valid in the super-adiabatic part of the convective zone where driving by the entropy is important. Indeed, the super-adiabatic layer is a place where the radiative losses of the eddies are important because of the optically thin layers. Assuming that the entropy (or equivalently the temperature) is diffusive (11.33) is no longer valid. Furthermore, departure from incompressible turbulence is the largest in that layer and, accordingly, the cross terms between S_R and S_S no longer vanish (see SG). Therefore, the passive scalar assumption is not valid in the super-adiabatic layers. To avoid this assumption, one needs to include the radiative losses in the modeling.

One other approximation concerns the spatial separation between the modes and the contributing eddies. This approximation is less valid in the super-adiabatic region where the turbulent Mach number is no longer small, in particular for high ℓ order modes. This spatial separation can however be avoided if the kinetic energy spectrum associated with the turbulent elements ($E(k)$) is properly coupled with the spatial dependence of the modes (work in progress).

The CoRoT mission, launched 27 December, is precise enough to detect solar-like oscillations with amplitudes as low as the solar p modes [67]. Furthermore, thanks to its long term (up to 150 days) and *continuous* observations, it is possible with CoRoT to resolve solar-like oscillations, and hence to measure not only the mode amplitudes but also *directly* the mode linewidths (see e.g. [68]). Similarly as in the case of the Sun, it is now possible with CoRoT to derive direct constraints on $\mathcal{P}$ for stars with different characteristics: evolutionary status, effective temperature, gravity, chemical composition, magnetic field, rotation, surface convection, etc. We emphasize below some physical processes and conditions that we expect to address thanks to the CoRoT data.

Some solar-like pulsators are young stars that show rather strong activity (e.g. HD 49933, HD 181420, HD 175726, HD 181906,...). A high level of activity is often linked to the presence of strong magnetic field. Effects of the magnetic field are not taken into account in the calculation of the mode excitation rates. A strong magnetic field can more or less inhibit convective transport (see e.g. [69, 70]) Furthermore, as shown by Jacoutot et al. [71], a strong magnetic field can significantly change the way turbulent kinetic energy is spatially distributed and leads to a less efficient driving of the acoustic modes. In that framework, the CoRoT target HD 175726 is probably an interesting case. Indeed, this star shows both a particularly high level of activity and solar-like oscillations with amplitudes significantly lower than expected [72].

Young and active stars rotate usually faster than the Sun. As shown recently by Belkacem et al. [73], the presence of rotation introduces additional sources of driving. However, in the case of a moderate rotator such as HD 49933, these additional sources of driving remain negligible compared to the Reynolds stress and the entropy source term. On the other hand, the presence of rotation has an indirect effect on mode driving through the modification of the mode eigenfunctions. An open issue is: will the CoRoT or the Kepler mission be able to test the expected effect of rotation (see [71])?

Solar-like oscillations have now been firmly detected in several red giant stars, from both Doppler velocity measurements (see the review by Bedding and Kjeldsen [74]) as well as from space based photometry measurements [75, 76]. More recently, detection of solar-like oscillations by CoRoT in a huge number of red giant stars has been announced by de Ridder et al. [77]. Why look at solar-like oscillations in red giant stars? Toward the end of their lives, stars like the Sun greatly expand to become giant stars. A consequence of this great expand, is the existence of a very dilute convective envelope. A low density favors a vigorous convection, hence higher Mach numbers (M_t). The theoretical models of stochastic excitation are strictly valid in a medium where M_t is—as in the Sun and α Cen A—rather small. Hence, the higher M_t, the more questionable the different approximations and the assumptions involved in the theory. Hence, red giant stars allow us to test the theory of mode driving by turbulent in more extreme conditions. Finally, most of theories of stochastic excitation are developed for radial modes only. Dolginov and Muslimov [12], GMK and Belkacem et al. [19] have considered the non-radial case. There are interesting applications of such non-radial formalisms, for instance the case of solar g modes [78], but also g modes in massive stars that can in principle be excited in their central convective zones [79].

Acknowledgements I am very grateful to Marie-Jo Goupil and Kévin Belkacem for their valuable comments and advise. I am indebted to J. Leibacher for his careful reading of the manuscript. I am grateful to the organizers of the CNRS school of St-Flour for their invitation and I thank the CNRS for the financial support.

References

1. Lighthill, M.J.: Proc. R. Soc. Lond. A **211**, 564 (1952)
2. Bedding, T.R., Kjeldsen, H.: Commun. Asteroseismol. **150**, 106 (2007)
3. Libbrecht, K.G.: APJ **334**, 510 (1988)
4. Baudin, F., Samadi, R., Goupil, M.-J., et al.: A&A **433**, 349 (2005)
5. Samadi, R., Belkacem, K., Goupil, M.J., Dupret, M.-A., Kupka, F.: A&A **489**, 291 (2008)
6. Unno, W., Kato, S.: PASJ **14**, 417 (1962)
7. Stein, R.F.: Sol. Phys. **2**, 385 (1967)
8. Ulrich, R.K.: ApJ **162**, 993 (1970)
9. Leibacher, J.W., Stein, R.F.: Astrophys. Lett. **7**, 191 (1971)
10. Deubner, F.L.: A&A **44**, 371 (1975)
11. Goldreich, P., Keeley, D.A.: APJ **212**, 243 (1977)
12. Dolginov, A.Z., Muslimov, A.G.: Ap&SS **98**, 15 (1984)
13. Balmforth, N.J.: MNRAS **255**, 639 (1992)
14. Goldreich, P., Murray, N., Kumar, P.: APJ **424**, 466 (1994)
15. Samadi, R., Goupil, M.J.: A&A **370**, 136 (2001) (SG3)
16. Chaplin, W.J., Houdek, G., Elsworth, Y., et al.: MNRAS **360**, 859 (2005)
17. Samadi, R., Nordlund, Å., Stein, R.F., Goupil, M.J., Roxburgh, I.: A&A **404**, 1129 (2003)
18. Belkacem, K., Samadi, R., Goupil, M.J., Kupka, F., Baudin, F.: A&A **460**, 183 (2006)
19. Belkacem, K., Samadi, R., Goupil, M.-J., Dupret, M.-A.: A&A **478**, 163 (2008)
20. Chaplin, W.J., Elsworth, Y., Isaak, G.R. et al.: MNRAS **298**, L7 (1998)
21. Chaplin, W.J., Basu, S.: Sol. Phys. **36** (2008)
22. Houdek, G. In: ESA Special Publication, Proceedings of SOHO 18/GONG 2006/HELAS I, Beyond the spherical Sun, Published on CDROM, vol. 624, p. 28.1 (2006)
23. Kjeldsen, H., Bedding, T.R., Arentoft, T. et al.: APJ **682**, 1370 (2008)
24. Cowling, T.G.: MNRAS **101**, 367 (1941)
25. Unno, W., Osaki, Y., Ando, H., Saio, H., Shibahashi, H.: Nonradial oscillations of stars, 2nd edn. University of Tokyo Press, Tokyo (1989))
26. Stein, R.F., Nordlund, Å.: ApJ **546**, 585 (2001)
27. Osaki, Y.: Progress of seismology of the sun and stars. In: Osaki, Y., Shibahashi, H. (eds.) Lecture Notes in Physics, vol. 367, p. 75. Springer, Berlin (1990)
28. Lesieur, M.: Turbulence in fluids. Academic Publishers, Kluwer (1997)
29. Houdek, G., Balmforth, N.J., Christensen-Dalsgaard, J., Gough, D.O.: Aap **351**, 582 (1999)
30. Batchelor, G.K.: The Theory of Homogeneous Turbulence. University Press (1970)
31. Gough, D.O.: APJ **214**, 196 (1977)
32. Samadi, R., Kupka, F., Goupil, M.J., Lebreton, Y., van't Veer-Menneret, C.: A&A **445**, 233 (2006)
33. Kolmogorov, A.N.: Dokl. Akad. Nauk SSSR **30**, 299 (1941)
34. Oboukhov, A.: Dokl. Akad. Sci. Nauk SSSR **32**, 22 (1941)
35. Spiegel, E.: J. Geophys. Res. **67**, 3063 (1962)
36. Musielak, Z.E., Rosner, R., Stein, R.F., Ulmschneider, P.: APJ **423**, 474 (1994)
37. Samadi, R., Nordlund, Å., Stein, R.F., Goupil, M.J., Roxburgh, I.: A&A **403**, 303 (2003)
38. Samadi, R., Goupil, M.J., Lebreton, Y.: A&A **370**, 147 (2001)
39. Georgobiani, D., Stein, R.F., Nordlund Å.: Solar MHD theory and observations: a high spatial resolution perspective. In: Leibacher, J., Stein, R.F., Uitenbroek, H. (eds.) Astronomical Society of the Pacific Conference Series, vol. 354, p. 109 (2006)

40. He, G.-W., Rubinstein, R., Wang, L.-P.: Phys. Fluids **14**, 2186 (2002)
41. Jacoutot, L., Kosovichev, A.G., Wray, A.A., Mansour, N.N.: APJ **682**, 1386 (2008)
42. Samadi, R., Georgobiani, D., Trampedach, R. et al.: A&A **463**, 297 (2007)
43. Belkacem, K., Samadi, R., Goupil, M.J., Kupka, F.: A&A **460**, 173 (2006)
44. Kupka, F., Robinson, F.J.: MNRAS **374**, 305 (2007)
45. Abdella, K., McFarlane, N.: J. Atm. Phys. **54**, 1850 (1997)
46. Gryanik, V., Hartmann, J.: J. Atmos. Sci. **59**, 2729 (2002)
47. Rosenthal, C.S., Christensen-Dalsgaard, J., Nordlund, Å., Stein, R.F., Trampedach, R.: A&A **351**, 689 (1999)
48. Böhm-Vitense, E.: Zeitschr. Astrophys. **46**, 108 (1958)
49. Nordlund, Å., Stein, R.F.: Stellar structure: theory and test of connective energy transport. In: Gimenez, A., Guinan, E.F., Montesinos, B. (eds.) Astronomical Society of the Pacific Conference Series, vol. 173, p. 91(1999)
50. Samadi, R., Ludwig, H., Belkacem, K., Goupil, M.-J., Dupret, M.: A&A vol. 509, A15 (2010) (astro-ph/0910.4027)
51. Benomar, O., Baudin, F., Campante, T.L., et al.: A&A **507**, L13 (2009)
52. Samadi, R., Ludwig, H., Belkacem, K., et al.: A&A **509**, A16 (2010) (astro-ph/0910.4037)
53. Stein, R., Georgobiani, D., Trampedach, R., Ludwig, H.-G., Nordlund, Å.: Sol. Phys. **220**, 229 (2004)
54. Ledoux, P., Walraven, T.: In: Flugge, S. (ed.) Handbuch der Physik, vol. 51, p. 353. Springer, New York (1958)
55. Goldreich, P., Keeley, D.A.: APJ **211**, 934 (1977)
56. Christensen-Dalsgaard, J., Frandsen, S.: Sol. Phys. **82**, 469 (1983)
57. Gough, D.: Problems of stellar convection. In: Spiegel, E., Zahn, J.-P. (eds.) Lecture Notes in Physics, vol. 71, p. 15. Springer (1976)
58. Balmforth, N.J.: MNRAS **255**, 632 (1992)
59. Xiong, D.R., Cheng, Q.L., Deng, L.: MNRAS **319**, 1079 (2000)
60. Houdek, G.: Commun. Asteroseismol. **157**, 137 (2008)
61. Nordlund, Å., Stein, R.F.: APJ **546**, 576 (2001)
62. Roca Cortés, T., Montañés, P., Pallé, P.L., et al.: Stellar structure: theory and test of connective energy transport. In: Gimenez, A., Guinan, E.F., Montesinos, B. (eds.) Astronomical Society of the Pacific Conference Series, vol. 173, p. 305 (1999)
63. Bohm-Vitense, E.: Introduction To Stellar Astrophysics, vol. 3. Cambridge University Press, Cambridge (1989)
64. Kjeldsen, H., Bedding, T.R., Butler, R.P., et al.: APJ **635**, 1281 (2005)
65. Fletcher, S.T., Chaplin, W.J., Elsworth, Y., Schou, J., Buzasi, D.: MNRAS **824** (2006)
66. Kjeldsen, H., Bedding, T.R.: Aap **293**, 87 (1995)
67. Michel, E., Baglin, A., Auvergne, M. et al.: Science **322**, 558 (2008)
68. Appourchaux, T., Michel, E., Auvergne, M., et al.: A&A **488**, 705 (2008)
69. Proctor, M.R.E., Weiss, N.O.: Rep. Prog. Phys. **45**, 1317 (1982)
70. Vögler, A., Shelyag, S., Schüssler, M., et al.: A&A **429**, 335 (2005)
71. Jacoutot, L., Kosovichev, A.G., Wray, A., Mansour, N.N.: APJ **684**, L51 (2008)
72. Mosser, B., Michel, E., Appourchaux, T., et al.: A&A **506**, 33 (2009)
73. Belkacem, K., Mathis, S., Goupil, M. J., Samadi, R.: A&A **508**, 345 (2009)
74. Bedding, T.R., Kjeldsen, H.: Memorie della Societa Astronomica Italiana, **77**, 384 (2006)
75. Barban, C., Matthews, J.M., de Ridder, J. et al.: A&A **468**, 1033 (2007)
76. de Ridder, J., Barban, C., Carrier, F., et al.: A&A **448**, 689 (2006)
77. de Ridder, J., Barban, C., Baudin, F., et al.: Nature **459**, 398 (2009)
78. Belkacem, K., Samadi, R., Goupil, M.J., et al.: A&A **494**, 191 (2009)
79. Samadi, R., Belkacem, K., Goupil, M.-J., Dupret M.-A., Brun, A., Noels, A.: Ap&SS **328**, 253 (2010)

Index

J.-P. Rozelot and C. Neiner (eds.), *The Pulsation of the Sun and the Stars*,   341
Lecture Notes in Physics 832, DOI: 10.1007/978-3-642-19928-8,
© Springer-Verlag Berlin Heidelberg 2011

Made in United States
Orlando, FL
22 March 2026

79558001R00201